LOGIC AND
WITH APPLICATIONS

FOURTH EDITION

MAI Publishing

LOGIC AND SET THEORY WITH APPLICATIONS

FOURTH EDITION

Philip Cheifetz
Nassau Community College

Frank Avenoso
College of Charleston

Kenneth Lemp
Nassau Community College

Jay Martin
Nassau Community College

Ellen Schmierer
Nassau Community College

Michael Steuer
Nassau Community College

Theresa Vecchiarelli
Nassau Community College

Cover Design: Eva Reck

ISBN: 0-916060-09-8

Printed in the United States of America
10 9 8 7 6 5 4 3 2 1

Preface

In the coming years, a solid foundation in logic and set theory will be essential to succeeding in an increasingly technological world. This textbook serves as an introduction to the basic ideas of these topics. *Logic and Set Theory with Applications* is targeted toward non-science students and prospective elementry school teachers who seek to improve their skills in logical thinking and organization of information. Minimal mathematical background is necessary.

The fourth edition represents a refined rewriting of the third edition. Chapter One introduces the student to the basic concepts of logic and is similar in content to the opening chapter of the third edition. Chapter Two introduces arguments and the testing of validity by TF methods. A new section on conditional proof strategy has been added. Chapter Three covers valid argument forms and formal proofs. Chapter Four includes some of the more innovative material of this new edition. Sections on fuzzy logic, flowcharting techniques and electronic circuits demonstrate a few of the many applications of logic. Chapter Five introduces set theory. Applications of set theory provide the content of Chapter Six, including probability and a section on networks.

Several variations are possible for those instructors not wishing to cover Chapters One through Six consecutively. For an earlier treatment of set theory, Chapters Five and Six may be covered before Chapters One through Four. (In this case, it may be desirable to defer coverage of Section 6.1: Testing the Validity of Arguments using Venn Diagrams, until after Chapter Three.) Instructors teaching a shorter course may omit any of the sections in Chapters Four or Six without loss of continuity. Chapter Four may be covered at any point after Chapter One.

Feedback from users of the third edition has prompted many changes. Each section concludes with two groups of exercises: one designed to be used as illustrative examples or in-class practice problems, the other containing homework problems. At the end of each chapter is a cumulative set of chapter review exercises. Following each chapter review is a sample exam covering all topics in that chapter. Answers to all in-class exercises, chapter reviews and sample exams appear at the end of the text as well as answers to all odd homework problem sets. Answers to the odd problems are given as well.

We would like to thank our colleagues at Nassau Community College for their ideas and suggestions. We are particularly grateful to Carmine DeSanto, Jerry Honig, Abe Mantell, Lilia Orlova, Rochelle Robert and David Sher.

Philip Cheifetz
Frank Avenoso
Kenneth Lemp
Jay Martin
Ellen Schmierer
Michael Steuer
Theresa Vecchiarelli

Table of Contents

CHAPTER ONE

STATEMENTS AND TRUTH TABLES

A unifying concept in mathematics is the proof of arguments. To determine if an argument is valid we must examine its component parts, that is, the sentences or statements that comprise the argument. Such sentences are called *TF* statements. In this chapter we will examine individual *TF* statements and the relationship between combinations of *TF* statements.

1.1 Introduction to Logic

TF Statements

A *TF statement* is a declarative sentence to which we can assign a truth value of either true or false, but not both at the same time. A *TF* statement is also referred to as a *simple statement.*

Example 1

The sentence "Thanksgiving is in November" is a *TF* statement since we can decide whether the sentence is true or false.

■

Example 2

The sentence "Lincoln was the sixteenth President of the United States" is a *TF* statement since we can decide whether the sentence is true or false.

■

Example 3

"Is that an Audi?" is not a *TF* statement since it is a question, not a declarative sentence.

■

Example 4

"Sit down" is not a *TF* statement. It is a command, not a declarative sentence.

■

TF sentences are often represented by *variables.* In this text, the variables used will be lowercase letters. As an illustration of our notation,

p: Precision is required in logic

would be read "p represents (or stands for) the statement 'Precision is required in logic.' "

Negations

To negate a *TF* statement is to change its truth value. A *TF* statement may be negated by using the word *not*. It may also be negated by placing the words *it is not true that* in front of the statement. The mathematical symbol for the word *not* is ~. This symbol is also used to mean *it is not true that.* When the symbol ~ is placed in front of a letter such as p, the new statement, $\sim p$ is read "not p."

Example 5

Let p: The capybara is a bird. Then $\sim p$: It is not true that the capybara is a bird, or simply "The capybara is not a bird."

■

Example 6

Let e: The emu is an extinct bird. Therefore $\sim e$: It is not true that the emu is an extinct bird, or simply "The emu is not an extinct bird." ■

The following table is called a *truth table*. The truth table indicates the instances when a *TF* statement is true and when it is false. Truth values are displayed within the table using the capital letters *T* and *F*. The truth table below shows us that the negation of a true statement is false, while the negation of a false statement is true.

p	$\sim p$
T	F
F	T

The truth table is used to summarize our previous results about negations. A common misconception is that the negation of a *TF* statement must be false. Many *TF* sentences that use the word *not* are true statements. Consider the false statement n: Three nickels equal one dime. Then $\sim n$: Three nickels do not equal one dime. Surely this is a true statement.

Connectives

A *connective* is a word or group of words used to join two *TF* statements. In our study of logic, the four basic connectives we will use are *and*, *or*, *if...then...*, and *...if and only if....* These connectives are referred to respectively as conjunction, disjunction, implication, and biconditional.

Simple and Compound Statements

If a *TF* statement does not contain a connective, it is called a *simple statement*. If two or more simple statements are joined by one or more connectives, the resulting statement is called a *compound statement*. We will investigate the truth values of compound statements for the remainder of this chapter.

Conjunctions

If we join two *TF* statements by the word *and*, we call the resulting compound statement the *conjunction* of the two statements. The symbol for the conjunction is $\wedge$. The statement $p \wedge q$ is read "*p and q*" or "*p* conjunction *q*."

Example 7

Let p: Hamm plays soccer. Let s: Hamm stars in television commercials. The conjunction "Hamm plays soccer and stars in television commercials" is written symbolically as $p \wedge s$.

■

Example 8

Let l: Letterman hosts a television show. Let h: Howard hosts a radio show. The conjunction "Letterman hosts a television show and Howard hosts a radio show" can be written symbolically as $l \wedge h$.

■

Example 9

Let c: Jilly likes coffee. Let d: Jilly likes doughnuts. Then $c \wedge \sim d$ represents the compound statement "Jilly likes coffee and Jilly does not like doughnuts", or simply "Jilly likes coffee but not doughnuts." Notice that the word *but* is sometimes used to mean *and*.

■

Let us examine the truth values of a conjunction. By the nature of the word "and", if a conjunctive statement is to be true, both of its component sentences must be true. If either of the component sentences is false, the conjunction is false. *Thus, the only time a conjunction is true is when p and q are both true.*

The following truth table summarizes the results about conjunctions. Notice that there are exactly four ways that the truth values of the two statements can combine.

p	q	$p \wedge q$
T	T	T
T	F	F
F	T	F
F	F	F

Disjunctions

If we join two *TF* statements by the word *or*, the resulting compound statement is called the *disjunction* of the two statements. The symbol for the disjunction is $\vee$. The statement $p \vee q$ is read "*p or q*" or "*p* disjunction *q*." Sometimes the word *either* is implied as the correlative of the word *or.*

Example 10

Let p: I will pass English. Let s: I will go to summer school. Then the disjunction $p \vee s$ represents the compound statement "Either I will pass English or I will go to summer school" or simply "I will pass English or go to summer school."

■

In the English language, the word "or" is often used in the exclusive sense. When we say "Either we will go to the movies or go out to dinner," we often mean that we will do *one* of these two activities, but not *both*. In logic, the "or" is inclusive. In the above example, we mean that we will go to the movies, or go to dinner, or do both. The following truth table summarizes our results about disjunctions. *Notice that a disjunction is false only when p and q are both false.*

p	q	$p \vee q$
T	T	T
T	F	T
F	T	T
F	F	F

Conditionals

Given two statements p and q, a compound statement of the form *"If p then q"* is called a *conditional* statement or an *implication.* The symbol for a conditional statement is $\rightarrow$. The statement $p \rightarrow q$ is read *"p* implies *q"* or "If *p* then *q*." The first statement, *p,* is called the *antecedent* or the left hand side (LHS) of the conditional statement, while *q* is called the *consequent* or the right hand side (RHS) of the conditional statement.

Example 11

Let j: I have a part-time job. Let c: I pay for my cell phone. The implication "If I have a part-time job then I pay for my cell phone" is symbolically written as $j \rightarrow c$, while $c \rightarrow j$ is read "If I pay for my cell phone then I have a part-time job."

■

Example 12

Let f: I order flounder. Let r: I go to a restaurant. The statement "I order flounder if I go to a restaurant" is symbolized as $r \rightarrow f$. Notice that regardless of where the antecedent is located in the English sentence, in the symbolization it must appear as the LHS of the conditional statement.

■

Example 13

Using the underlined letters, the symbolization of "The food will go bad if left unrefrigerated" is $u \to b$.

■

The truth table for a conditional statement is less intuitive. Mathematicians feel that *the only time a conditional statement is false is when the LHS is true and the RHS is false.* The following truth table shows the truth values for a conditional statement.

p	q	$p \to q$
T	T	T
T	F	F
F	T	T
F	F	T

Example 14

Let p: Pigs fly, and let q: Two plus two equals four. The statement $p \to q$ is read "If pigs fly then two plus two equals four." This is a true statement since its LHS is false and its RHS is true. It corresponds to the third row of the conditional truth table above. The sentence $q \to p$ is read "If two plus two equals four then pigs fly." This is a false statement. It corresponds to the second row of the table.

■

Biconditionals

Given two statements p and q, a compound statement of the form $(p \to q) \wedge (q \to p)$ is called a *biconditional* statement. Because of this cumbersome notation, we denote the biconditional of p and q as $p \leftrightarrow q$. The statement $p \leftrightarrow q$ is read *"p if and only if q."*

Example 15

Let r: Roses are red, and let v: Violets are blue. Then $r \leftrightarrow v$ is read "Roses are red if and only if violets are blue."

■

The truth table for the biconditional is shown below.

p	q	$p \to q$	$q \to p$	$(p \to q) \wedge (q \to p)$
T	T	T	T	T
T	F	F	T	F
F	T	T	F	F
F	F	T	T	T

Noting that the heading of the last column is the definition of $p \leftrightarrow q$, our results may be summarized in the truth table below.

p	q	$p \leftrightarrow q$
T	T	T
T	F	F
F	T	F
F	F	T

Notice that for a biconditional to be true the LHS and the RHS must have the same truth value. That is, *for a biconditional to be true both sides of the biconditional must be true, or both sides must be false.* Notice that unlike a conditional statement, the order in the biconditional is unimportant. This can be seen from the truth table for the biconditional.

Example 16

Let s: I study, and g: I get a good grade. Then $s \leftrightarrow g$ is read as "I will study if and only if I get a good grade." The biconditional may also be read as "I get a good grade if and only if I study." ■

The Role of Parentheses in Compound Statements

Often, compound statements have more than one connective. These connectives can sometimes be grouped in more than one way, and different groupings may convey different meanings. Consider the statement "Emeril will buy apples or bananas and cherries." This statement has three simple ideas: a: Emeril will buy apples, b: [Emeril will buy] bananas and c: [Emeril will buy] cherries. The statement contains an "or" as well as an "and" connective. Therefore, a reasonable question to ask is, "Does it matter whether we think of this statement as the disjunction $a \vee (b \wedge c)$, or as the conjunction $(a \vee b) \wedge c$?"

If the statement is written as the disjunction $a \vee (b \wedge c)$, Emeril may buy apples, or buy both bananas and cherries, or buy all three fruits. If the statement is written as the conjunction $(a \vee b) \wedge c$, Emeril will buy apples or bananas (or both) and definitely buy cherries. The following table summarizes the possible ways Emeril could fill a shopping cart, depending on where the parentheses are placed.

Disjunction $a \vee (b \wedge c)$ What's in the Cart?	Conjunction $(a \vee b) \wedge c$ What's in the Cart?
apples	apples and cherries
bananas and cherries	bananas and cherries
apples and bananas and cherries	apples and bananas and cherries

The results are not the same. This is because there is ambiguity in the English statement as it was written. However, a mathematical statement must not have any ambiguity. It must have one and only one meaning. In an English sentence, the ambiguity is often remedied by using a comma to separate ideas. In order to construct compound statements that are non-ambiguous in logic, parentheses are often required to express the correct English meaning. Thus, if we wrote "Emeril will buy apples, or bananas and cherries," we mean $a \vee (b \wedge c)$, while if we wrote "Emeril will buy apples or bananas, and cherries," we mean $(a \vee b) \wedge c$.

Example 17

Is the following statement a conjunction or disjunction?

"I will order the soup or the oysters, and the rib steak."

Solution

It is a conjunction. A symbolization could be $(s \vee o) \wedge r$. ■

Example 18

Is the statement a conditional or a conjunction?

"I will go to the night club, and if you are not there I will leave."

Solution

It is a conjunction. A symbolization might be $n \wedge (\sim y \rightarrow l)$. ■

Example 19

Is this statement a conditional or a disjunction?

"If the team comes in last place, then they will fire their manager or trade for a better second baseman."

Solution

This is a conditional statement. A symbolization is $l \rightarrow (m \vee s)$. ■

Truth Table Summary

The truth value of a compound statement depends on the truth value of its individual simple sentences, and the connective used.

The following is a summary of the truth tables for the connectives we have discussed.

p	q	$p \wedge q$	$p \vee q$	$p \rightarrow q$	$p \leftrightarrow q$
T	T	T	T	T	T
T	F	F	T	F	F
F	T	F	T	T	F
F	F	F	F	T	T

There are four sentences that stress the main points found in the truth tables for conjunction, disjunction, conditional and biconditional statements. They are:

$p \wedge q$ is true only when both parts are true.

$p \vee q$ is false only when both parts are false.

$p \rightarrow q$ is false only if the LHS is true and the RHS is false.

$p \leftrightarrow q$ is true only when the LHS and the RHS are the same.

Example 20

Suppose p is a statement that we know is true and q is a statement that we know is false. Then

$p \vee q$ is true

$p \wedge q$ is false

$p \rightarrow q$ is false

$p \leftrightarrow q$ is false

■

In-Class Exercises and Problems for Section 1.1

In-Class Exercises

In Exercises 1–15, use the underlined letter to write the sentences in symbolic form:

1. I like to swim and play baseball.
2. I will watch the movie if it stars Susan Lucci.
3. Either I will redo my history homework or study logic.

4. I will buy you candy if and only if it is both your birthday and Valentines day.
5. It is not the case that Humpty Dumpty had a great fall.
6. If you like Colorado, then you like to ski and climb mountains.
7. It's not true that doctors and lawyers are rich.
8. They go to the Super Bowl, and if they win they celebrate.
9. If they go to the Super Bowl and celebrate, then they won.
10. Solving a crossword puzzle is time-consuming but enjoyable.
11. If she has a bronze medal she did not win but if she has a silver medal, she came in second.
12. If the Senate and House pass the bill, then it will become a law and the police will enforce it.
13. If freezing follows either snow or rain, the roads will become icy.
14. She will take calculus or statistics but if she takes economics then she will not take history.
15. The rich dark Turkish coffee was poured into the large green ceramic mug. (Choose your own symbols)

In Exercises 16–21, let p: She has red hair. Let q: She wears glasses. Let r: She is tall. Let s: She plays the flute. Create sentences for each of the symbolizations. For example, if the symbolization is $p \to q$, the sentence would be "If she has red hair then she wears glasses."

16. $q \wedge \sim r$
17. $\sim (s \to q)$
18. $\sim p \vee r$
19. $q \to (p \wedge s)$
20. $p \leftrightarrow (q \wedge r)$
21. $s \vee (p \to q)$

In Exercises 22–36, enter true, false, or can't be determined because of insufficient information.

22. If $\sim p$ is true and q is true then $p \wedge q$ must be _______.
23. If p is false and q is false then $p \to q$ must be _______.
24. If p is false and q is true then $p \leftrightarrow q$ must be _______.

25. If $\sim p$ is true then $p \vee q$ must be _______.

26. If $p \to q$ is false then q is _______.

27. If $p \leftrightarrow q$ is false and $\sim p$ is false then q is _______.

28. If $p \vee q$ is false then $p \to q$ is _______.

29. If $p \to (q \to r)$ is false then $p \to q$ is _______.

30. If $p \leftrightarrow (\sim q \vee r)$ is false and p is true then $r \to s$ is _______.

31. If $\sim p$ is true, the truth value of $q \to p$ is _______.

32. If q is true the truth value of $q \vee p$ is _______.

33. If $q \to \sim p$ is false, then p must be _______.

34. If $q \to \sim p$ is true then q must be _______.

35. If $\sim p \leftrightarrow q$ and $\sim q \wedge r$ are both true, then $w \to p$ is _______.

36. If $\sim (p \vee q)$ is true then the value of $q \to p$ is _______.

Problems

In Problems 37–51, classify the sentences as simple *TF* statements, conjunctions, disjunctions, conditionals, or biconditionals. Then symbolize them using the underlined letters.

37. Black bears hibernate.

38. Bill Clinton was a president or I'm a monkey's uncle.

39. The vegetables are hard if and only if they are not steamed.

40. Nine times ten equals ninety and six divided by two equals three.

41. If you sing, then you like music.

42. Salt is sweet and mice do not have 12 legs.

43. Her eyes are blue but her hair is brown.

44. Weeds grow quickly if they are not attended.

45. Many commuters do not take the Long Island Rail Road into Manhattan.

46. If you want to go to law school, you must take the LSAT exam.

47. If you do not smoke a pipe then you don't need a pipe lighter.

48. Scott will buy a dog if and only if it doesn't shed.

49. If the salary is not a positive number, then the computer writes an error message.

50. Either you will be able to write a computer program if you do well in logic or you will become a computer operator.

51. Computers are speedy and accurate but they can't reflect.

In Problems 52–61, each sentence is an example of a compound statement. Symbolize each sentence, using appropriate letters as symbols for the *TF* statements.

52. Economics equilibrium is achieved if and only if supply is equal to demand.

53. If today is Sunday, we will either go on a picnic or go to the game.

54. You eat at Beefy Delight and not at Charlie's Crab House if you like steak but not lobster.

55. It is not true that if a woman swims fast then she can run fast.

56. If a woman cannot swim fast then she can run fast.

57. I will go out with him if and only if he has a good job and he likes to dance.

58. Averi will not grow tall if she doesn't eat her spinach but only eats her dessert.

59. If she gets a new car she will withdraw her savings, and if she buys a new boat she will get a loan or not quit her weekend job.

60. If you are married and filing jointly then you will use Table X, or if you are married and not filing jointly then you will use Table Y.

61. The computer will perform Routine-A if the salary is greater than one hundred dollars and not greater than one thousand dollars, or if the employee has worked overtime.

62. Fill in the following truth table.

p	q	$p \wedge q$	$p \vee q$	$p \rightarrow q$	$p \leftrightarrow q$
T	T				
T	F				
F	T				
F	F				

In Problems 63–82, fill in each blank with either true or false.

63. If an implication is false, then its LHS must be _____ and its RHS must be _____.

64. If a conjunction is true then both its parts must be _______.

65. If an implication is false, its RHS must be _______.

66. If a disjunction is false, both its parts must be _______.

67. If a biconditional is true and its LHS is false then its RHS is _______.

68. If p is true and q is false then $p \vee q$ must be _______.

69. If p is true and q is false then $p \rightarrow q$ must be _______.

70. If p is true and q is false then $p \leftrightarrow q$ must be _______.

71. If p is false and q is false then $p \leftrightarrow q$ must be _______.

72. If p is false and q is false then $p \rightarrow q$ must be _______.

73. If p is false then $p \rightarrow q$ must be _______.

74. If p is true then $p \vee q$ must be _______.

75. If q is true then $p \rightarrow q$ must be _______.

76. If q is false then $p \wedge q$ must be _______.

77. If $p \leftrightarrow q$ is false and p is true then q is _______.

78 If $p \leftrightarrow q$ is true and p is true then q is _______.

79. If q is false and $p \rightarrow q$ is true then p must be _______.

80. Suppose the truth value of r is unknown, but s is true. The truth value of $r \rightarrow s$ is _______.

81. Suppose the truth value of r is unknown, but s is true. The truth value of $r \vee s$ is _______.

82. Suppose the truth value of r is unknown, but s is false. The truth value of $r \wedge s$ is _______.

1.2 Evaluating *TF* Statements

We have seen that the truth value of a compound statement depends on the truth value of its individual simple sentences and the particular connective used. Sometimes we wish to know the truth value when we combine more than two simple statements into a compound statement. The strategy we use to obtain the truth value is to consider the truth values two at a time.

Example 1

Suppose we are told that p and q are true but r is false. Determine the truth value of the compound statement $(p \to q) \wedge r$.

Solution

We begin by noticing that the compound statement is a conjunction. For a conjunction to be true, both of its components must be true. Since r is known to be false, the compound statement $(p \to q) \wedge r$ is false for the given truth values.

■

Example 2

Let a and b be false and c be true. What is the truth value of $(a \to c) \vee b$?

Solution

The statement $(a \to c) \vee b$ is a disjunction. Disjunctions are true if either side is true. Since b is not true, the truth value of the disjunction will depend on $(a \to c)$. Since the LHS of this implication is false and the RHS is true, $(a \to c)$ is true. Therefore, $(a \to c) \vee b$ is true for the given truth values.

■

The simplest way to resolve the truth value of the compound statement is to resolve the truth value of its components. First, replace each simple sentence with its given truth value. Then, resolve each component, working from the inside out. The next three examples show how this is done.

Example 3

Suppose r: The computer has 512 megabytes of RAM, d: The computer has a read-write disk drive, and c: The computer has a cordless mouse. Assume r, d, and c are all true. Find the truth value of $\sim[(r \wedge \sim d) \vee c]$.

Solution

First, replace each statement with its truth values. Then resolve each component, working outward from the innermost statements.

$$\sim[(r \wedge \sim d) \vee c]$$
$$\sim[(T \wedge \sim T) \vee T]$$
$$\sim[(T \wedge F) \vee T]$$
$$\sim[F \vee T]$$
$$\sim[T]$$
$$F$$

Therefore, the statement $\sim[(r \wedge \sim d) \vee c]$ is false for the given truth values.

■

Example 4

If p is true while q and r are false, find the truth value of $(p \rightarrow q) \rightarrow \sim r$.

Solution

$$(p \rightarrow q) \rightarrow \sim r$$
$$(T \rightarrow F) \rightarrow \sim F$$
$$F \rightarrow T$$
$$T$$

Therefore, $(p \rightarrow q) \rightarrow \sim r$ is a true statement for the given truth values.

■

Example 5

Let c: She gets a new cat, w: She withdraws money from her savings, b: She buys a new boat, and q: She quits her part-time job. Suppose c and w are true while b and q are false. What is the truth value of the compound statement "If she gets a new cat she will withdraw money from her savings, but if she buys a new boat she will not quit her part-time job"?

Solution

First, symbolize the compound statement as $(c \rightarrow w) \wedge (b \rightarrow \sim q)$. Then replace each statement with its truth value and then resolve each component, working outward from the innermost statements.

$$(c \to w) \wedge (b \to \sim q)$$
$$(T \to T) \wedge (F \to \sim F)$$
$$T \wedge (F \to T)$$
$$T \wedge T$$
$$T$$

Therefore, the compound statement is true for the given truth values. ■

Example 6

Suppose p and q are both true. What are the truth vaues of $p \to (q \vee r)$ and $p \to (q \wedge r)$?

Solution

The truth value of r is not given, yet we are able to determine the truth value of $p \to (q \vee r)$. The RHS is a disjunction, and we know that the disjunction is true since one of its parts is true. The only way an implication is false is if the LHS is true and the RHS is false. Therefore, since both the LHS and RHS of the implication are true, $p \to (q \vee r)$ is true. In the implication $p \to (q \wedge r)$, the truth value of the RHS is unknown, since it is a conjunction. Therefore, we can not determine the truth value of $p \to (q \wedge r)$. ■

In-Class Exercises and Problems for Section 1.2

In-Class Exercises

In Exercises 1–15 determine whether the compound statements are true, false or can't be determined because of insufficient information. Assume that p and q are true and r and s are false.

1. $\sim s \to q$
2. $\sim s \to r$
3. $p \leftrightarrow q$
4. $\sim q \to \sim r$
5. $q \vee z$
6. $p \wedge r$
7. $p \vee y$
8. $\sim p \leftrightarrow s$

9. $(q \to s) \to (a \vee p)$

10. $r \vee (\sim s \vee n)$

11. $q \wedge (\sim p \vee b)$

12. $s \to (p \wedge e)$

13. $\sim [p \to (q \wedge s)] \wedge r$

14. $(p \wedge s) \leftrightarrow \sim (q \vee \sim r)$

15. $[(w \vee p) \wedge (\sim q \vee s)] \to [(p \leftrightarrow r) \vee \sim q]$

For Exercises 16–25, answer true, false or can't be determined because of insufficient information.

16. Suppose $(p \wedge s) \to r$ is false. What is the truth value of *r*?

17. What is the truth value of *r* if $(s \vee \sim r) \vee (p \to q)$ is false?

18. If $(s \vee \sim r) \wedge (p \to q)$ is true, what is the truth value of *s*?

19. Suppose $(p \leftrightarrow \sim r) \to (s \wedge \sim p)$ and *p* are both true. What is the truth value of *r*?

20. If $(a \wedge \sim b) \leftrightarrow (\sim c \to d)$ and *b* are both true, what is the truth value of *c*?

21. What is the truth value of *m* if $(m \to n) \vee (\sim p \wedge r)$ is true and *r* is false?

22. Suppose *r* is false. Then what is the truth value of the statement $\sim (p \vee \sim r) \to (r \leftrightarrow w)$?

23. What is the truth value of *a* if $\sim [(c \to \sim a) \vee (\sim p \leftrightarrow s)]$ is true?

24. If $\sim [(p \vee \sim r) \wedge (\sim s \to b)]$ and *p* are both false, what is the truth value of *r*?

25. If $[(\sim r \to s) \vee \sim d] \leftrightarrow (w \wedge p)$ is true and *p* is false, what is the truth value of *d*?

Problems

For Problems 26–45, determine whether the compound statements are true, false or can't be determined because of insufficient information. Assume that *p* and *q* are true and *r* and *s* are false.

26. $q \vee \sim s$

27. $(p \wedge q) \to s$

28. $s \rightarrow (r \vee q)$

29. $(r \vee s) \leftrightarrow q$

30. $q \rightarrow \sim (p \wedge s)$

31. $\sim (q \rightarrow r)$

32. $\sim q \rightarrow r$

33. $(z \wedge s) \rightarrow p$

34. $p \rightarrow (m \rightarrow q)$

35. $(p \rightarrow q) \leftrightarrow (\sim p \vee q)$

36. $[p \rightarrow (q \wedge r)] \leftrightarrow [(p \rightarrow q) \wedge (p \rightarrow r)]$

37. $\sim [(\sim s \vee p) \wedge (q \vee \sim r)]$

38. $[(s \vee \sim r) \wedge (p \rightarrow \sim q)] \rightarrow (\sim p)$

39. $(p \rightarrow q) \wedge z$

40. $z \rightarrow (\sim r \rightarrow q)$

41. $(\sim r \vee p) \rightarrow (s \wedge q)$

42. $\sim [(\sim r \rightarrow s) \wedge (p \wedge q)]$

43. $(p \rightarrow r) \rightarrow \sim s$

44. $[(m \leftrightarrow n) \wedge c] \rightarrow p$

45. $[(a \vee p) \wedge (r \rightarrow k)] \leftrightarrow (c \vee s)$

For Problems 46–55, answer true, false or can't be determined because of insufficient information.

46. If $(c \rightarrow \sim d) \wedge b$ is true, and c is true, what is the truth value for d?

47. What is the truth value of $m \rightarrow (a \vee p)$ if m is false?

48. If $(n \vee \sim p) \leftrightarrow (c \wedge g)$ is true and p is false, what is the truth value of c?

49. Let $(a \vee \sim b) \rightarrow (\sim b \vee c)$ be true. What is the truth value of a?

50. If $(r \rightarrow \sim s) \rightarrow (a \wedge \sim w)$ is false and $\sim w$ is true, then what is the truth value of a?

51. Suppose p is false. Find the truth value of $\sim p \leftrightarrow [(p \rightarrow r) \wedge p]$.

52. What is the truth value of $(\sim m \rightarrow \sim n) \vee (a \wedge \sim b)$ if we know that n is false?

53. If $(a \wedge \sim b) \vee (p \wedge \sim r)$ is true and $\sim a$ is true, what is the truth value of r?

54. What is the truth value of $\sim [(\sim c \rightarrow d) \wedge (a \rightarrow \sim b)]$ given that a is true and b is false?

55. If both $(\sim p \wedge \sim c) \leftrightarrow (\sim d \wedge p)$ and p are false, what is the truth value of c?

1.3 Truth Tables with Two Simple Statements

Introduction

More often than not, we do not know the truth values of the components of a compound statement. We can, however, construct a truth table that lists all the possible combinations of truth values for the compound statement. We have already seen this technique when we constructed truth tables for negations, conjunctions, disjunctions, implications, and biconditionals.

When the truth table for negation was constructed for a single simple statement, the truth table only had two rows, one row for true and one for false. When we constructed truth tables for compound statements that contained two simple statements, the truth tables had four rows, one for each possible combination of truth values.

Example 1

Construct a truth table for $(p \vee q) \rightarrow p$.

Solution

We begin by noticing that there are only two simple statements involved, p and q. Therefore, the truth table will have four rows. We will have a column for each of the simple statements, one column for $p \vee q$ and finally one column for our entire statement $(p \vee q) \rightarrow p$.

p	q	$p \vee q$	$(p \vee q) \rightarrow p$
T	T		
T	F		
F	T		
F	F		

Now we fill in the table by rows.

p	q	$p \vee q$	$(p \vee q) \rightarrow p$
T	T	T	T
T	F	T	T
F	T	T	F
F	F	F	T

In this example, the compound statement $(p \vee q) \rightarrow p$ is true except when p is false and q is true.

■

Example 2

Construct a truth table for the statement $(r \vee \sim s) \rightarrow (\sim r \vee s)$.

Solution

We begin by noticing that there are only two simple statements involved, r and s. Therefore, the truth table will have four rows. We will have a column for each of the simple statements, one column for each of the negations, one for $(r \vee \sim s)$, one for $(\sim r \vee s)$, and finally, one for the entire compound statement $(r \vee \sim s) \rightarrow (\sim r \vee s)$.

r	s	$\sim r$	$\sim s$	$r \vee \sim s$	$\sim r \vee s$	$(r \vee \sim s) \rightarrow (\sim r \vee s)$
T	T					
T	F					
F	T					
F	F					

Now we fill in the table by rows, using the truth values for each component as we proceed from left to right. We obtain:

r	s	$\sim r$	$\sim s$	$r \vee \sim s$	$\sim r \vee s$	$(r \vee \sim s) \rightarrow (\sim r \vee s)$
T	T	F	F	T	T	T
T	F	F	T	T	F	F
F	T	T	F	F	T	T
F	F	T	T	T	T	T

Using the truth table, it is now a simple matter to determine the conditions under which the compound statement is true. In this case, $(r \vee \sim s) \rightarrow (\sim r \vee s)$ is always true, except for the case when r is true and s is false.

■

Many times the column for the negation of simple *TF* statements is omitted from the truth table, as is shown in the next example.

Example 3

Construct a truth table for $(p \to \sim q) \to (q \to \sim p)$.

Solution

Once again the truth table will have four rows, but this time we will not include a column for the negations of p and q. Thus, the truth table will have only five columns.

p	q	$p \to \sim q$	$q \to \sim p$	$(p \to \sim q) \to (q \to \sim p)$
T	T	F	F	T
T	F	T	T	T
F	T	T	T	T
F	F	T	T	T

In this case, no matter what the truth values of the simple statements, the compound statement $(p \to \sim q) \to (q \to \sim p)$ is always true. ■

Compact Truth Tables

Rather than constructing a column in a truth table for every *TF* statement in a complicated compound statement, and then a column for each sub-statement, we often construct a more compact truth table that yields the same result.

Example 4

Determine when $(p \to q) \leftrightarrow (q \vee \sim p)$ is true and when it is false.

Solution

We begin by noting that the truth table will have four rows since there are two simple sentences. Previously, the truth table would have looked like this:

p	q	$p \to q$	$q \vee \sim p$	$(p \to q) \leftrightarrow (q \vee \sim p)$
T	T			
T	F			
F	T			
F	F			

Instead of the above truth table, we will use the compact form.

p	q	$p \to q$	$\leftrightarrow$	$q \vee \sim p$
T	T			
T	F			
F	T			
F	F			

Notice how the table is slightly different than in previous examples. We have already entered the truth values for the statements p and q. Now, we will resolve the third and fifth columns.

p	q	$p \to q$	$\leftrightarrow$	$q \vee \sim p$
T	T	T		T
T	F	F		F
F	T	T		T
F	F	T		T

Finally, we resolve the fourth column.

p	q	$p \to q$	$\leftrightarrow$	$q \vee \sim p$
T	T	T	T	T
T	F	F	T	F
F	T	T	T	T
F	F	T	T	T

The fourth column is the resolution of the compound statement and it is always true. Therefore, $(p \to q) \leftrightarrow (q \vee \sim p)$ is always true. ■

Tautologies

One of the main uses of truth tables is to determine if a compound statement is *always* true. Statements that are always true, regardless of the truth values of the simple statements involved are called *tautologies*. The compound statement we examined in Example 3, $(p \to \sim q) \to (q \to \sim p)$, is a tautology.

Example 5

Determine whether or not the statement $\sim (z \vee x) \leftrightarrow (\sim z \wedge \sim x)$ is a tautology.

Solution

z	x	$\sim z$	$\sim x$	$(z \vee x)$	$\sim(z \vee x)$	$\sim z \wedge \sim x$	$\sim(z \vee x) \leftrightarrow (\sim z \wedge \sim x)$
T	T	F	F	T	F	F	T
T	F	F	T	T	F	F	T
F	T	T	F	T	F	F	T
F	F	T	T	F	T	T	T

The statement $\sim(z \vee x) \leftrightarrow (\sim z \wedge \sim x)$ is a tautology since it is always true.

■

Example 6

Use a compact truth table to determine whether or not the biconditional statement in Example 5 is a tautology.

Solution

We construct a compact truth table for $\sim(z \vee x) \leftrightarrow (\sim z \wedge \sim x)$ using the steps below.

z	x	$z \vee x$	$\sim(z \vee x)$	$\leftrightarrow$	$(\sim z \wedge \sim x)$
T	T	T	F	T	F
T	F	T	F	T	F
F	T	T	F	T	F
F	F	F	T	T	T

Again, we see that $\sim(z \vee x) \leftrightarrow (\sim z \wedge \sim x)$ is a tautology.

■

In-Class Exercises and Problems for Section 1.3

In-Class Exercises

In Exercises 1–10, construct a truth table to determine if the given compound statement is a tautology.

1. $p \wedge \sim q$
2. $p \rightarrow (p \wedge q)$
3. $(\sim p \wedge q) \vee \sim(\sim q \rightarrow \sim p)$
4. $(\sim p \rightarrow q) \rightarrow (p \vee q)$

5. $\sim(p \to q) \to (p \wedge \sim q)$
6. $[p \wedge (p \vee q)] \leftrightarrow p$
7. $[p \wedge (\sim p \vee q)] \to q$
8. $[p \vee (\sim p \wedge q)] \leftrightarrow [(p \vee \sim p) \wedge (p \vee q)]$
9. $[\sim q \wedge (\sim p \to q)] \vee [(q \leftrightarrow \sim p) \wedge p]$
10. $[(p \to q) \wedge (q \to p)] \to (\sim p \leftrightarrow \sim q)$

Problems

In Problems 11–30, use a truth table to determine if the given compound statement is a tautology.

11. $(p \wedge q) \to p$
12. $(p \wedge q) \to (q \vee p)$
13. $q \to (p \vee q)$
14. $(q \to p) \to (p \to q)$
15. $\sim(p \to q) \to (\sim p \to \sim q)$
16. $(p \to q) \to (\sim p \vee q)$
17. $[p \wedge (p \to q)] \leftrightarrow q$
18. $[p \wedge (p \to q)] \to q$
19. $[(\sim p \vee \sim q) \wedge q] \to p$
20. $(p \to \sim q) \to \sim(p \wedge q)$
21. $\sim(\sim p \to q) \leftrightarrow (p \to \sim q)$
22. $\sim(p \vee q) \leftrightarrow (\sim p \vee \sim q)$
23. $(p \leftrightarrow q) \leftrightarrow [(p \to q) \wedge (q \to p)]$
24. $[(p \to q) \wedge q] \to p$
25. $[(p \to q) \wedge \sim p] \to \sim q$
26. $\sim(p \to q) \leftrightarrow (\sim p \vee q)$
27. $[(q \to p) \to p] \leftrightarrow (\sim q \vee p)$
28. $\sim(p \leftrightarrow q) \leftrightarrow [(p \wedge \sim q) \vee (q \wedge \sim p)]$
29. $\sim(p \wedge q) \leftrightarrow (p \to \sim q)$
30. $(p \wedge q) \leftrightarrow \sim(p \to \sim q)$

1.4 Truth Tables with Three Simple Statements

As we discussed earlier, when only two variables are involved in the construction of a compound sentence, four different sets of truth values must be considered. This results in a truth table that has four rows. If a compound statement involves three simple statements, there will be exactly eight ways that the truth values of these variables can combine. This will result in a truth table that has eight rows, as shown below.

p	q	r
T	T	T
T	T	F
T	F	T
T	F	F
F	T	T
F	T	F
F	F	T
F	F	F

Observe the *TF* entries in each column. The first column has four entries of *T* followed by four entries of *F*. The second column has the truth values alternating in groups of two. The last column has the truth values alternating, beginning with *T*. If we fill in a truth table for a compound statement that contains three simple statements in this way, we are assured of accounting for all possible truth values

Example 1

Construct a truth table for $p \to (q \wedge r)$.

Solution

We begin by listing a column for each simple statement, as well as for each compound component. We then assign truth values to the simple sentences, entering our truth values by columns. The first column will have four entries of *T* followed by four entries of *F*. The second column will have the truth values alternating in groups of two, beginning with *T*. The last column has the truth values alternating, beginning with *T*.

p	q	r	$q \wedge r$	$p \to (q \wedge r)$
T	T	T		
T	T	F		
T	F	T		
T	F	F		
F	T	T		
F	T	F		
F	F	T		
F	F	F		

We now fill in the truth values for $q \wedge r$, by considering the truth values of q and r in each row.

p	q	r	$q \wedge r$	$p \to (q \wedge r)$
T	T	T	T	
T	T	F	F	
T	F	T	F	
T	F	F	F	
F	T	T	T	
F	T	F	F	
F	F	T	F	
F	F	F	F	

Finally, we resolve the truth values of the entire compound statement by combining the truth values of the first column with the truth values in the fourth column using the conditional connective. For example, in the first row of the truth table, $p \to (q \wedge r)$ corresponds to $T \to T$, and is therefore true. In the second row, $p \to (q \wedge r)$ corresponds to $T \to F$, which is false. We continue to resolve the the compound statement row by row.

p	q	r	$q \wedge r$	$p \rightarrow (q \wedge r)$
T	T	T	T	T
T	T	F	F	F
T	F	T	F	F
T	F	F	F	F
F	T	T	T	T
F	T	F	F	T
F	F	T	F	T
F	F	F	F	T

We see that the statement $p \rightarrow (q \wedge r)$ is true five out of eight times. ■

Example 2

For what values of p, q, and r is the statement $(p \vee q) \vee (p \rightarrow r)$ true?

Solution

We construct a truth table and include all the components. We will need columns for the three simple statements p, q, and r, as well as the statements $p \vee q$, $p \rightarrow r$, and $(p \vee q) \vee (p \rightarrow r)$. Therefore, the truth table will have eight rows and six columns. The completed table is shown below.

p	q	r	$p \vee q$	$p \rightarrow r$	$(p \vee q) \vee (p \rightarrow r)$
T	T	T	T	T	T
T	T	F	T	F	T
T	F	T	T	T	T
T	F	F	T	F	T
F	T	T	T	T	T
F	T	F	T	T	T
F	F	T	F	T	T
F	F	F	F	T	T

The statement $(p \vee q) \vee (p \rightarrow r)$ is always true, regardless of the truth values of the variables. Therefore, it is a tautology. ■

Example 3

Determine if the statement $(p \wedge q) \leftrightarrow r$ is a tautology.

Solution

We construct a truth table with eight rows and five columns.

p	q	r	$p \wedge q$	$(p \wedge q) \leftrightarrow r$
T	T	T	T	T
T	T	F	T	F
T	F	T	F	F
T	F	F	F	T
F	T	T	F	F
F	T	F	F	T
F	F	T	F	F
F	F	F	F	T

The statement $(p \wedge q) \leftrightarrow r$ is not a tautology. ■

Example 4

Use a truth table to determine if $[p \wedge (q \vee r)] \leftrightarrow [(p \wedge q) \vee (p \wedge r)]$ is a tautology.

Solution

We construct a truth table as shown below. Notice that aside from the three columns needed for our variables p, q and r, only three more columns are required in this compact table: one for the LHS of the biconditional, one for its RHS, and one for the symbol $\leftrightarrow$ itself which appears between these two sides.

p	q	r	$p \wedge (q \vee r)$	$\leftrightarrow$	$(p \wedge q) \vee (p \wedge r)$
T	T	T			
T	T	F			
T	F	T			
T	F	F			
F	T	T			
F	T	F			
F	F	T			
F	F	F			

To determine the truth values of $p \wedge (q \vee r)$ we first resolve the truth values of $(q \vee r)$.

p	q	r	$p \wedge (q \vee r)$	$\leftrightarrow$	$(p \wedge q) \vee (p \wedge r)$
T	T	T	T		
T	T	F	T		
T	F	T	T		
T	F	F	F		
F	T	T	T		
F	T	F	T		
F	F	T	T		
F	F	F	F		

Next, resolve the truth values of $p \wedge (q \vee r)$.

p	q	r	$p \wedge (q \vee r)$		$\leftrightarrow$	$(p \wedge q) \vee (p \wedge r)$
T	T	T	T	T		
T	T	F	T	T		
T	F	T	T	T		
T	F	F	F	F		
F	T	T	F	T		
F	T	F	F	T		
F	F	T	F	T		
F	F	F	F	F		

The boxed truth values are those of the LHS of the biconditional. We now determine the truth values of $(p \wedge q)$ and of $(p \wedge r)$.

p	q	r	$p\wedge(q\vee r)$		$\leftrightarrow$	$(p\wedge q)\vee(p\wedge r)$	
T	T	T	T	T		T	T
T	T	F	T	T		T	F
T	F	T	T	T		F	T
T	F	F	F	F		F	F
F	T	T	F	T		F	F
F	T	F	F	T		F	F
F	F	T	F	T		F	F
F	F	F	F	F		F	F

Next, we use the truth values of $(p\wedge q)$ and of $(p\wedge r)$ to determine the truth values of the disjunction connecting these statements.

p	q	r	$p\wedge(q\vee r)$		$\leftrightarrow$	$(p\wedge q)\vee(p\wedge r)$		
T	T	T	T	T		T	T	T
T	T	F	T	T		T	T	F
T	F	T	T	T		F	T	T
T	F	F	F	F		F	F	F
F	T	T	F	T		F	F	F
F	T	F	F	T		F	F	F
F	F	T	F	T		F	F	F
F	F	F	F	F		F	F	F

The fourth and seventh columns of truth values are those of the LHS and the RHS, respectively, of the biconditional statement. Using these boxed truth values, we finally resolve the truth values of the biconditional itself, recalling that a biconditional is true if and only if its LHS and RHS have the same truth value.

p	q	r	$p \wedge (q \vee r)$	$\leftrightarrow$	$(p \wedge q) \vee (p \wedge r)$
T	T	T	$T \quad T$	T	$T \quad T \quad T$
T	T	F	$T \quad T$	T	$T \quad T \quad F$
T	F	T	$T \quad T$	T	$F \quad T \quad T$
T	F	F	$F \quad F$	T	$F \quad F \quad F$
F	T	T	$F \quad T$	T	$F \quad F \quad F$
F	T	F	$F \quad T$	T	$F \quad F \quad F$
F	F	T	$F \quad T$	T	$F \quad F \quad F$
F	F	F	$F \quad F$	T	$F \quad F \quad F$

The truth values in the sixth column reveal that the statement $[p \wedge (q \vee r)] \leftrightarrow [(p \wedge q) \vee (p \vee r)]$ is always true. Hence, it is a tautology.

■

Example 5

Determine if $[(p \wedge q) \rightarrow r] \rightarrow [(p \rightarrow r) \wedge (q \rightarrow r)]$ is a tautology.

Solution

We will use a compact truth table. The numbers at the bottom of the columns refer to the order in which we resolved the truth values of the component statements.

p	q	r	$(p \wedge q) \rightarrow r$	$\rightarrow$	$[(p \rightarrow r)$	$\wedge$	$(q \rightarrow r)]$
T	T	T	$T \quad T$	T	T	T	T
T	T	F	$T \quad F$	T	F	F	F
T	F	T	$F \quad T$	T	T	T	T
T	F	F	$F \quad T$	F	F	F	T
F	T	T	$F \quad T$	T	T	T	T
F	T	F	$F \quad T$	F	T	F	F
F	F	T	$F \quad T$	T	T	T	T
F	F	F	$F \quad T$	T	T	T	T
1	2	3	4 5	9	6	8	7

The columns numbered 5 and 8 contain the truth values of the LHS and RHS respectively. The column numbered 9 contains the truth values of the entire conditional statement. Since this conditional statement is not always true, it is not a tautology.

■

In-Class Exercises and Problems for Section 1.4

In-Class Exercises

For Exercises 1–10, construct a truth table to determine if the given statement is a tautology.

1. $[p \vee (q \wedge r)] \leftrightarrow [(p \vee q) \wedge (p \vee r)]$
2. $[p \wedge (q \vee r)] \leftrightarrow [(p \wedge q) \vee r]$
3. $[\sim p \rightarrow \sim (q \wedge \sim r)] \leftrightarrow [\sim q \vee (r \vee p)]$
4. $[p \rightarrow (q \vee r)] \leftrightarrow [(p \rightarrow q) \vee (p \rightarrow r)]$
5. $[\sim p \vee (\sim q \vee \sim r)] \vee [\sim p \rightarrow \sim (q \wedge \sim r)]$
6. $[p \vee (q \rightarrow r)] \rightarrow [(p \vee q) \rightarrow (p \vee r)]$
7. $[\sim (p \vee q) \wedge r] \rightarrow [(\sim q \rightarrow r) \vee p]$
8. $[p \rightarrow (q \rightarrow r)] \leftrightarrow [(p \rightarrow q) \rightarrow r)]$
9. $\{p \vee [q \wedge \sim (r \vee p)]\} \leftrightarrow \sim (\sim q \vee r)$
10. $\{[q \wedge \sim (\sim p \wedge r)] \vee [(\sim q \wedge p) \vee (\sim q \wedge \sim r)]\} \leftrightarrow (p \vee \sim r)$

Problems

For Problems 11–20, construct a truth table to determine if the given statement is a tautology.

11. $[p \wedge (q \vee r)] \leftrightarrow [(p \wedge q) \vee (p \wedge r)]$
12. $(\sim p \vee q) \vee (\sim q \vee r)$
13. $[p \rightarrow (q \wedge r)] \leftrightarrow [(p \rightarrow q) \wedge (p \rightarrow r)]$
14. $[(r \leftrightarrow \sim p) \wedge q] \vee \sim (q \wedge \sim r)$
15. $[p \wedge (q \rightarrow r)] \rightarrow [(p \wedge q) \rightarrow (p \wedge r)]$
16. $[(p \rightarrow q) \rightarrow r] \vee (r \rightarrow p)$

17. $[p \to (q \to r)] \to [(p \to q) \to r)]$

18. $\{r \wedge \sim [p \vee (r \to \sim q)]\} \wedge (\sim p \to \sim r)$

19. $[p \to (\sim q \vee r)] \leftrightarrow [\sim p \to (q \to r)]$

20. $\sim [p \to (q \to r)] \leftrightarrow [\sim p \to (\sim q \to \sim r)]$

For Problems 21–30, fill in the blanks.

21. We have seen that if a compound statement involves two simple statements, its truth table will have four rows. If a compound statement involves three simple statements, its truth table will have ________ rows.

22. Notice that $2^2 = 4$, and $2^3 = 8$. Using similar reasoning, if a compound statement involves 4 simple statements, its truth table will have ________ rows.

23. In constructing a truth table for a compound statement that involves two variables, we can ensure that all possible arrangements of truth values will be accounted for by filling in the first column with two entries of *T* followed by ________ entries of *F*. The second column will have alternating values of *T* and ________.

24. When constructing a truth table that involves three simple statements we can ensure that all possible arrangements of truth values will be accounted for by filling in the first column of truth values with four entries of *T* followed by ________ entries of *F*. The second column will contain entries of *T* and *F* alternating ________ at a time. The third column will contain entries of *T* and *F* alternating ______________________________.

25. When a truth table has four simple statements we can ensure that all possible arrangements of truth values will be accounted for by filling in the first column of truth values with eight entries of *T* followed by ______________________________. The second column will have entries of *T* and *F* alternating four at a time. The third column will have entries of *T* and *F* alternating ________ at a time. The last column will have entries of *T* and *F* alternating ________ at a time.

26. Compound statements that are always true, regardless of the truth values of the variables involved, are called ___________.

27. The compound statement $[p \wedge (q \vee r)] \leftrightarrow [(p \wedge q) \vee (p \wedge r)]$ is a tautology. Hence, $[a \wedge (b \vee c)] \leftrightarrow$ ___________ is a tautology.

28. The statement $[p \rightarrow (q \wedge r)] \leftrightarrow [(\sim p \vee (q \wedge r)]$ is a tautology. Therefore, the statement $[\sim a \rightarrow (b \wedge \sim c)] \leftrightarrow$ ___________ is a tautology.

29. The statement $[\sim p \vee (\sim q \wedge r)] \leftrightarrow [(\sim p \vee \sim q) \wedge (\sim p \vee r)]$ is a tautology. Hence, $[m \vee (l \wedge \sim w)] \leftrightarrow$ ___________ is a tautology.

30. The statement $[p \rightarrow (q \rightarrow r)] \leftrightarrow [(p \rightarrow q) \rightarrow r]$ is *not* a tautology. Therefore, $[(\sim a \rightarrow b) \rightarrow \sim c)] \leftrightarrow$ ___________ is *not* a tautology.

1.5 Equivalences–Part I

Introduction

Two compound statements that always have the same truth value, regardless of the truth values of the variables involved, are said to be logically equivalent. When two such statements are joined by the symbol $\Leftrightarrow$, the resulting expression is referred to as an *equivalence.*

Equivalences allow us to express the same thought in two different, but logically equivalent ways. Equivalences are used extensively in the remainder of this chapter and in many of the chapters that follow. While there are infinitely many equivalences, a few are used with such great frequency as to warrant special attention. A list of the frequently used equivalences is summarized on page 53.

Double Negation Equivalence

The *double negation equivalence* is expressed symbolically as $\sim(\sim p) \Leftrightarrow p$. This equivalence states that the negation of a negation of a statement is equivalent to the original statement.

Example 1

Let d: I like to dance. Then $\sim d$: I don't like to dance. The statement "It is not true that I don't like to dance," symbolized as $\sim(\sim d)$, is logically equivalent to the statement "I like to dance." The truth table verifies this relationship.

d	$\sim d$	$\sim(\sim d)$
T	F	T
F	T	F

Since the columns for d and $\sim(\sim d)$, are identical, these statements are equivalent. That is, $\sim(\sim d) \Leftrightarrow d$.

■

Commutative Equivalence

The *commutative equivalences,* which are expressed symbolically as $(p \wedge q) \Leftrightarrow (q \wedge p)$ or $(p \vee q) \Leftrightarrow (q \vee p)$, pertain only to the *conjunction* and *disjunction* connectives.

This equivalence tells us that the order in which the statements occur in a conjunction or in a disjunction is immaterial. You encountered commutative relationships when dealing with addition and multiplication of real numbers. The order in which you add or multiply two numbers has no effect on the resulting sum or product.

Example 2

Verify that $(p \wedge q) \Leftrightarrow (q \wedge p)$ is an equivalence.

Solution

p	q	$p \wedge q$	$q \wedge p$
T	T	T	T
T	F	F	F
F	T	F	F
F	F	F	F

Since $p \wedge q$ and $q \wedge p$ always have the same truth value, these two conjunctions are logically equivalent. You will be asked to verify the disjunctive form of the commutative equivalence in the exercises.

■

Example 3

Let e: The Earth revolves around the sun. Let m: The moon revolves around the Earth. Write a statement equivalent to "The Earth revolves around the sun and the moon revolves around the Earth," using the commutative equivalence.

Solution

The original conjunction can be expressed symbolically as $e \wedge m$. Using the commutative equivalence we obtain $m \wedge e$. Therefore, an equivalent statement is "The moon revolves around the Earth and the Earth revolves around the sun."

■

Associative Equivalence

The *associative equivalences* are expressed symbolically as $[(p \wedge q) \wedge r] \Leftrightarrow [p \wedge (q \wedge r)]$ or $[(p \vee q) \vee r] \Leftrightarrow [p \vee (q \vee r)]$. Like the commutative equivalences, the associative equivalences pertain only to the *conjunction* and *disjunction* connectives. Notice that the order in which the statements appear is unchanged—only the order in which the operations are performed is altered.

You encountered associative relationships when dealing with addition and multiplication, since the way in which we group three numbers when adding or when multiplying has no effect on the resulting sum or product. For example, $(2+3)+4 = 2+(3+4)$ and $(2 \times 3) \times 4 = 2 \times (3 \times 4)$.

Example 4

Verify that $[(p \vee q) \vee r] \Leftrightarrow [p \vee (q \vee r)]$ is an equivalence.

Solution

We will construct a compact truth table for the disjunctive form of the associative equivalence. The numbers you see at the bottom of each column indicate the order in which these columns were filled in with truth values. Notice that column 5 is the resolution of the left hand side of the equivalence and column 7 is the resolution of the right hand side of the equivalence.

p	q	r	$[(p$	$\vee$	$q)$	$\vee$	$r]$		$[p$	$\vee$	$(q$	$\vee$	$r)]$
T	T	T		T		T				T		T	
T	T	F		T		T				T		T	
T	F	T		T		T				T		T	
T	F	F		T		T				T		F	
F	T	T		T		T				T		T	
F	T	F		T		T				T		T	
F	F	T		F		T				T		T	
F	F	F		F		F				F		F	
1	2	3		4		5				7		6	

Observing the entries in columns 5 and 7, we notice that they are identical. Therefore, $[(p \vee q) \vee r] \Leftrightarrow [p \vee (q \vee r)]$ is an equivalence. ■

Example 5

Write a statement equivalent to "In the summer, either I play golf or baseball, or I go sailing" using the associative equivalence.

Solution

The statement can be symbolized as $(g \vee b) \vee s$. Using the associative equivalence, this is equivalent to $g \vee (b \vee s)$. In words, we would say "In the summer, I play golf, or I either play baseball or go sailing." ■

Distributive Equivalence

The *distributive equivalences* involve two different connectives, and it has two forms. One form tells us that a conjunction distributes over a disjunction. This form can be expressed symbolically as $[p \wedge (q \vee r)] \Leftrightarrow [(p \wedge q) \vee (p \wedge r)]$

The other form of the distributive equivalence tells us that a disjunction distributes over a conjunction. It is expressed as $[p \vee (q \wedge r)] \Leftrightarrow [(p \vee q) \wedge (p \vee r)]$

Example 6

Show that $p \wedge (q \vee r)$ is logically equivalent $(p \wedge q) \vee (p \wedge r)$.

Solution

We construct a compact truth table. The order in which the columns were completed is again indicated by the numbers you see at the bottom of each column.

p	q	r	$[(p$	$\wedge$	$(q \vee r)]$		$[(p \wedge q)$	$\vee$	$(p \wedge r)]$
T	T	T		T	T		T	T	T
T	T	F		T	T		T	T	F
T	F	T		T	T		F	T	T
T	F	F		F	F		F	F	F
F	T	T		F	T		F	F	F
F	T	F		F	T		F	F	F
F	F	T		F	T		F	F	F
F	F	F		F	F		F	F	F
1	2	3		5	4		6	8	7

Observing the entries in columns 5 and 8, we notice that they always have the same truth value. Therefore, the statements $p \wedge (q \vee r)$ and $(p \wedge q) \vee (p \wedge r)$ are logically equivalent.

■

Example 7

Write a statement equivalent to "She will sleep late and eat either lunch or brunch."

Solution

The statement can be symbolized as $s \wedge (l \vee b)$. We may rewrite $s \wedge (l \vee b)$ as $(s \wedge l) \vee (s \wedge b)$ by using the distributive equivalence. In English, we would then say "She will sleep late and eat lunch, or she will sleep late and eat brunch."

■

Example 8

Write an equivalent statement for "They will practice and win the gold medal, or they will practice and win the silver medal."

Solution

The sentence can be symbolized as $(p \wedge g) \vee (p \wedge s)$. Using the distributive equivalence we can rewrite this as $p \wedge (g \vee s)$. In English, we would say, "They will practice and either win the gold medal or win the silver medal."

■

DeMorgan's Equivalence

DeMorgan's equivalences are expressed symbolically as $\sim (p \vee q) \Leftrightarrow (\sim p \wedge \sim q)$ or $\sim (p \wedge q) \Leftrightarrow (\sim p \vee \sim q)$. This means that the negation of a disjunction is a *conjunction* with each of its components negated, and that the negation of a conjunction is a *disjunction* with each of its components negated.

Example 9

Express the negation of "I will drive or I will take the bus" as a conjunction.

Solution

The disjunction can be symbolized as $d \vee b$ so its negation would be symbolized as $\sim (d \vee b)$. DeMorgan's equivalence allows us to rewrite this as $\sim d \wedge \sim b$. In English, we would say "I will not drive *and* I will not take the bus."

■

Example 10

Express the negation of "I do homework and I study" as a disjunction.

Solution

Using DeMorgan's equivalence, $\sim(h \wedge s)$ is equivalent to $\sim h \vee \sim s$. This is the English sentence "I don't do homework *or* I don't study." ■

Example 11

Write a sentence equivalent to "I don't drink and I don't smoke."

Solution

DeMorgan's equivalence allows us to write $\sim d \wedge \sim s$ as $\sim(d \vee s)$. The corresponding English sentence is "It is not true that I drink or smoke."

■

In-Class Exercises and Problems for Section 1.5

In-Class Exercises

For Exercises 1–10, complete each statement by applying the indicated equivalence.

1. $\sim(\sim p) \Leftrightarrow$	(Double Negation)
2. $(w \wedge \sim r) \Leftrightarrow$	(Commutative)
3. $[p \wedge (q \vee r)] \Leftrightarrow$	(Distributive)
4. $\sim[\sim(\sim p \rightarrow s)] \Leftrightarrow$	(Double Negation)
5. $[g \vee (s \vee \sim p)] \Leftrightarrow$	(Associative)
6. $[(\sim c \vee d) \wedge (\sim c \vee \sim b)] \Leftrightarrow$	(Distributive)
7. $[(s \rightarrow r) \wedge p] \Leftrightarrow$	(Commutative)
8. $[(\sim a \wedge b) \wedge \sim c] \Leftrightarrow$	(Associative)
9. $[\sim b \wedge (\sim p \vee q)] \Leftrightarrow$	(Distributive)
10. $\sim w \Leftrightarrow$	(Double Negation)

For Exercises 11–20, write a statement that is logically equivalent to the given statement using DeMorgan's Equivalence.

11. $\sim(p \vee \sim q)$
12. $\sim(w \wedge \sim s)$
13. $\sim(\sim p \vee \sim s)$
14. $b \wedge \sim r$
15. $\sim a \wedge \sim b$
16. $q \vee \sim s$
17. $\sim[m \vee (a \rightarrow b)]$
18. $\sim[(p \leftrightarrow q) \wedge \sim r]$
19. $[p \vee \sim(a \rightarrow c)]$
20. $\sim[\sim(w \wedge r) \wedge (p \rightarrow q)]$

For Exercises 21–25, write a statement equivalent to the given statement using the indicated equivalence.

21. It is not true that I will take piano lessons and not go to class. (DeMorgan's)
22. Going around the block and parallel parking, doesn't a driver make. (Commutative)
23. We can paint the town red or we can paint it blue and green. (Distributive)
24. The swan's long neck glistened in the noonday sun. (Double Negation)
25. I will have milk but I don't want any cookies. (DeMorgan's)

Problems

For Problems 26–37, state the name of the equivalence that is illustrated in each statement.

26. $(a \vee \sim b) \Leftrightarrow (\sim b \vee a)$
27. $[\sim m \vee (p \wedge \sim s)] \Leftrightarrow [(\sim m \vee p) \wedge (\sim m \vee \sim s)]$
28. $(p \rightarrow q) \Leftrightarrow \sim[\sim(p \rightarrow q)]$
29. $\sim(c \wedge \sim d) \Leftrightarrow (\sim c \vee d)$
30. $[(r \vee s) \vee p] \Leftrightarrow [(r \vee (s \vee p)]$
31. $[(r \wedge s) \vee p] \Leftrightarrow [(r \vee p) \wedge (s \vee p)]$
32. $[(p \rightarrow w) \vee r] \Leftrightarrow [r \vee (p \rightarrow w)]$
33. $\sim[s \vee (r \rightarrow p)] \Leftrightarrow [\sim s \wedge \sim(r \rightarrow p)]$
34. $(\sim m \vee n) \Leftrightarrow \sim(m \wedge \sim n)$
35. $\sim[\sim(\sim s \leftrightarrow w)] \Leftrightarrow (\sim s \leftrightarrow w)$
36. $\{[(a \rightarrow b) \wedge c] \wedge d\} \Leftrightarrow [(a \rightarrow b) \wedge (c \wedge d)]$
37. $[\sim(c \rightarrow d) \vee e] \Leftrightarrow \sim[(c \rightarrow d) \wedge \sim e]$

For Problems 38–43, match each numbered statement with its lettered equivalent.

38. $\sim(w \wedge r)$	a. $\sim(r \vee \sim w)$
39. $\sim(r \vee w)$	b. $(\sim r \wedge w) \vee (\sim r \wedge s)$

40. $\sim[\sim(r \wedge w)]$ c. $(\sim r \wedge w) \wedge s$

41. $\sim r \wedge (w \vee s)$ d. $\sim(r \wedge w)$

42. $\sim r \wedge (w \wedge s)$ e. $\sim r \wedge \sim w$

43. $\sim r \wedge w$ f. $r \wedge w$

For Problems 44–53, write a statement equivalent to the given statement using the indicated equivalence.

44. It's not true that I drink and drive. (DeMorgan's)

45. I will either buy you flowers or take you out for dinner. (Commutative)

46. The euro is the currency of the European Union. (Double Negation)

47. In the winter, either I go skiing or ice skating, or I go tobogganing. (Associative)

48. Before you enter the restaurant you must wear shoes and either a jacket or a tie. (Distributive)

49. We will not go out for dinner and we will not stay home and cook. (DeMorgan's)

50. It is not true that pearls are not found in clams. (Double Negation)

51. The raspberry is plump and the banana is ripe. (Commutative)

52. I will eat pasta and chicken or pasta and shrimp. (Distributive)

53. It is not true that doctors advise us not to eat right or not to exercise. (DeMorgan's)

For Problems 54–58, use a truth table to verify each equivalence.

54. $(p \vee q) \Leftrightarrow (q \vee p)$

55. $[(p \wedge q) \wedge r] \Leftrightarrow [p \wedge (q \wedge r)]$

56. $[p \vee (q \wedge r)] \Leftrightarrow [(p \vee q) \wedge (p \vee r)]$

57. $\sim(p \wedge q) \Leftrightarrow (\sim p \vee \sim q)$

58. $\sim(p \vee q) \Leftrightarrow (\sim p \wedge \sim q)$

1.6 Equivalences–Part II

Conditional Equivalence

The *conditional equivalence* is expressed symbolically as $(p \to q) \Leftrightarrow (\sim p \vee q)$, that is, $p \to q$ and $\sim p \vee q$ are logically equivalent. The conditional equivalence allows us to express a conditional statement as a disjunction to which it is logically equivalent. Notice that the left side of the disjunction is the negation of the LHS of the original conditional statement. Although the equivalence of $p \to q$ and $\sim p \vee q$ is not immediately obvious, the following truth table will convince us that they are.

p	q	$p \to q$		$\sim p \vee q$
T	T	T		T
T	F	F		F
F	T	T		T
F	F	T		T
1	2	3		4

Since columns 3 and 4 always have the same truth values, $p \to q$ and $\sim p \vee q$ are logically equivalent statements.

Example 1

Let p: We get pasta, and n: I want a napkin. Write a disjunctive statement equivalent to "If we get pasta then I want a napkin."

Solution

The statement "If we get pasta then I want a napkin" is expressed symbolically as $p \to n$. We may write a conditional statement as an equivalent disjunction provided we negate the LHS of the conditional statement. So, $p \to n$ is equivalent to $\sim p \vee n$, which can be translated into "We don't get pasta or I want a napkin."

■

Contrapositive Equivalence

The *contrapositive equivalence* is expressed symbolically as $(p \to q) \Leftrightarrow (\sim q \to \sim p)$. It allows us to express one conditional statement as another conditional statement. The contrapositive of a conditional statement is formed by interchanging its LHS with its RHS, and negating both sides of the new implication. Thus, the contrapositive of $p \to q$ is $\sim q \to \sim p$. It is critical to realize that one can only form the contrapositive of a *conditional* statement.

Example 2

Write a statement logically equivalent to "If I work overtime I will arrive to the party late," using the contrapositive equivalence.

Solution

The implication can be symbolized as $w \rightarrow l$. Its contrapositive is symbolized as $\sim l \rightarrow \sim w$. The English translation would be "If I don't arrive to the party late, then I didn't work overtime."

■

Example 3

Write the contrapositive of the statement "If I don't exercise, then I gain weight."

Solution

The implication can be symbolized as $\sim e \rightarrow w$. Its contrapositive is symbolized as $\sim w \rightarrow e$. The English translation would be "If I don't gain weight, then I exercised," which is logically equivalent to the original statement.

■

The Negation of a Conditional Statement

The *conditional negation equivalence* is expressed symbolically as $\sim(p \rightarrow q) \Leftrightarrow (p \wedge \sim q)$. In order to see why this is so, we will apply both the conditional equivalence and DeMorgan's equivalence. As we have learned, the conditional equivalence states that $p \rightarrow q$ is equivalent to $\sim p \vee q$. Negating $p \rightarrow q$ is therefore equivalent to negating $\sim p \vee q$. Using DeMorgan's equivalence, $\sim(\sim p \vee q)$ is logically equivalent to $p \wedge \sim q$. Summarizing our results,

$$\begin{aligned}(p \rightarrow q) &\Leftrightarrow (\sim p \vee q)\\ \sim(p \rightarrow q) &\Leftrightarrow \sim(\sim p \vee q)\\ \sim(p \rightarrow q) &\Leftrightarrow (p \wedge \sim q)\end{aligned}$$

Therefore, negation of an implication is the *conjunction* of its LHS with the negation of its RHS.

Example 4

Use a compact truth table to show that the negation of $p \rightarrow q$ is logically equivalent to $p \wedge \sim q$.

Solution

p	q	$[\sim$	$(p \to q)]$		$p \wedge \sim q$
T	T	F	T		F
T	F	T	F		T
F	T	F	T		F
F	F	F	T		F
1	2	4	3		5

Since columns 4 and 5 are identical, $\sim(p \to q)$ is equivalent to $p \wedge \sim q$.

■

Example 5

Find the negation of "If China vetoes the resolution there will not be an invasion."

Solution

The given statement may be symbolized as $c \to \sim i$. Since the negation of an implication is the *conjunction* of its LHS with the negation of its RHS we have $\sim(c \to \sim i) \Leftrightarrow (c \wedge i)$. In English, this translation would be "China vetoes the resolution and there will be an invasion."

■

In-Class Exercises and Problems for Section 1.6

In-Class Exercises

In Exercises 1–10, use the conditional equivalence to complete each statement.

1. $(c \to k) \Leftrightarrow$
2. $(p \vee \sim r) \Leftrightarrow$
3. $(\sim p \to \sim s) \Leftrightarrow$
4. $(\sim w \vee \sim z) \Leftrightarrow$
5. $(\sim b \to a) \Leftrightarrow$
6. $[\sim m \to (a \wedge b)] \Leftrightarrow$
7. $[p \vee (c \wedge d)] \Leftrightarrow$
8. $[\sim (p \wedge r) \to k] \Leftrightarrow$

9. $[\sim a \rightarrow (r \wedge s)] \Leftrightarrow$

10. $[\sim (w \wedge h) \vee (a \wedge \sim c)] \Leftrightarrow$

In Exercises 11–15, use the contrapositive equivalence to complete each statement.

11. $(a \rightarrow \sim h) \Leftrightarrow$

12. $(\sim c \rightarrow s) \Leftrightarrow$

13. $(\sim p \rightarrow \sim q) \Leftrightarrow$

14. $[p \rightarrow (w \wedge a)] \Leftrightarrow$

15. $[\sim (m \vee n) \rightarrow b] \Leftrightarrow$

In Exercises 16–21, use the conditional negation equivalence to complete each statement.

16. $\sim (m \rightarrow b) \Leftrightarrow$

17. $\sim (\sim s \rightarrow w) \Leftrightarrow$

18. $(a \wedge \sim d) \Leftrightarrow$

19. $(\sim k \wedge c) \Leftrightarrow$

20. $\sim [(r \vee s) \rightarrow w] \Leftrightarrow$

21. $[p \wedge (w \vee q)] \Leftrightarrow$

In Exercises 22–28, match each numbered statement to its lettered equivalent.

22. $d \rightarrow \sim k$	a. $\sim d \rightarrow \sim k$
23. $k \rightarrow d$	b. $d \wedge \sim k$
24. $\sim (d \rightarrow \sim k)$	c. $\sim d \rightarrow k$
25. $\sim k \rightarrow \sim d$	d. $\sim k \rightarrow d$
26. $\sim k \rightarrow d$	e. $d \wedge k$
27. $\sim (d \rightarrow k)$	f. $k \vee \sim d$
28. $\sim d \rightarrow k$	g. $\sim d \vee \sim k$

In Exercises 29–33, write a statement that is logically equivalent to the given statement using the indicated equivalence.

29. I do homework or I don't pass. (Conditional Equivalence)

30. If I win the lottery, I'll quit my job. (Contrapositive Equivalence)

31. You'll get a fortune cookie if you go to the Chinese restaurant. (Conditional Equivalence)
32. Either the bird gets up early or it doesn't catch the worm. (Conditional Equivalence)
33. I'll fit into my jeans if I go on a diet. (Contrapositive Equivalence)

Problems

In Problems 34–43, match each numbered statement with its lettered equivalent.

34. $\sim b \rightarrow \sim a$	a. $\sim a \rightarrow b$
35. $a \vee \sim b$	b. $a \wedge \sim b$
36. $\sim(a \rightarrow b)$	c. $a \wedge b$
37. $a \rightarrow \sim b$	d. $\sim a \rightarrow \sim b$
38. $\sim a \wedge \sim b$	e. $\sim b \rightarrow a$
39. $\sim a \vee \sim b$	f. $b \rightarrow a$
40. $\sim(a \rightarrow \sim b)$	g. $\sim a \vee \sim b$
41. $a \vee b$	h. $a \rightarrow b$
42. $\sim a \rightarrow \sim b$	i. $a \rightarrow \sim b$
43. $\sim a \rightarrow b$	j. $\sim(\sim a \rightarrow b)$

In Problems 44–53, state the equivalence illustrated in each statement.

44. $(\sim q \vee \sim r) \Leftrightarrow (q \rightarrow \sim r)$
45 $(p \rightarrow q) \Leftrightarrow (\sim q \rightarrow \sim p)$
46. $(\sim p \rightarrow q) \Leftrightarrow (\sim q \rightarrow p)$
47. $(p \vee s) \Leftrightarrow (\sim p \rightarrow s)$
48. $(\sim p \wedge \sim s) \Leftrightarrow \sim(\sim p \rightarrow s)$
49. $\sim(p \rightarrow \sim r) \Leftrightarrow (p \wedge r)$
50. $[(q \wedge \sim p) \rightarrow \sim s] \Leftrightarrow [\sim(q \wedge \sim p) \vee \sim s]$
51. $\sim[q \rightarrow (\sim p \vee s)] \Leftrightarrow [q \wedge \sim(\sim p \vee s)]$

52. $[q \rightarrow (\sim p \vee s)] \Leftrightarrow [\sim q \vee (\sim p \vee s)]$

53. $[q \rightarrow (\sim p \vee s)] \Leftrightarrow [\sim (\sim p \vee s) \rightarrow \sim q]$

In Problems 54–58, write a statement that is

a. logically equivalent to the given statement using the contrapositive equivalence.

b. logically equivalent to the given statement using the conditional equivalence.

c. the negation of the given statement, expressed as a conjunction.

54. If revenue exceeds cost I will make a profit.

55. If I buy a digital camera I will post your picture on the web.

56. We will go out for dinner if we don't go to the movies.

57. If you take medicine then you'll feel better.

58. A cord of wood will come in handy if it keeps snowing.

1.7 Converses, Inverses, and Negations of Biconditionals

Introduction

Conditional statements are used extensively in logic, flowcharts, and computer programming. They are also used as a basis for solving thought questions on the SAT, LSAT, MCAT, and Graduate Record Examination. In the previous section, we introduced and proved the contrapositive equivalence. In so doing, we showed that given the conditional $p \rightarrow q$, a new conditional, $\sim q \rightarrow \sim p$, called its contrapositive, is logically equivalent to $p \rightarrow q$.

Given the conditional statement $p \rightarrow q$, two other associated conditional statements may be formed. Neither of these, as we shall see, is logically equivalent to the original conditional $p \rightarrow q$.

The Converse

We define the *converse* of the conditional statement $p \rightarrow q$ as the conditional statement $q \rightarrow p$. In words, we would say that to form the converse of a conditional statement, we simply interchange its LHS with its RHS.

Example 1

Write the converse of $r \to \sim s$.

Solution

To write the converse, we interchange the LHS with the RHS of the conditional statement. Therefore, the converse of $r \to \sim s$ is $\sim s \to r$.

■

Example 2

Write an English sentence that is the converse of the statement "If he called me then he liked me."

Solution

We first symbolize the given sentence as $c \to l$. Now we interchange the LHS with the RHS to produce $l \to c$. In English, this translates into "If he liked me then he called me."

■

Example 3

Is a conditional statement equivalent to its converse?

Solution

To answer this question, we will use the short form of a truth table to determine if $p \to q$ and $q \to p$ are equivalent.

p	q	$p \to q$		$q \to p$
T	T	T		T
T	F	F		T
F	T	T		F
F	F	T		T
1	2	3		4

Since columns 3 and 4 are not the same for *all* values of p and q, $p \to q$ and $q \to p$ are not logically equivalent. That is, a conditional statement and its converse are *not* equivalent.

■

The Inverse

We define the *inverse* of the conditional statement $p \to q$ as the conditional statement $\sim p \to \sim q$. Notice that to form the inverse of a conditional statement, we simply negate both the LHS and the RHS of the given conditional statement.

Example 4

Write, in English, the inverse of "If I don't pay the toll I will get a summons."

Solution

The implication is symbolized as $\sim p \rightarrow s$. If we negate the LHS and RHS of this conditional, we obtain $\sim(\sim p) \rightarrow \sim s$. Using the double negation equivalence, this is equivalent to $p \rightarrow \sim s$. Therefore, the English sentence would be "If I pay the toll then I don't get a summons."

■

Example 5

Is a conditional statement equivalent to its inverse?

Solution

We will again use a truth table to see if $p \rightarrow q$ and $\sim p \rightarrow \sim q$ are logically equivalent.

p	q	$p \rightarrow q$		$\sim p \rightarrow \sim q$
T	T	T		T
T	F	F		T
F	T	T		F
F	F	T		T
1	2	3		4

Since $p \rightarrow q$ and $\sim p \rightarrow \sim q$ are not the same for all values of p and q, a conditional statement and its inverse are *not* equivalent.

■

Students often erroneously assume that the negation of a conditional $p \rightarrow q$ is logically equivalent to its inverse $\sim p \rightarrow \sim q$. While this may seem reasonable, a truth table will show that it is *not* true.

Example 6

Construct a truth table for a conditional statement, its negation and its inverse to determine if the negation of a conditional statement is logically equivalent to its inverse.

Solution

p	q	$(p \rightarrow q)$	$\sim(p \rightarrow q)$	$\sim p \rightarrow \sim q$
T	T	T	F	T
T	F	F	T	T
F	T	T	F	F
F	F	T	F	T
1	2	3	4	5

Notice that the statements $\sim(p \to q)$ and $\sim p \to \sim q$ do not always have the same truth value, so they are not equivalent. Therefore, the negation of a conditional statement *is not* logically equivalent to its inverse.

■

Example 7

In words, what is the negation of the sentence "If you use Microsoft Word you can format documents easily"?

Solution

First, we symbolize the sentence as $w \to d$. The negation of $w \to d$ is symbolized as $\sim(w \to d)$ and is equivalent to $w \wedge \sim d$. In English, this is "You use Microsoft Word *and* can't format documents easily."

■

The Negation of a Biconditional Statement

The *biconditional negation equivalence* can be expressed as $\sim(p \leftrightarrow q) \Leftrightarrow [(p \wedge \sim q) \vee (q \wedge \sim p)]$. Recall that in Section 1.1, we defined $p \leftrightarrow q$ as $(p \to q) \wedge (q \to p)$. Note that a biconditional statement is a conjunction of an implication and its converse. To negate the statement $p \leftrightarrow q$, we actually negate the conjunctive statement $(p \to q) \wedge (q \to p)$. We therefore apply DeMorgan's equivalence, followed by the negation of the conditional equivalence. Symbolically, we write

$$\begin{aligned} \sim(p \leftrightarrow q) &\Leftrightarrow \sim[(p \to q) \wedge (q \to p)] \\ &\Leftrightarrow \sim(p \to q) \vee \sim(q \to p) \quad \text{(DeMorgan's)} \\ &\Leftrightarrow (p \wedge \sim q) \vee (q \wedge \sim p) \quad \text{(Conditional Negation)} \end{aligned}$$

Example 8

Write, in English, the negation of "I will board the plane if and only if they screen the luggage."

Solution

We can symbolize the sentence to be negated as $b \leftrightarrow s$. Its negation would be $\sim(b \leftrightarrow s)$, which is equivalent to $\sim[(b \to s) \wedge (s \to b)]$. If we apply DeMorgan's equivalence, our result is symbolized as $\sim(b \to s) \vee \sim(s \to b)$. Now using the conditional negation, we obtain $(b \wedge \sim s) \vee (s \wedge \sim b)$. Translating back to English, we have "I board the plane and they do not screen the luggage, or they screen the luggage and I don't board the plane."

■

In-Class Exercises and Problems for Section 1.7

In-Class Exercises

For Exercises 1–10, write the associated converse or inverse as indicated.

1. $r \to \sim s$ (Converse)
2. $\sim q \to \sim p$ (Converse)
3. $\sim p \to q$ (Inverse)
4. $m \to \sim r$ (Inverse)
5. $(m \vee p) \to \sim s$ (Converse)
6. $d \to \sim (b \wedge p)$ (Inverse)
7. $\sim (p \leftrightarrow q) \to (c \vee d)$ (Converse)
8. $(p \vee \sim q) \to \sim (p \wedge a)$ (Converse)
9. $\sim (p \leftrightarrow q) \to (c \vee d)$ (Inverse)
10. $(r \wedge \sim s) \to \sim (w \vee c)$ (Inverse)

In Exercises 11–13, express the negation of each biconditional statement as the disjunction of two conjunctions.

11. $\sim (p \leftrightarrow \sim r)$
12. $\sim (\sim w \leftrightarrow s)$
13. $\sim (\sim m \leftrightarrow \sim n)$

In Exercises 14–18, write the inverse and the converse of each given conditional statement.

14. If you live in a glass house then you shouldn't throw stones.
15. You should bring a present if you go to the birthday party.
16. If you can't stand the heat, then you stay out of the kitchen.
17. You don't take your umbrella if it's not raining.
18. The stars can't be seen if the sun is shining.

In Exercises 19–23, use the biconditional negation equivalence to negate each sentence.

19. You should buy a new computer if and only if you have owned yours for more than two years.

20. I will buy you a dozen roses if and only if I don't get fired.
21. You won't fall off the horse if and only if you take riding lessons.
22. We play bridge if and only if we have four players.
23. You can see the stars if and only if the sun is not shining.

Problems

In Problems 24–33, symbolize each statement. Then express its inverse, and its converse, both symbolically and in English.

24. If you like warm weather you will like Puerto Rico.
25. Melissa cannot drive if she is too sleepy.
26. If Doris is angry she squints.
27. If Sam does not get time and a half he does not work overtime.
28. If it rained the streets are wet.
29. If you live alone your food bills are low.
30. Weeds grow quickly if they are unattended.
31. If you swim, then you like water.
32. If the movie receives a one star rating, it will not be a hit at the box office.
33. If you do not do well on the MCATs you will not get into medical school.
34. Express the contrapositive of Problem 24 as an English sentence.
35. Express the contrapositive of Problem 30 as an English sentence.
36. Express the contrapositive of Problem 33 as an English sentence.

For Problems 37–40, use the biconditional negation equivalence to negate each sentence.

37. The waiter will get a tip if and only if his service is adequate.
38. Sharon won't work if and only if she wins the lottery.
39. We will use the motor if and only if we don't pick up the oars.
40. The rain will turn to snow if and only if the temperature drops below the freezing point.

Frequently Used Equivalences

1. Double Negation Equivalence	$\sim(\sim p) \Leftrightarrow p$
2. Commutative Equivalences	$(p \wedge q) \Leftrightarrow (q \wedge p)$
And statements or statements	$(p \vee q) \Leftrightarrow (q \vee p)$
3. Associative Equivalences	$[(p \wedge q) \wedge r] \Leftrightarrow [p \wedge (q \wedge r)]$
	$[(p \vee q) \vee r] \Leftrightarrow [p \vee (q \vee r)]$
4. Distributive Equivalences	$[p \wedge (q \vee r)] \Leftrightarrow [(p \wedge q) \vee (p \wedge r)]$
	$[p \vee (q \wedge r)] \Leftrightarrow [(p \vee q) \wedge (p \vee r)]$
5. DeMorgan's Equivalences	$\sim(p \vee q) \Leftrightarrow (\sim p \wedge \sim q)$
	$\sim(p \wedge q) \Leftrightarrow (\sim p \vee \sim q)$
6. Conditional Equivalence	$(p \rightarrow q) \Leftrightarrow (\sim p \vee q)$
7. Biconditional Equivalence	$(p \leftrightarrow q) \Leftrightarrow [(p \rightarrow q) \wedge (q \rightarrow p)]$
8. Contrapositive Equivalence	$(p \rightarrow q) \Leftrightarrow (\sim q \rightarrow \sim p)$
9. Conditional Negation Equivalence	$\sim(p \rightarrow q) \Leftrightarrow (p \wedge \sim q)$
10. Biconditional Negation Equivalence	$\sim(p \leftrightarrow q) \Leftrightarrow [(p \wedge \sim q) \vee (q \wedge \sim p)]$

Chapter 1 Review

For questions 1–10, match each numbered statement with its lettered equivalent.

1. $\sim(\sim p \to \sim q)$	a. $\sim p \vee \sim q$
2. $\sim(\sim p \vee q)$	b. $q \to p$
3. $p \vee q$	c. $\sim p \to q$
4. $\sim(p \wedge q)$	d. $p \wedge \sim q$
5. $\sim p \vee q$	e. $(p \wedge \sim q) \vee (p \wedge r)$
6. $\sim p \to \sim q$	f. $\sim p \wedge q$
7. $p \vee (\sim q \wedge r)$	g. $p \vee (\sim q \vee r)$
8. $(p \vee \sim q) \vee r$	h. $\sim(\sim p \to q)$
9. $\sim p \wedge \sim q$	i. $p \to q$
10. $p \wedge (\sim q \vee r)$	j. $(p \vee \sim q) \wedge (p \vee r)$

For questions 11–23, name the equivalence illustrated.

11. $(p \vee q) \Leftrightarrow (q \vee p)$
12. $[p \wedge (q \wedge r)] \Leftrightarrow [(p \wedge q) \wedge r]$
13. $[p \vee (q \wedge r)] \Leftrightarrow [(p \vee q) \wedge (p \vee r)]$
14. $(\sim p \leftrightarrow q) \Leftrightarrow [(\sim p \to q) \wedge (q \to \sim p)]$
15. $\sim(p \to q) \Leftrightarrow (p \wedge \sim q)$
16. $\sim(\sim p \wedge q) \Leftrightarrow (p \vee \sim q)$
17. $(\sim p \to q) \Leftrightarrow (p \vee q)$
18. $(\sim p \vee \sim q) \Leftrightarrow \sim(p \wedge q)$
19. $(p \vee \sim q) \Leftrightarrow (\sim p \to \sim q)$
20. $[(p \to q) \vee (r \wedge s)] \Leftrightarrow \{[(p \to q) \vee r] \wedge [(p \to q) \vee s]\}$
21. $\sim[\sim(p \to q)] \Leftrightarrow (p \to q)$
22. $(\sim p \to q) \Leftrightarrow (\sim q \to p)$
23. $[(\sim p \to q) \vee r] \Leftrightarrow [r \vee (\sim p \to q)]$

For questions 24–43 express the negation of each given statement as a conjunction or as a disjunction if possible. Simplify each statement using equivalences where appropriate. Your answer should not involve any conditional or biconditional symbols.

24. I play field hockey but not ice hockey.
25. Latesha will smile if she is happy.
26. Margorie will play the drums if and only if Juan is ill.
27. I will take statistics or calculus.
28. We will play poker if and only if I don't have a business trip.
29. The computer is cheap, but the flat screen monitor is expensive.
30. If the answer to part three is true then the answer to part four is also true.
31. If I do a chemistry experiment and it smells like rotten eggs then I used sulfur dioxide.
32. A lawn will grow if and only if it is seeded and watered.
33. If I have the salad and main course I will not be able to eat dessert.
34. If it is a sentence it must have a subject and a verb.
35. Either you stay in school or you do not become an accountant.
36. Computers are fast and accurate but they can't think.
37. In her psychology class she will study Skinner or Freud.
38. The small brown house on top of the windswept hill is where the old Polish man lived for the last thirty years.
39. I'm going to stay in New York or if I leave I will relocate to Los Angeles.
40. If I stay in school I will study chemistry and work in the lab.
41. Native Americans hunted buffalo for their skins and their meat or they were cold and hungry.
42. Working after school is not fun, but I make money.
43. If a flag is waved and a parade is in progress then there is sure to be a band playing.

For questions 44–53, write a statement that is logically equivalent to the given statement using the indicated equivalence.

44. She wants to be a doctor or a lawyer.
(Conditional Equivalence)

45. It is not true that they will visit and stay for two weeks.
(DeMorgan's Equivalence)

46. If she has a manicure then she will have a pedicure.
(Contrapositive Equivalence)

47. The dog will not bite if you feed her.
(Conditional Equivalence)

48. Money is both a necessity and an evil.
(Commutative Equivalence)

49. Calculators have made statistics easier but you have to do the homework.
(DeMorgan's Equivalence)

50. If a number is even and prime then the number is two.
(Contrapositive Equivalence)

51. It is not true that the enemy will be defeated if they are not supplied with missiles.
(Conditional Negation Equivalence)

51. You use the Internet, and either your research skills will be enhanced or you will learn a great deal.
(Distributive Equivalence)

53. I will go to the play if and only if the tickets are not expensive.
(Biconditional Equivalence)

For questions 54–58, express the sentence "I will wear my hat and gloves if it is cold outside" symbolically using the underlined letters. First symbolically, and then as an English sentence, write its:

54. Contrapositive

55. Conditional Equivalence

56. Negation

57. Inverse

58. Converse

Sample Exam: Chapter 1

1. A statement logically equivalent to "It is not true that I like apples and bananas" is
   a. I don't like apples and I don't like bananas.
   b. If I like apples then I like bananas.
   c. I like apples or I like bananas.
   d. If I don't like apples then I don't like bananas.
   e. None of these.

2. The contrapositive of "I will buy you flowers or take you out for dinner if I get a raise" is
   a. If I bought you flowers or took you out for dinner then I got a raise.
   b. If I didn't buy you flowers and I didn't take you out for dinner then I didn't get a raise.
   c. If I didn't get a raise then I didn't buy you flowers and I didn't take you out for dinner.
   d. I didn't get a raise or I will buy you flowers or take you out for dinner.
   e. None of these.

3. The negation of $\sim(p \vee q) \rightarrow r$ is
   a. $(p \vee q) \rightarrow \sim r$ b. $(p \vee q) \vee \sim r$
   c. $(\sim p \wedge \sim q) \wedge \sim r$ d. $\sim r \rightarrow (p \vee q)$ e. None of these

4. A statement logically equivalent to $\sim[\sim a \vee (b \rightarrow c)]$ is
   a. $a \vee \sim(b \rightarrow c)$ b. $a \wedge (b \wedge \sim c)$
   c. $a \wedge \sim(b \vee \sim c)$ d. $a \rightarrow \sim(b \rightarrow c)$ e. None of these

5. If $(p \wedge q) \leftrightarrow (\sim r \vee s)$ is true and r is false then the truth value of p is
   a. True b. False c. Indeterminable

6. Let $p \# q$ be defined as $\sim p \rightarrow q$. Then $\sim(p \# q)$ is
   a. $p \rightarrow q$ b. $p \rightarrow \sim q$
   c. $p \wedge \sim q$ d. $\sim p \wedge \sim q$ e. None of these

7. Is $[(p \vee q) \rightarrow r] \leftrightarrow [(\sim p \wedge \sim q) \rightarrow \sim r]$ a tautology?
   a. Yes b. No c. Cannot be determined

8. Given that $[(\sim p \rightarrow q) \wedge \sim p] \rightarrow q$ is a tautology, if p is false, then q is
   a. True b. False c. Indeterminable

9. If the converse of a statement is "If you like tennis then you will go to the U.S. Open," then the inverse of the original statement is
   a. If you do not go to the U.S. Open, then you do not like tennis.
   b. If you don't like tennis, then you will not go to the U.S. Open.
   c. You go to the U.S. Open if you like tennis.
   d. If you go to the U.S. Open, then you like tennis.
   e. None of these.

10. A statement equivalent to "If you go to culinary school, then you like to cook" is
    a. If you like to cook then you go to culinary school.
    b. Either you do not go to culinary school or you like to cook.
    c. If you do not go to culinary school then you do not like to cook.
    d. If you do not go to culinary school then you like to cook.
    e. None of these.

11. Is $p \rightarrow (r \wedge s)$ equivalent to $(\sim r \vee \sim s) \rightarrow \sim p$?
    a. Yes   b. No   c. Cannot be determined

12. The negation of "We will win if we try" is
    a. We will not win if we don't try.
    b. If we don't win we did not try.
    c. We win and we don't try.
    d. We try but we don't win.
    e. None of these

13. A symbolization of "The grass will look beautiful if it is watered and mowed" is
    a. $g \rightarrow (w \wedge m)$ b. $(g \rightarrow w) \wedge m$
    c. $(w \wedge m) \rightarrow g$ d. $w \wedge (m \rightarrow g)$ e. None of these

14. The statement $\sim [k \rightarrow (g \vee h)]$ is equivalent to
    a. $\sim k \rightarrow \sim (g \vee h)$ b. $\sim k \vee (g \vee h)$
    c. $k \wedge (g \vee h)$ d. $\sim (g \vee h) \rightarrow \sim k$ e. None of these

For questions 15–20, assume that *a* and *b* are true and *c* and *d* are false. Determine the truth value of each statement.

15. $(a \wedge b) \rightarrow \sim c$
    a. True   b. False   c. Indeterminable

16. $\sim c \leftrightarrow (a \wedge d)$
    a. True   b. False   c. Indeterminable

17. $\sim[c \rightarrow \sim(\sim a \vee \sim s)]$

a. True b. False c. Indeterminable

18. $\sim(a \vee b) \rightarrow s$

a. True b. False c. Indeterminable

19. $s \rightarrow [(a \vee c) \wedge \sim d]$

a. True b. False c. Indeterminable

20. $r \rightarrow [(d \rightarrow w) \leftrightarrow (a \wedge c)]$

a. True b. False c. Indeterminable

CHAPTER TWO

ARGUMENTS

Arguments form the foundation for all reasoning in mathematics and indeed for all the sciences. In this chapter, we will examine the concept of valid and invalid arguments and the process by which we test them for validity.

2.1 Testing for Validity Using Truth Tables

Introduction

An *argument* is a collection of *TF* statements, called *premises*, which are always assumed to be true, followed by a final statement, called the *conclusion* of the argument. The conclusion is often preceded by the word "therefore."

To show that an argument is *valid*, we must prove that, *based on the truth of its premises*, its conclusion *must* be true. It is crucial to understand that for an argument to be valid it is not sufficient to find one, two, or even several cases for which true premises lead to a true conclusion. For the argument to be valid, the conclusion must *always* be true if the premises are true. If true premises lead to a conclusion that is false even in one instance, the argument is *invalid*.

Consider the argument below that consists of two premises, followed by a conclusion.

We will go to the game or study.
We will not study.
Therefore, we will go to the game.

In order to determine if an argument is valid or invalid, it is usually helpful to express the argument symbolically. When we do so, it is customary to list all the premises above a horizontal line, and to place the conclusion below the line. This format is shown below.

$$\begin{array}{c} g \vee s \\ \sim s \\ \hline g \end{array}$$

Constructing Truth Tables for Validity

Example 1

Determine whether the argument below is valid or invalid.

We will go to the game or study.
We will not study.
Therefore, we will go to the game.

Solution

We write the argument symbolically.

$$\begin{array}{c} g \vee s \\ \sim s \\ \hline g \end{array}$$

We next construct a truth table whose headings are the simple statements involved, the premises, and the conclusion, as shown below.

		Premise 1	Premise 2	Conclusion
g	s	$g \vee s$	$\sim s$	g
T	T	T	F	T
T	F	T	T	T
F	T	T	F	F
F	F	F	T	F

To test the validity of the argument, we need only concern ourselves with those cases for which all the premises are true. In the table above, this occurs only in row two. This row is boxed in the table below.

		Premise 1	Premise 2	Conclusion
g	s	$g \vee s$	$\sim s$	g
T	T	T	F	T
T	F	T	T	T
F	T	T	F	F
F	F	F	T	F

In this row, the conclusion is also true. Therefore, the argument is valid, since whenever all the premises are true, the conclusion is true. ■

Example 2

Determine whether the argument below is valid or invalid.

If I lift weights I will become stronger.
I became stronger.
Therefore, I lifted weights.

Solution

We first symbolize the argument.

$$\begin{array}{l} w \rightarrow s \\ \underline{s \qquad} \\ w \end{array}$$

Now we construct a truth table whose headings are the simple statements, premises, and conclusion.

		Premise 1	Premise 2	Conclusion
s	w	$w \to s$	s	w
T	T	T	T	T
T	F	T	T	F
F	T	F	F	T
F	F	T	F	F

We are only concerned with those cases in which all the premises are true. In the table above, this occurs in the first and second rows. However, in this example the conclusion is not always true when both premises are true, as can be seen in row two. *If there is any case in which true premises lead to a false conclusion, the argument is invalid.* Therefore, this argument is invalid.

		Premise 1	Premise 2	Conclusion
s	w	$w \to s$	s	w
T	T	T	T	T
T	F	T	T	F
F	T	F	F	T
F	F	T	F	F

■

Example 3

Determine whether the following argument is valid or invalid.

If I don't go on a date then I will watch the game or take a nap.
I don't go on a date or I wash my car.
I don't wash my car.
Therefore, I take a nap.

Solution

The argument is symbolized below.

$$\sim d \to (g \vee n)$$
$$\sim d \vee w$$
$$\underline{\sim w}$$
$$n$$

Construct a truth table whose headings are the simple statements, premises, and the conclusion. Since four variables are involved, this truth table will have $2^4 = 16$ rows. The premises are all true only in rows 10, 12, and 14. The conclusion is not always true in these three cases. Therefore, the argument is invalid.

				Premise 1	Premise 2	Premise 3	Conclusion
d	g	n	w	$\sim d \to (g \vee n)$	$\sim d \vee w$	$\sim w$	n
T	T	T	T	T	T	F	T
T	T	T	F	T	F	T	T
T	T	F	T	T	T	F	F
T	T	F	F	T	F	T	F
T	F	T	T	T	T	F	T
T	F	T	F	T	F	T	T
T	F	F	T	T	T	F	F
T	F	F	F	T	F	T	F
F	T	T	T	T	T	F	T
F	T	T	F	T	T	T	T
F	T	F	T	T	T	F	F
F	T	F	F	T	T	T	F
F	F	T	T	T	T	F	T
F	F	T	F	T	T	T	T
F	F	F	T	F	T	F	F
F	F	F	F	F	T	T	F

■

Example 4

Is the following argument valid or invalid?

$$\begin{array}{l} p \to q \\ \sim q \vee r \\ \underline{\sim r} \\ \sim p \end{array}$$

Solution

Since the argument involves three variables, our truth table will have $2^3 = 8$ rows.

			Premise 1	Premise 2	Premise 3	Conclusion
p	q	r	$p \to q$	$\sim q \vee r$	$\sim r$	$\sim p$
T	T	T	T	T	F	F
T	T	F	T	F	T	F
T	F	T	F	T	F	F
T	F	F	F	T	T	F
F	T	T	T	T	F	T
F	T	F	T	F	T	T
F	F	T	T	T	F	T
F	F	F	T	T	T	T

There is only one case in where all the premises are true. This occurs in row eight. Since the conclusion is also true in this case, the argument is valid.

■

In-Class Exercises and Problems for Section 2.1

In-Class Exercises

For Exercises 1–10, use a truth table to determine whether the argument is valid or invalid.

1. $w \to x$
$\sim w$
$\overline{}$
$\sim x$

2. $c \to d$
$e \to d$
$\overline{}$
$c \to e$

3. $m \to p$
$\sim m \vee \sim p$
$\overline{}$
p

4. $\sim (r \wedge s)$
s
$\overline{}$
$\sim r$

5. $\sim q \to \sim p$

$\sim (q \to r)$

$p \vee r$

6. $\sim s$

$s \vee q$

$p \leftrightarrow q$

$\sim p$

7. $p \to r$

$r \vee q$

$q \wedge p$

---

r

8. $\sim (m \wedge n)$

$\sim m \to \sim c$

c

$\sim n$

9. $p \to q$

$q \leftrightarrow r$

$\sim r$

$\sim p$

10. $\sim m \vee s$

m

$(s \vee q) \to w$

$\sim w$

Problems

For Problems 11–20, construct a truth table to determine whether the argument is valid or invalid.

11. $p \to q$

$\sim p \vee \sim q$

$\sim p$

12. $p \to q$

$q \to r$

$p \to r$

13. $\sim (\sim p \vee w)$

$\sim s \to w$

$p \to s$

14. $\sim (\sim p \vee w)$

$\sim s \to w$

$p \to s$

15. $\sim (\sim p \vee w)$

$\sim s \to w$

$p \to s$

16. $p \to q$

$q \to r$

p

---

r

17.
$$\begin{array}{l} p \rightarrow q \\ q \rightarrow p \\ \hline p \leftrightarrow q \end{array}$$

18.
$$\begin{array}{l} \sim(p \vee \sim r) \\ r \leftrightarrow w \\ \hline \sim w \end{array}$$

19.
$$\begin{array}{l} \sim a \wedge b \\ b \rightarrow \sim c \\ c \vee \sim a \\ \hline c \end{array}$$

20.
$$\begin{array}{l} \sim(a \rightarrow \sim b) \\ \sim a \vee c \\ c \rightarrow \sim d \\ \hline d \rightarrow a \end{array}$$

2.2 Counterexamples for Invalid Arguments

Introduction

We know that if an argument is valid, its conclusion must be true based on the truth of the premises. If an argument is invalid, there must be at least one case in which the conclusion is false, even though the premises are true. A *counterexample* is an illustration that shows that the argument is not valid. To produce a counterexample, we must state the truth values of the variables such that all the premises are true, yet the conclusion is false.

Example 1

Is the given argument valid or invalid? If it is invalid, produce a counterexample.

If I lift weights I will become stronger.
I became stronger.
Therefore, I lifted weights.

Solution

You may recognize this invalid argument from Example 2 of the previous section. The symbolized argument is expressed as:

$$\begin{array}{l} w \rightarrow s \\ s \\ \hline w \end{array}$$

Reproducing the truth table associated with this argument, we notice that when s is true and w is false, the premises are both true but the conclusion is false.

		Premise 1	Premise 2	Conclusion
s	w	$w \rightarrow s$	s	w
T	T	T	T	T
T	F	T	T	F
F	T	F	F	T
F	F	T	F	F

■

Counterexample Charts

Row two in the truth table in Example 1 provides us with our counterexample, that is, a set of truth values for which all the premises are true, but the conclusion is false. We display this result in the following counterexample chart.

s	w
T	F

Example 2

Consider the following argument. Is the argument valid or invalid? If it is invalid, produce a counterexample chart.

If I lift weights I will become stronger.
I didn't become stronger.
Therefore, I didn't lift weights.

Solution

The argument is expressed symbolically as

$$\begin{array}{l} w \rightarrow s \\ \sim s \\ \hline \sim w \end{array}$$

Constructing a truth table whose headings are the simple statements, premises, and conclusion, we obtain:

		Premise 1	Premise 2	Conclusion
s	w	$\sim s$	$w \rightarrow s$	$\sim w$
T	T	F	T	F
T	F	F	T	T
F	T	T	F	F
F	F	T	T	T

No counterexample can be produced since the conclusion is always forced to be true whenever the premises are true. This is exhibited in the fourth row of the truth table. Hence, the argument is valid. ■

Example 3

Is the following argument valid or invalid? If it is invalid, produce a counterexample chart.

$$\begin{array}{l} w \to j \\ j \vee b \\ \sim b \\ \hline w \end{array}$$

Solution

We construct the truth table whose headings are the simple statements, premises, and conclusion.

			Premise 1	Premise 2	Premise 3	Conclusion
b	j	w	$w \to j$	$j \vee b$	$\sim b$	w
T	T	T	T	T	F	T
T	T	F	T	T	F	F
T	F	T	F	T	F	T
T	F	F	T	T	F	F
F	T	T	T	T	T	T
F	T	F	T	T	T	F
F	F	T	F	F	T	T
F	F	F	T	F	T	F

Noting the entries in the sixth row, we observe that when b is false, j is true, and w is false, the argument has true premises with a false conclusion and is therefore invalid. The truth values of the variables that provide us with our counterexample are displayed in the counterexample chart below.

b	j	w
F	T	F

■

Example 4

Is the following argument valid or invalid? If it is invalid, produce a counterexample chart.

$$\begin{array}{l} p \to q \\ (\sim p \vee q) \to r \\ \hline p \end{array}$$

Solution

First, the truth table is constructed.

			Premise 1	Premise 2	Conclusion
p	q	r	$p \to q$	$(\sim p \vee q) \to r$	p
T	T	T	T	T	T
T	T	F	T	F	T
T	F	T	F	T	T
T	F	F	F	T	T
F	T	T	T	T	F
F	T	F	T	F	F
F	F	T	T	T	F
F	F	F	T	F	F

Observing the entries in rows five and seven, we note that there are two instances where true premises produce a false conclusion. Therefore, we have two equally correct counterexamples.

p	q	r
F	T	T
F	F	T

Either of the two counterexamples is sufficient to show that the argument is invalid.

■

In-Class Exercises and Problems for Section 2.2

In-Class Exercises

For Exercises 1–10, use a truth table to determine whether the argument is valid or invalid. If the argument is invalid produce a counterexample chart.

1. $\sim(c \to g)$
   $\underline{\sim g}$
   c

2. $p \to q$
   $\underline{q \vee \sim r}$
   $\sim r$

3. $r \to (w \vee \sim s)$
   $\underline{\sim(w \vee \sim s)}$
   r

4. $(p \wedge r) \vee s$
   $\underline{\sim s}$
   $\sim r$

5. $(b \to k) \wedge r$
   $\underline{\sim r \vee b}$
   k

6. $(q \vee r) \leftrightarrow \sim p$
   $\underline{\sim p \wedge r}$
   q

7. $\sim a$
   $\sim a \to b$
   $\underline{c \to b}$
   c

8. $\sim b \to d$
   $\sim b \vee c$
   $\underline{c \to d}$
   d

9. $b \wedge \sim a$
   $a \vee (b \to c)$
   $\underline{c \vee d}$
   d

10. $p \vee (q \wedge s)$
    r
    $\underline{r \to \sim p}$
    s

Problems

For Problems 11–26, construct a truth table to determine whether the argument is valid or invalid. If the argument is invalid produce a counterexample chart.

11. $\sim r \to s$
    $\underline{s}$
    r

12. $\sim(w \to r)$
    $\underline{w \leftrightarrow s}$
    s

13. $\sim(c \wedge \sim k)$
$k \rightarrow p$
$\overline{\quad p \quad}$

14. $(p \rightarrow q) \rightarrow r$
$\sim q \rightarrow \sim p$
$\overline{\quad r \quad}$

15. $\sim a \rightarrow \sim b$
$\sim b \rightarrow c$
$\overline{\sim c \rightarrow a}$

16. $\sim(n \leftrightarrow p) \vee c$
$\sim c$
$\overline{\sim p \rightarrow \sim n}$

17. $\sim(n \leftrightarrow p) \vee c$
$\sim c$
$\overline{\sim p \rightarrow \sim n}$

18. $(q \rightarrow p) \wedge c$
$\sim p$
$\overline{\quad q \quad}$

19. $\sim a$
$\sim m \vee a$
$\sim m \rightarrow p$
$\overline{p \vee a}$

20. $p \wedge r$
$\sim p \rightarrow q$
$\overline{\quad q \quad}$

21. r
$\overline{r \vee (s \rightarrow g)}$

22. $\sim(p \vee s)$
$\sim p \rightarrow q$
$\overline{\quad q \quad}$

23. $p \rightarrow q$
$\sim s \vee q$
p
$\overline{\quad s \quad}$

24. $p \rightarrow q$
$q \rightarrow r$
r
$\overline{\quad p \quad}$

25. $p \rightarrow q$
$s \rightarrow r$
$p \vee s$
$\overline{q \vee r}$

26. $\sim p \vee q$
$q \rightarrow r$
$r \rightarrow s$
$\overline{p \rightarrow s}$

2.3 Testing for Validity Using the *TF* Method–The Direct Approach

You have seen that as the number of variables involved in an argument increases, the truth table used for testing validity becomes unwieldy. Another technique for testing the validity of an argument, called the *TF* method, makes the process easier. With this method, we assume all the premises are true and try to arrive at the truth value of the conclusion.

Example 1

Determine whether the following argument is valid or invalid.

We will go to the game or study.
We will not study.
Therefore, we will go to the game.

Solution

By assumption, each premise must be true. This is illustrated below.

$$\begin{array}{ll} g \vee s & T \\ \underline{\sim s} & T \\ g & \end{array}$$

Notice that since the first premise is a disjunction and is assumed true, we can be sure that g and s are not both false. Unfortunately, we cannot be sure whether only g is true, only s is true, or they are both true. Examining our second premise however, we note that there is only one truth value of s such that $\sim s$ is true. In order for $\sim s$ to be true, s must be false. Now we consider the first premise again. Since we know that s is false, our assumption that $g \vee s$ is true requires that g must be true. Note that the conclusion of the argument is g. Hence, the conclusion is forced to be true when all the premises are true. Thus, the argument is valid.

■

In general, if a disjunction is true, and the negation of one side of this disjunction is also true, the other side of the disjunction must also be true. This kind of pattern recognition will be expanded upon in Chapter Three.

Example 2

Determine whether the following argument is valid or invalid.

If I traveled at a speed of over 70 miles per hour I received a summons.
I received a summons.
Therefore, I traveled at a speed of over 70 miles per hour.

Solution

First we represent the argument symbolically.

$$\begin{array}{l} s \rightarrow r \\ \underline{r \qquad} \\ \quad s \end{array}$$

By assumption, each premise must be true.

$$\begin{array}{ll} s \rightarrow r & T \\ \underline{r \qquad} & T \\ \quad s & \end{array}$$

The first premise is an implication. There are three different sets of truth values for r and s that make this conditional statement true. Thus we cannot, at this point, determine truth values for s and r. We know only that s cannot be true while r is false. The assumption that our second premise is true however, requires that r be true. But if r is true, the implication $s \rightarrow r$ is true *regardless* of the truth value of s. Therefore, the conclusion, which is also s, is not forced to be true. Consequently, this argument is invalid. ■

The fact that the argument in Example 2 is invalid implies that based on the truth of the premises the conclusion is not forced to be true. Notice that when r is true and s is false, each premise is true, yet the conclusion is not true. The counterexample chart is shown below.

r	s
T	F

Example 3

Determine if the argument is valid or invalid.

If six plus six is twelve then three times three is nine.
Three times three is nine.
Therefore, six plus six is twelve.

Solution

Using the letters s and r we symbolize the argument.

$$\begin{array}{l} s \rightarrow r \\ \underline{r \qquad} \\ \quad s \end{array}$$

Notice that the argument has exactly the same form as the argument in Example 2, so the argument must be invalid. Recall that for an argument to be valid, we must be able to show that, ***based on the truth of its premises,*** the conclusion must be true. We do not dispute the fact that six plus six is twelve, but rather that one cannot conclude this fact from the given premises.

■

Example 4

Determine if the argument is valid or invalid.

If six plus six is twelve then three times three is nine.
Six plus six is twelve.
Therefore, three times three is nine.

Solution

We first represent the argument symbolically.

$$\begin{array}{ll} s \rightarrow r & T \\ \underline{s \quad\quad} & T \\ \;\; r & \end{array}$$

The first premise is an implication. Again, we know that s cannot be true while r is false. The assumption that our second premise is true however, requires that s must be true. Thus, r is forced to be true in order for $s \rightarrow r$ to be true. Since r is the conclusion, and based on our reasoning, it was forced to be true, this argument is valid. In this case, we are able to show, based on the truth of the premises only, that three times three is nine.

■

In general, arguments may have many premises. The next example will show us how to deal with an argument that has three premises. This same procedure can be used with arguments involving any number of premises.

Example 5

Determine whether the argument below is valid or invalid.

$$\begin{array}{l} \sim p \vee q \\ r \rightarrow p \\ \underline{\sim q \quad\quad} \\ \;\sim r \end{array}$$

Solution

We begin our reasoning by noting that the third premise, $\sim q$, must be true. Note that the first premise involves q. Since $\sim p \vee q$ is assumed true and $\sim q$ is true, $\sim p$ must be true and thus p is false. Since $r \to p$ is a premise, it too must be true. But since p is false, r must be false. Hence, the conclusion, $\sim r$, is forced to be true. The argument is valid.

■

Example 6

Is the following argument valid or invalid? If it is invalid, produce a counterexample chart.

If my salary increases by five percent I will go to Rome.
Either I go to Rome or I buy a computer.
I didn't buy a computer.
Therefore, my salary didn't increase by five percent.

Solution

First, we represent the argument symbolically.

$$\begin{array}{l} s \to r \\ r \vee c \\ \underline{\sim c} \\ \sim s \end{array}$$

The premises $\sim c$ and $r \vee c$ are both true, forcing r to be true. Since $s \to r$ is true, and r is true, s can either be true or false; the implication is true regardless of the truth value of s. The conclusion, $\sim s$, is not forced to be true. Therefore, the argument is invalid.

A counterexample will provide us with truth values for c, r, and s such that every premise is true yet the conclusion is false. We already know that c must be false and r must be true in order for the last two premises to be true. If s is true, the conclusion of the argument will be false, and the first premise will still be true. The counterexample chart

c	r	s
F	T	T

provides us with a set of truth values for the variables such that every premise will be true, yet the conclusion will be false.

■

Example 7

Is the following argument valid or invalid? If it is invalid, produce a counterexample chart.

If I don't go on a date then I will watch the game or take a nap.
I don't go on a date or I wash my car.
I don't wash my car.
Therefore, I take a nap.

Solution

First, symbolize the argument.

$$\begin{array}{l} \sim d \rightarrow (g \vee n) \\ \sim d \vee w \\ \underline{\sim w \qquad\qquad} \\ \qquad n \end{array}$$

Both $\sim w$ and $\sim d \vee w$ are true, so $\sim d$ must be true. The first premise must be true, so $g \vee n$ must be true. However, there are three ways for a disjunction to be true. The statement $g \vee n$ is true when g and n are both true, when g is true and n is false and when g is false and n is true. Thus, the conclusion, n, is not forced be true. The argument is invalid. The counterexample chart for this argument is shown below.

d	g	n	w
F	T	F	F

■

In-Class Exercises and Problems for Section 2.3

In-Class Exercises

For Exercises 1–6, answer true, false or can't be determined because of insufficient information.

1. If $m \wedge \sim n$ is a premise, what is the truth value of $\sim m \rightarrow s$?
2. If $\sim m \vee n$ is a premise, what is the truth value of $n \rightarrow p$?
3. If $\sim r \vee w$ and r are both premises, what is the truth value of $r \leftrightarrow w$?
4. If $\sim (a \vee \sim b)$ is a premise, what is the truth value of $\sim a \leftrightarrow \sim b$?
5. If $\sim (d \rightarrow \sim c)$ is a premise, what is the truth value of $d \vee p$?
6. If $w \wedge \sim s$ is a premise, what is the truth value of $\sim w \vee \sim s$?

For Exercises 7–11, determine whether each statement is true or false.

7. If $a \rightarrow \sim b$ is true, then b must be true.

8. If $\sim a \wedge b$ is true, then b must be false.

9. If $\sim m \vee p$ is true, then p must be true.

10. If $\sim (q \vee p)$ is true, then p and q must be false.

11. If $\sim (w \rightarrow \sim s)$ is true, then s must be true.

For Exercises 12–21, use the *TF* method to determine if each argument is valid or invalid. If it is invalid, produce a counterexample chart.

12. $(p \wedge q) \vee r$
$\sim r$
___
p

13. $(s \vee d) \rightarrow \sim k$
k
___
$\sim s$

14. $a \rightarrow (b \vee c)$
$\sim a \wedge c$
___
b

15. $(m \leftrightarrow n) \rightarrow r$
$\sim r \wedge m$
___
$n \vee c$

16. $(\sim r \vee s) \rightarrow \sim q$
$\sim s \vee p$
q
___
p

17. $w \vee \sim b$
$b \wedge z$
$(\sim w \vee z) \rightarrow a$
___
$a \vee p$

18. $(p \wedge s) \vee e$
$\sim e$
$p \rightarrow w$
___
$w \rightarrow e$

19. $(c \wedge g) \rightarrow \sim m$
$m \wedge g$
$a \rightarrow c$
___
$\sim a$

20. $(m \rightarrow n) \leftrightarrow q$
$\sim q$
$(\sim n \vee w) \rightarrow s$
___
$a \rightarrow s$

21. $\sim (p \vee \sim s) \rightarrow r$
$\sim r \wedge s$
$p \rightarrow w$
$c \rightarrow w$
___
c

For Exercises 22–26, express each argument symbolically. Then use the *TF* method to determine if it is valid or invalid. If it is invalid, produce a counterexample chart.

22. The streets are not wet. If it rains the streets get wet. Therefore, it did not rain.
23. If the Rangers win I will go to the playoffs. Either the Rangers win or the Islanders win. The Islanders do not win. Therefore, I will not go to the playoffs.
24. If I quit my job I cannot pay for insurance. I quit my job or I get a scholarship. I didn't get a scholarship. Therefore, I cannot pay for insurance.
25. We will play basketball if and only if we have five players. We don't have five players but Michael Jordan is a great player. Therefore, if Michael Jordan is a great player, we will play basketball.
26. He is a man or a mouse. If he is a mouse he squeaks and eats cheese. He does not squeak. Therefore, he is a man.

Problems

For Problems 27–36, answer either true, false, or can't be determined because of insufficient information.

27. If $\sim p$ is true and $\sim p \rightarrow q$ is a premise, then q must be _______.
28. If $r \rightarrow s$ is a premise and s is false, then r must be _______.
29. If $p \leftrightarrow q$ and $\sim q$ are both premises, p must be _______.
30. If $p \rightarrow q$ is false, p is _______ and q is _______.
31. If $\sim (p \wedge q)$ is a premise, then p must be _______.
32. Suppose $\sim (p \leftrightarrow \sim q)$ and q are premises of an argument. Then $\sim p$ must be _______.
33. If $\sim p \wedge q$ is a premise, then $\sim p$ is _______.
34. If $s \wedge r$ is a premise, then $\sim (s \rightarrow r)$ is _______.
35. Let $p \wedge q$ be a premise. Then $r \rightarrow q$ is _______.
36. Suppose $\sim p \vee \sim q$ is a premise and p is a premise as well. Then q is _______.

For Problems 37–46, answer yes or no.

37. If $p \to q$ is true, must p be true? _______.

38. If $p \wedge q$ is true, must p be true? _______.

39. If q is true, must $p \to q$ be true? _______.

40. If p is true, must $p \vee q$ be true? _______.

41. If q is true, must $p \wedge q$ be true? _______.

42. If $p \to q$ is true, must p and q both be true? _______.

43. If $p \leftrightarrow q$ is false, must p and q both be false? _______.

44. If p is false, must $p \to q$ be true? _______.

45. If $p \vee q$ is true, must p be true? _______.

46. If $p \to q$ is true and p is false, must q be false? _______.

For Problems 47–54, use the *TF* method to determine whether each argument is valid or invalid. If it is invalid, produce a counterexample chart.

47.
$$\begin{array}{l} r \to (p \vee q) \\ r \wedge \sim p \\ \hline \sim q \to p \end{array}$$

48.
$$\begin{array}{l} s \vee \sim (d \wedge p) \\ d \wedge p \\ \hline \sim s \end{array}$$

49.
$$\begin{array}{l} w \to s \\ \sim w \to \sim p \\ c \wedge \sim s \\ \hline \sim p \end{array}$$

50.
$$\begin{array}{l} a \to (\sim a \vee b) \\ a \\ b \to \sim c \\ \hline \sim c \end{array}$$

51.
$$\begin{array}{l} \sim (w \vee \sim p) \\ p \to a \\ a \vee b \\ \hline b \end{array}$$

52.
$$\begin{array}{l} l \to (m \vee p) \\ \sim (m \vee p) \\ \sim l \to \sim w \\ \hline \sim w \end{array}$$

53.
$$\begin{array}{l} s \vee r \\ \sim r \vee g \\ g \to q \\ \sim s \\ \hline q \end{array}$$

54.
$$\begin{array}{l} \sim (a \to \sim w) \\ c \leftrightarrow d \\ w \to (b \wedge c) \\ d \to (c \vee p) \\ \hline p \end{array}$$

For Problems 55–59, express each argument symbolically. Then use the *TF* method to determine if the argument is valid or invalid. If it is invalid, produce a counterexample chart.

55. If I take the subway to work, I spend money. If I don't take the subway to work, then I get some exercise. I walk to work and do not spend money. Therefore, I get some exercise.
56. If today is Sunday then I do not have to go to work. If the weather is not nice then I can't play golf. Today is Sunday and the weather is not nice. Therefore, I can't play golf but I do not have to go to work.
57. If it is not the case that the dish is hot and spicy then Nick will order it. The dish is not hot. Therefore, Nick will order it.
58. If I read it in the newspaper or see it on the evening news then the story is true. If I don't see it on the evening news, then it is not a big story. It is a big story or it is a developing story. It is not a developing story. Therefore, the story is true.
59. The pizza does not have anchovies or it does have onions. If the pizza has onions I will have indigestion. I don't have indigestion. Therefore, the pizza has anchovies.

2.4 Testing for Validity Using the *TF* Method–Equivalences

Introduction

When testing the validity of an argument, it is often useful to replace a statement used in that argument with another, logically equivalent statement. As we discussed in Chapter One, if two statements are equivalent their truth values are always the same, regardless of the truth values of the variables involved. That is, they are either both true or both false. In Sections 1.5, 1.6, and 1.7 we examined a number of equivalences. For instance, we saw that the statement $p \rightarrow q$ is equivalent to both $\sim p \vee q$ and $\sim q \rightarrow \sim p$. Similarly, $\sim (p \vee q)$ is equivalent to $\sim p \wedge \sim q$, and $\sim (p \rightarrow q)$ is equivalent to $p \wedge \sim q$.

Be aware, however, that the fact that two statements are equivalent does not by itself reveal any information about the individual truth values of the variables that are involved in statements.

Using the Conditional Equivalence

Recall that the conditional equivalence allows us to express any conditional statement of the form $p \rightarrow q$ as a disjunction of the form $\sim p \vee q$, to which it is logically equivalent. Symbolically, the conditional equivalence is written as $(p \rightarrow q) \Leftrightarrow (\sim p \vee q)$. In the following example, the conditional equivalence will prove useful in determining whether an argument is valid.

Example 1

Test the following argument for validity. If it is invalid, produce a counterexample chart.

$$\begin{array}{c} (p \vee \sim q) \rightarrow r \\ \underline{\sim p \rightarrow \sim q} \\ r \end{array}$$

Solution

A close inspection of the argument reveals that the variables in the second premise appear in the LHS of the first premise. We note that the second premise is an implication and the LHS of the first premise is a disjunction. Since the conditional equivalence can be used to express any implication as a disjunction, we suspect that this equivalence may be helpful. Applying the conditional equivalence to the second premise, we express the implication as $p \vee \sim q$. Notice that we still do not know the individual truth values for *p* and *q*, but it is of no consequence, since $p \vee \sim q$ is exactly the LHS of the first premise. Hence, *r* must be true. The argument is valid.

■

Example 2

Test the following argument for validity. If it is invalid, produce a counterexample chart.

$$\begin{array}{c} (p \rightarrow q) \rightarrow r \\ r \rightarrow \sim s \\ \underline{\sim p \vee q} \\ \sim s \end{array}$$

Solution

If we apply the conditional equivalence to the third premise, $\sim p \vee q$, we may rewrite it as $p \rightarrow q$, which is the LHS of the first premise, $(p \rightarrow q) \rightarrow r$. Since $\sim p \vee q$ is true, $p \rightarrow q$ is also true. Notice again that it is not necessary to know the individual truth values for p and q. We know $(p \rightarrow q) \rightarrow r$ is true and its LHS, $p \rightarrow q$, is also true. Therefore, we can conclude that r must be true. Since both $r \rightarrow \sim s$ and r are true, $\sim s$ is forced to be true. Thus the argument is valid. ■

Using the Contrapositive Equivalence

The contrapositive equivalence states that any conditional statement of the form $p \rightarrow q$ is logically equivalent to the conditional statement $\sim q \rightarrow \sim p$. Symbolically, the contrapositive equivalence is expressed as $(p \rightarrow q) \Leftrightarrow (\sim q \rightarrow \sim p)$. In the next example, the contrapositive equivalence will be useful in determining whether an argument is valid.

Example 3

Test the following argument for validity.

$$\begin{array}{l} p \rightarrow (\sim q \rightarrow r) \\ (\sim p \vee s) \rightarrow w \\ \underline{\sim (\sim r \rightarrow q)} \\ \quad w \end{array}$$

Solution

The conditional statement within the parentheses in the third premise is the contrapositive of the RHS of the first premise $p \rightarrow (\sim q \rightarrow r)$. Since $\sim (\sim r \rightarrow q)$ is true, the contrapositive equivalence tells us that $\sim (\sim q \rightarrow r)$ must also be true, so the RHS of $p \rightarrow (\sim q \rightarrow r)$ must be false. Therefore, p is false, so $\sim p$ is true. This means that the LHS of $(\sim p \vee s) \rightarrow w$ must be true. But since the LHS of this true conditional statement is true, its RHS must also be true. Therefore, the argument is valid. ■

Using DeMorgan's Equivalence

As we discussed in Section 1.5, DeMorgan's equivalences allow us to express $\sim (p \vee q)$ as $(\sim p \wedge \sim q)$ and $\sim (p \wedge q)$ as $(\sim p \vee \sim q)$. We can express these two equivalences symbolically by writing

$\sim(p \vee q) \Leftrightarrow (\sim p \wedge \sim q)$ and $\sim(p \wedge q) \Leftrightarrow (\sim p \vee \sim q)$. DeMorgan's equivalence will be used in the next two examples.

Example 4

Test the following argument for validity. If it is invalid, produce a counterexample chart.

$$\begin{array}{l} \sim(p \wedge q) \rightarrow r \\ \underline{\sim p \vee \sim q} \\ \quad r \end{array}$$

Solution

The second premise, $\sim p \vee \sim q$, is equivalent to $\sim(p \wedge q)$ by DeMorgan's equivalence. Consequently, $\sim(p \wedge q)$ is true. The first premise has $\sim(p \wedge q)$ as its LHS. Therefore, *r* must be true, and the argument is valid. Again, notice that we still do not know, nor do we need to know, the truth values for *p* and *q*.

■

Example 5

Test the following argument for validity. If it is invalid, produce a counterexample chart.

$$\begin{array}{l} \sim p \rightarrow (q \wedge \sim r) \\ \sim q \vee r \\ \underline{p \rightarrow (s \wedge h)} \\ \quad h \end{array}$$

Solution

The appearance of the variables *q* and *r* in both the first and second premises may provide a hint as to how we should proceed. The RHS of the first premise is the negation of the second premise, because $\sim(\sim q \vee r)$ is equivalent to $q \wedge \sim r$ by DeMorgan's equivalence. Since $\sim q \vee r$ is true, $q \wedge \sim r$ must be false. In order for the first premise to be true, ~*p* must be false, making *p* true. Turning to the third premise, $s \wedge h$ must be true since *p* is true. The only way a conjunction can be true is if each of its components is true. Therefore, *h* must be true. The argument is valid.

■

Using the Conditional Negation Equivalence

When we negate the conditional statement $p \to q$, we obtain $p \wedge \sim q$. If both of these statements appear in an argument, it is important to recognize that one is the negation of the other.

Example 6

Test the following argument for validity. If it is invalid, produce a counterexample chart.

$$\begin{array}{l} r \to (d \wedge \sim e) \\ r \vee s \\ \underline{d \to e} \\ \quad s \end{array}$$

Solution

The negation of the third premise, $d \to e$, is $d \wedge \sim e$, which is the RHS of the first premise. Since $d \to e$ is true, $d \wedge \sim e$ must be false. Since $r \to (d \wedge \sim e)$ is true but its RHS is false, r must be false. However, $r \vee s$ is true, so s must be true. The argument is valid. ■

Example 7

Test the following argument for validity. If it is invalid, produce a counterexample chart.

$$\begin{array}{l} (p \wedge \sim q) \to (r \vee s) \\ \underline{\sim (p \to q)} \\ \quad r \end{array}$$

Solution

The conditional negation equivalence tells us that the second premise is exactly the LHS of the first premise. Since $\sim (p \to q)$ is true, so is $p \wedge \sim q$. Therefore, $r \vee s$ must be true. However, there are three ways a disjunction can be true, so r is not necessarily true. Since the argument is clearly invalid, we now produce a counterexample chart. Notice that since $\sim (p \to q)$ is true, $p \to q$ must be false. This means that p is true and q is false. We know that the RHS of the first premise must be true. This suggests we construct the counterexample chart as follows:

p	q	r	s
T	F	F	T

With these truth values, every premise is true, but the conclusion is false. The argument is invalid.

■

In-Class Exercises and Problems for section 2.4

In-Class Exercises

For Exercises 1–10, use your knowledge of equivalences to test the validity of each argument. If the argument is invalid, produce a counterexample chart.

1. $\sim p \vee s$
   $\sim p \to (q \wedge r)$
   $\sim q \vee \sim r$
   $\overline{\quad s \quad}$

2. $\sim p \vee s$
   $\sim q \to r$
   $(q \vee r) \leftrightarrow s$
   $\overline{\quad \sim p \quad}$

3. $(\sim r \to q) \to s$
   $\sim q \to r$
   $s \to (r \wedge w)$
   $\overline{\quad \sim w \quad}$

4. $\sim p \vee \sim q$
   $(p \to \sim q) \to a$
   $a \to \sim b$
   $\overline{\quad b \to d \quad}$

5. $(m \to \sim n) \to (q \vee n)$
   $\sim q \wedge r$
   $n \to \sim m$
   $\overline{\quad \sim n \vee p \quad}$

6. $a \wedge b$
   $(m \wedge \sim n) \to \sim b$
   $(\sim m \vee n) \to p$
   $\overline{\quad p \vee r \quad}$

7. $(\sim p \to r) \leftrightarrow b$
   $b \to (c \to d)$
   $\sim d$
   $p \vee r$
   $\overline{\quad w \to \sim c \quad}$

8. $\sim (w \to \sim s)$
   $s \leftrightarrow (p \vee \sim q)$
   $(\sim p \to \sim q) \to r$
   $\sim r \vee p$
   $\overline{\quad p \quad}$

9.
$$\begin{array}{l} (p \wedge \sim q) \to s \\ (s \vee w) \to (\sim r \vee m) \\ \sim (p \to q) \\ c \to (r \wedge \sim m) \\ \hline \quad \sim c \end{array}$$

10.
$$\begin{array}{l} (r \to \sim w) \to (\sim s \vee p) \\ (s \to p) \to b \\ q \wedge (w \to \sim r) \\ \hline \quad b \end{array}$$

Problems

For Problems 11–20, test each argument for validity. If the argument is invalid, produce a counterexample chart.

11.
$$\begin{array}{l} q \to w \\ q \vee (d \wedge c) \\ \sim d \vee \sim c \\ \hline \quad w \end{array}$$

12.
$$\begin{array}{l} \sim d \to c \\ (\sim c \to d) \to w \\ w \to (c \wedge s) \\ \hline \quad \sim s \end{array}$$

13.
$$\begin{array}{l} g \wedge m \\ (b \wedge \sim e) \to \sim m \\ (\sim b \vee e) \to q \\ \hline \quad c \vee q \end{array}$$

14.
$$\begin{array}{l} q \to w \\ \sim d \to c \\ (d \vee c) \leftrightarrow w \\ \hline \quad q \end{array}$$

15.
$$\begin{array}{l} d \to (\sim b \to \sim e) \\ \sim c \wedge b \\ (e \to b) \vee c \\ \hline \quad \sim d \end{array}$$

16.
$$\begin{array}{l} g \to \sim m \\ \sim (q \wedge d) \to g \\ \sim q \vee \sim d \\ \hline \quad m \to p \end{array}$$

17.
$$\begin{array}{l} \sim m \vee d \\ \sim (q \to r) \to p \\ p \to (m \wedge \sim d) \\ (w \vee p) \vee (q \wedge \sim r) \\ \hline \quad w \end{array}$$

18.
$$\begin{array}{l} b \leftrightarrow (q \vee m) \\ \sim q \to m \\ b \to (p \to c) \\ \sim c \\ \hline \quad s \to \sim p \end{array}$$

19. $m \vee q$

$\sim(s \to \sim w)$

$(\sim q \to \sim r) \to m$

$w \leftrightarrow (q \vee \sim r)$

---

q

20. $(\sim w \vee q) \to \sim(m \to \sim s)$

$(s \to \sim m) \wedge \sim r$

$\sim b \to (w \to q)$

$b \vee p$

2.5 Testing for Validity Using the *TF* Method–Conditional Conclusions

Introduction

If an argument contains a conditional conclusion, we can use a clever bit of reasoning to determine if the argument is valid or invalid. Recall that the *only* way a conditional statement can be false is if its LHS is true and its RHS is false. Therefore, if we can show that when the LHS of a conditional statement is true its RHS must also be true, the conditional statement must be true. This line of reasoning is the basis for the technique known as the *conditional proof.*

The Conditional Proof Strategy

When an argument contains a conditional conclusion, we can employ the method of conditional proof to test its validity. Note that if the LHS of a conditional conclusion is false, the argument is valid, since when the LHS of an implication is false, the implication must be true, thus making the argument valid. The only condition we have to consider is the case when the LHS of the conclusion is true.

The conditional strategy is to *assume* the LHS of the conclusion is true and show that under this assumption, the RHS of the conditional statement *must* be true, thus making the argument valid.

Example 1

Use a conditional strategy to show that the following argument is valid.

$a \to b$

$b \to \sim c$

$a \to \sim c$

Solution

Since the conclusion, $a \to \sim c$, is a conditional statement, we can assume that a, its LHS, is true. We now proceed as we did previously.

Since the first premise is true and since a is assumed to be true, b must be true. The second premise, $b \to \sim c$, is true and since b is true, $\sim c$ must be true. Thus, when the LHS of the conclusion is true, its RHS must be true as well. Therefore $a \to \sim c$ is true and the argument is valid.

■

If we are testing for validity in an argument that contains a conditional conclusion, the conditional strategy does not have to be the first step in our reasoning. We can assume the LHS of the conclusion is true at any step of the reasoning process.

Example 2

Use a conditional strategy to show that the following argument is valid.

$$\begin{array}{l} \sim p \vee q \\ \sim r \to s \\ \sim q \\ \underline{\sim p \to (s \to \sim w)} \\ \quad w \to r \end{array}$$

Solution

Since the conclusion of this argument is a conditional statement, we try to use the conditional strategy. If we assume w is true, we can proceed no further. While w appears in the premise $\sim p \to (s \to \sim w)$, we cannot reason further since the truth values of $\sim p$ and s are as yet unknown. Rather, we choose to begin with the first and third premises. Since $\sim q$ is true q is false. The first premise is a disjunction and one of its parts is false, so $\sim p$ must be true. The fourth premise $\sim p \to (s \to \sim w)$ has $\sim p$ as its LHS, so $s \to \sim w$ must be true. Now use the fact that the argument has a conditional conclusion. We assume its LHS, w, is true. If w is true $\sim w$ is false so s must be false. Now we turn our attention to the second premise. We know $\sim r \to s$ must be true and s is false. This means that $\sim r$ must be false, so r is true. But r is the RHS of the conclusion. Therefore we have shown that whenever the LHS of the conclusion is true, its RHS must be true. Therefore, the argument is valid.

■

Example 3

Use a conditional strategy to test the following argument for validity.

$$\begin{array}{c} p \to \sim w \\ \sim p \to (s \vee q) \\ \hline w \to s \end{array}$$

Solution

Since the conclusion of this argument is a conditional statement, we can try to use the conditional strategy. We assume w is true and consider the first premise $p \to \sim w$. Since premises are true and since w is assumed true, $\sim w$ must be false. Therefore, p must be false. The second premise, $\sim p \to (s \vee q)$ is true, and since p is false, we know $\sim p$ is true. Therefore $(s \vee q)$ is true. Since the disjunction is true, at least one of its parts must be true, but we can't be sure that s, the RHS of the conclusion is true. Since s is not *forced* to be true, we try to construct a counterexample chart for this argument, i.e. a set of truth values that makes every premise true and the conclusion false. Since we have assumed w is true, in order for the conclusion to be false, s must be false. The counterexample chart is shown below.

p	q	s	w
F	T	F	T

■

It is worth restating the strategy for a conditional approach. To use this approach, we must be presented with an argument whose conclusion is a conditional statement. If so, we can assume that the LHS of the conclusion is true and try to show that the RHS must be true. If we can, the argument is valid.

There are two common mistakes students make when trying to use this strategy. The first is assuming the LHS of *any* conclusion is true. The conditional proof strategy can be used only if the conclusion is a conditional statement. The other error is assuming the *entire* conclusion is true. This assumption is incorrect, since you would be assuming what you want to prove.

In-Class Exercises and Problems for Section 2.5

In-Class Exercises

For Exercises 1–12, test each argument for validity using the conditional proof technique. If the argument is invalid, produce a counterexample chart.

1. $p \to (q \wedge r)$
   $\underline{\qquad\qquad}$
   $p \to r$

2. $\sim a \to \sim b$
   $\sim a \vee c$
   $\underline{\qquad\qquad}$
   $b \to c$

3. $w \to \sim s$
   $b \to \sim w$
   $\underline{\qquad\qquad}$
   $s \to b$

4. $r \vee q$
   $\sim b \vee \sim q$
   $\underline{\qquad\qquad}$
   $\sim r \to \sim b$

5. $\sim (h \vee p) \vee w$
   $s \to \sim w$
   $\underline{\qquad\qquad}$
   $h \to \sim s$

6. $\sim k \to (p \wedge c)$
   $\sim c \vee a$
   $\underline{\qquad\qquad}$
   $\sim k \to a$

7. $r \to \sim p$
   $\sim r \vee (q \to s)$
   q
   $\underline{\qquad\qquad}$
   $p \to \sim s$

8. $(p \wedge a) \to m$
   $\sim m \vee (c \wedge d)$
   p
   $\underline{\qquad\qquad}$
   $a \to c$

9. b
   $\sim v \vee s$
   $(q \vee r) \to x$
   $b \to (x \to v)$
   $\underline{\qquad\qquad}$
   $q \to s$

10. $\sim p \to \sim q$
    $(p \wedge r) \to w$
    $(r \vee s) \to q$
    $\underline{\qquad\qquad}$
    $r \to w$

11. $\sim d \to p$
    $p \to (\sim m \to n)$
    $\sim d$
    $(m \vee d) \to r$
    $\underline{\qquad\qquad}$
    $\sim n \to (r \wedge s)$

12. $(d \wedge a) \to z$
    d
    $a \wedge (b \to p)$
    $z \to (x \vee c)$
    $\underline{\qquad\qquad}$
    $\sim p \to c$

For Exercises 13–16, express each argument symbolically. Then determine if the argument is valid or invalid. If it is invalid, produce a counterexample chart.

13. If I get married, I go on a honeymoon. If I go to Anguilla or St. Barts then I will be a June bride. If I do not go to Hawaii, then I will not be a June bride. If I go on a honeymoon, I go to Anguilla. Therefore, if I get married, I go to Hawaii.

14. It is not the case that I buy a flat screen television or go on a cruise. If I go on a cruise or buy a computer then if get a raise I buy a flat screen television. Therefore, if I buy a computer then I get a raise.

15. If I lift weights, then I will be healthy and gain muscle mass. If I do not exercise, I will not be healthy. I either lift weights or do not hire a personal trainer. Therefore, if I hire a personal trainer then either I exercise or overeat.

16. You are not part of the in-crowd or you either do Sudoku number puzzles or surf the web. If you do Sudoku number puzzles then you have a lot of patience. You do not surf the web. Therefore, if you are part of the in-crowd you do not have a lot of patience.

Problems

For Problems 17–28, test each argument for validity using the conditional proof technique. If the argument is invalid, produce a counterexample chart.

17. $\sim s \vee \sim p$

$\overline{s \to \sim p}$

18. $(a \wedge \sim b) \vee c$

$\overline{\sim(a \to \sim b) \to c}$

19. $k \vee \sim(a \vee \sim b)$

$b \to c$

$\overline{\sim k \to (c \vee n)}$

20. $\sim d \vee \sim x$

$b \vee x$

$\overline{\sim b \to \sim d}$

21. $\sim(a \vee m)$

$(m \vee d) \to (j \to a)$

$\overline{d \to j}$

22. $(s \to \sim w) \to \sim p$

$s \to a$

$\overline{p \to (a \wedge q)}$

23.
$$\begin{array}{l} \sim(\sim p \vee s) \\ p \to (q \to l) \\ \underline{q \vee m} \\ \sim l \to m \end{array}$$

24.
$$\begin{array}{l} d \to \sim a \\ r \to (a \wedge \sim b) \\ \underline{d \vee \sim q} \\ q \to \sim r \end{array}$$

25.
$$\begin{array}{l} (p \wedge a) \to m \\ \sim m \vee (c \wedge d) \\ \underline{p} \\ a \to \sim c \end{array}$$

26.
$$\begin{array}{l} (p \vee \sim q) \to \sim r \\ \underline{\sim m \to q} \\ r \to m \end{array}$$

27.
$$\begin{array}{l} n \to \sim g \\ (p \vee q) \to r \\ \sim w \to \sim r \\ \underline{\sim g \to p} \\ n \to w \end{array}$$

28.
$$\begin{array}{l} \sim q \\ (a \vee b) \to q \\ \sim c \vee n \\ \underline{c \to \sim (b \vee w)} \\ w \to n \end{array}$$

For Problems 29–31, express each argument symbolically. Then determine if the argument is valid or invalid. If it is invalid, produce a counterexample chart.

29. If the comptroller mails out dividend checks, it is June 30 or December 31. If it is December 31, then tax statements are not mailed. The comptroller mails monthly statements or tax statements. The comptroller is not mailing monthly statements. Therefore, if the comptroller is mailing dividend checks then it is not June 30.

30. If I speak Spanish and Portuguese, I will take Italian lessons. Either I do not take Italian lessons or I visit Rome and Venice. I speak Spanish. Therefore, if I speak Portuguese, I visit Rome.

31. If I do not take singing lessons then I cannot be a thespian. If I take singing lessons and play the piano, I can be a rock singer. If I play the piano or the guitar, I can be a thespian. Therefore, if I play the piano, I can be a rock singer.

2.6 Testing for Validity Using the *TF* Method–The Indirect Approach

Introduction

We have seen that an argument can be either valid or invalid. There are no other possibilities. If an argument is invalid, it can be shown that the conclusion may be false even though the premises are true.

If we are unable to directly deduce an argument's validity because there does not seem to be a relationship between the premises, we must resort to other methods.

Indirect Approach

One of the alternate methods for testing for validity is the *indirect approach*. It is important to understand the strategy for an indirect approach. If we assume that the argument is *invalid*, we are assuming that it is possible for the conclusion to be *false* while every premise is true. If we are able to provide truth values for the variables such that the conclusion is false while all premises are true, then we will have proven that our argument is invalid. If we are unable to do this, because of any contradiction to the laws of logic, our assumption must have been incorrect, and thus the argument must be valid.

Example 1

Use an indirect approach to show that the following argument is valid.

$$\begin{array}{c} p \to q \\ q \to r \\ \hline p \to r \end{array}$$

Solution

Using the indirect approach, we begin by assuming the conclusion is *false*. The conclusion is an implication. The only way it can be false is if p is true and r is false. The first premise, $p \to q$, must be true, and since we deduced p was true from our indirect approach, q must be true. The second premise, $q \to r$, is true, and since q was deduced true, r must be true as well. However, our assumption led us to the fact that if the conclusion was assumed false, r had to be false. We now have r is true and r is false, clearly a contradiction. Therefore, the assumption must have been incorrect. Since the assumption was that the argument was invalid, and this led to a contradiction, the argument must be valid.

■

In the above example, it would be impossible to produce a counterexample chart, since the argument was valid. If our indirect approach does not lead to a contradiction, we would strongly suspect that the argument was indeed invalid, and we would try to produce a counterexample chart.

Example 2

Use the indirect approach to test the argument below for validity.

$$\begin{array}{l} p \to q \\ r \to q \\ \hline p \vee \sim r \end{array}$$

Solution

Using the indirect approach, we assume that the conclusion is *false*. The conclusion is a disjunction. The only way a disjunction can be false is if both of its components are false. This means that p is false and $\sim r$ is false, which implies that r is true. Since the second premise, $r \to q$, is assumed true, and r is true, q must be true. We know p is false and q is true. This confirms the fact that the first premise, $p \to q$, is indeed true. We have *not* produced a contradiction. That is, it is possible to have true premises and a false conclusion. Therefore, our argument is invalid. We now produce a counterexample chart based on the discussion above.

p	q	r
F	T	T

■

Example 3

Test the given argument for validity.

$$\begin{array}{l} m \vee \sim n \\ n \\ r \to \sim p \\ m \to (\sim p \to q) \\ \hline \sim q \to \sim r \end{array}$$

Solution

It appears that the direct *TF* approach is the method of choice. We begin with the first two premises. Knowing that n and $m \vee \sim n$ are both true implies that m must be true. Since m is the antecedent of the premise $m \to (\sim p \to q)$, we know that $\sim p \to q$ must be true. There are three ways for $\sim p \to q$ to be true, so we are at an impasse. This suggests considering an indirect approach. We therefore assume the conclusion, $\sim q \to \sim r$, is *false*. Since it is an implication, $\sim q$ must be true and $\sim r$ must be false. We know both n and m are true from our attempt at a direct approach. We also know $\sim p \to q$ is true. Since $\sim q$ was assumed true, q is false. The truth value of $\sim p$ must be false in order for the premise $\sim p \to q$ to remain true. Applying this fact to the third premise, $r \to \sim p$, r must be false. This means $\sim r$ is true. However, $\sim r$ was assumed false. Since $\sim r$ cannot be both true and false, we have arrived at a contradiction. Therefore, the argument is valid.

■

Example 4

Test the given argument for validity.

$$\begin{array}{l} d \to r \\ (b \wedge c) \to d \\ \underline{a \wedge b} \\ \quad r \end{array}$$

Solution

Since we can begin our reasoning with the conjunction $a \wedge b$, it again appears that the direct *TF* approach is the method of choice. The conjunction $a \wedge b$ is true, so both a and b must be true. We try to incorporate the truth value of b into the second premise. But we can go no further since we do not know the truth value of c or d. Once again, this suggests considering an indirect approach. Therefore, we assume the conclusion, r, is *false*. For the first premise to remain true, d must be false. If the second premise $(b \wedge c) \to d$, is to remain true, $b \wedge c$ must be false. Since b is true, c is forced to be false. Using the indirect method, we have assigned truth values to all the variables such that the premises are true and the conclusion is false, and produced no contradictions in doing so. Hence, argument is invalid. Constructing a counterexample chart using the truth values obtained, we write:

a	b	c	d	r
T	T	F	F	F

■

In-Class Exercises and Problems for Section 2.6

In-Class Exercises

For Exercises 1–10, use the indirect approach to test each argument for validity. If the argument is invalid, produce a counterexample chart.

1.
$$\begin{array}{l} (\sim m \to p) \to r \\ \underline{\sim r \lor q} \\ q \end{array}$$

2.
$$\begin{array}{l} a \to b \\ \underline{m \leftrightarrow a} \\ \sim m \lor b \end{array}$$

3.
$$\begin{array}{l} r \lor (\sim p \to s) \\ \underline{\sim r \to (\sim s \lor p)} \\ p \lor r \end{array}$$

4.
$$\begin{array}{l} p \\ \sim p \lor (\sim a \to b) \\ \underline{\sim a \lor c} \\ c \end{array}$$

5.
$$\begin{array}{l} (p \lor q) \to r \\ \sim r \lor s \\ \underline{\sim p \to w} \\ s \to w \end{array}$$

6.
$$\begin{array}{l} (c \lor d) \leftrightarrow g \\ a \to (b \lor c) \\ \underline{a} \\ \sim d \to g \end{array}$$

7.
$$\begin{array}{l} w \land a \\ \sim a \lor (c \to s) \\ \underline{w \to (\sim s \lor n)} \\ c \to n \end{array}$$

8.
$$\begin{array}{l} w \to s \\ (s \lor r) \to q \\ \underline{\sim q \lor p} \\ p \lor (w \to r) \end{array}$$

9.
$$\begin{array}{l} r \land a \\ r \to (\sim p \lor s) \\ \sim s \lor (r \to w) \\ \underline{p \leftrightarrow a} \\ w \end{array}$$

10.
$$\begin{array}{l} \sim p \lor s \\ p \\ c \to (r \lor w) \\ \underline{w \to (s \to r)} \\ c \to r \end{array}$$

In Exercises 11–13, express each argument symbolically. Then determine if it is valid or invalid. If it is invalid, produce a counterexample chart.

11. If the dog barks, then she does not get a treat. If the dog rolls over, then she gets petted and praised. The dog barks or rolls over. Therefore, if you give the dog a treat, she gets praised.

12. If you e-mail the invitations then you get quick responses. Either you do not get quick responses, or you have time to plan for the party and call the caterers two weeks before the party. If you do not have to make telephone calls to guests then you have time to plan the party. Therefore, if you e-mail the invitations, you do not have to make telephone calls to the guests.

13. If Kathy drives her new SUV, then she has exorbitant gasoline bills. If she has exorbitant gasoline bills, she cannot save money for a cruise. If Kathy wins Lotto, she can save money for a cruise and drive her new SUV. Therefore, Kathy's husband pays for her gasoline or she does not win the Lotto.

Problems

For Problems 14–21, use the indirect approach to test each argument for validity. If the argument is invalid, produce a counterexample chart.

14. $s \rightarrow (c \vee m)$
$c \wedge h$
$\overline{\quad\quad\quad\quad}$
$s \rightarrow \sim b$

15. $p \vee q$
$\sim r \vee \sim q$
$\overline{\quad\quad\quad\quad}$
$\sim p \rightarrow \sim r$

16. $(a \vee b) \rightarrow c$
$\overline{\quad\quad\quad\quad}$
$b \rightarrow (c \vee d)$

17. $g \rightarrow (w \wedge x)$
$\overline{\quad\quad\quad\quad}$
$g \rightarrow x$

18. $a \rightarrow d$
$\sim d \vee p$
$p \rightarrow \sim c$
$\overline{\quad\quad\quad\quad}$
$(a \wedge b) \rightarrow c$

19. $p \rightarrow k$
$\sim p \rightarrow (w \wedge s)$
$\sim s \vee p$
$\overline{\quad\quad\quad\quad}$
k

20. $p \to (d \to \sim z)$

$p \to q$

$c \vee d$

$\sim q$

$z \vee c$

21. $\sim p \to \sim n$

$r \to \sim s$

n

$\sim p \vee (\sim s \to q)$

$q \vee r$

In Problems 22–25, express each given argument symbolically. Then determine if it is valid or invalid. If it is invalid, produce a counterexample chart.

22. If you pay your taxes and obey the law then you are a good citizen. If you are a good citizen, you drive within the speed limit. If you do not drive within the speed limit, then either you are not a good citizen or you didn't obey the law. Therefore, if you pay your taxes, you drive within the speed limit.

23. If you check your luggage then if you go on a long trip you must arrive at the airport two hours before the flight departs. You check your luggage or you leave your car in long-term parking. It is not the case that you go on a long trip and leave your car in long-term parking. Therefore, you arrive at the airport two hours before the flight departs or you do not go on a long trip.

24. If Grandma makes lasagna or ziti then Grandpa is happy. If Grandpa is not happy or Grandma makes meatballs, then Grandpa complains. Either Grandpa does not complain or Grandma makes ziti. Therefore, if Grandma makes meatballs, Grandpa is happy.

25. If I buy a new car, we carpool to work. If we carpool to work then I pick you up but you eat breakfast at home. If you take the train to work then either you do not eat breakfast at home or you wake up an hour earlier. Either I buy a new car and you wake up an hour earlier, or I do not buy a new car but you take the train to work. Therefore, either I pick you up or you wake up an hour earlier

Chapter 2 Review

For questions 1–10, answer true, false or can't be determined.

1. Let s be a premise. What is the truth value of $s \vee w$?
2. Suppose $p \leftrightarrow q$ is true and p is a premise. What is the truth value of $\sim q \rightarrow w$?
3. If $\sim p \vee (r \wedge s)$ is a premise and $w \rightarrow s$ is false, what is the truth value of p?
4. If both $p \rightarrow q$ and $(p \rightarrow q) \rightarrow \sim r$ are premises, what is the truth value of r ?
5. If $p \wedge (q \vee r)$ is a premise, what is the truth value of $p \wedge q$?
6. Suppose $(r \vee w) \wedge s$ and $s \rightarrow \sim w$ are both premises, what is the truth value of r?
7. If both $s \leftrightarrow (q \vee r)$ and $\sim s$ are premises, what is the truth value of $r \rightarrow p$?
8. If $(r \vee s) \rightarrow \sim p$ and $\sim r \rightarrow s$ are both premises, what is the truth value of p?
9. Suppose $\sim (s \rightarrow w) \rightarrow (q \vee a)$ and $s \wedge \sim w$ are premises. What is the truth value of a?
10. Let $r \rightarrow s$ and $\sim r \vee \sim s$ be premises. What is the truth value of r?

For question 11–20, use the *TF* method to determine if each argument is valid or invalid. If it is invalid, produce a counterexample chart.

11. $w \rightarrow \sim p$
$\sim s \rightarrow p$
$\sim (\sim r \vee s)$
$\overline{\quad \sim w \quad}$

12. $q \wedge \sim a$
$a \vee \sim p$
$\sim p \rightarrow (q \vee b)$
$\overline{\quad b \quad}$

13. $\sim p$
$\sim (p \leftrightarrow w)$
$s \rightarrow w$
$\overline{\quad s \quad}$

14. $\sim p \vee c$
$q \rightarrow \sim c$
p
$\sim q \rightarrow (w \wedge r)$
$\overline{\quad s \rightarrow r \quad}$

15. $b \to (a \wedge \sim c)$
$\sim a \vee p$
$p \to d$
b
$\overline{d \wedge \sim c}$

16. $(m \vee n) \to p$
$\sim p$
$n \leftrightarrow a$
$\overline{m \vee a}$

17. $(w \vee s) \to \sim q$
$q \wedge p$
$a \to w$
$\sim a \to (b \vee p)$
$\overline{b}$

18. $s \wedge w$
$(d \to h) \leftrightarrow \sim s$
$\sim a \vee h$
$a \vee (b \wedge c)$
$\overline{c \vee z}$

19. $b \wedge c$
$(p \vee q) \to \sim w$
$(\sim p \vee r) \to (s \vee \sim c)$
$(b \vee d) \to w$
$\overline{s}$

20. a
$\sim r \vee (w \wedge s)$
$(a \vee b) \to r$
$(m \wedge \sim s) \leftrightarrow \sim w$
$\overline{m}$

In questions 21–28, use your knowledge of equivalences to test the validity of each argument. If the argument is invalid, produce a counterexample chart.

21. $(s \to w) \to r$
$r \to (a \wedge b)$
$\sim s \vee w$
$\overline{c \to b}$

22. $a \wedge r$
$(p \wedge \sim q) \to \sim r$
$(\sim p \vee q) \to w$
$\overline{w \vee k}$

23. $(\sim w \to p) \wedge \sim q$
$b \to (\sim q \vee s)$
$(\sim p \to w) \to b$
$\overline{s}$

24. $b \to (a \wedge \sim c)$
$\sim a \vee c$
$p \to \sim b$
$\overline{p}$

25. $\sim(a \to b)$
$c \leftrightarrow (\sim b \vee p)$
$(w \vee a) \to c$
___
w

26. $m \to n$
$(\sim m \vee n) \to (\sim a \wedge s)$
$(a \vee \sim s) \vee w$
$(w \vee b) \to c$
___
c

27. $(a \wedge b) \to (\sim p \vee q)$
$(p \to q) \leftrightarrow w$
$\sim(a \to \sim b)$
$c \to \sim(w \vee s)$
___
$\sim c$

28. $\sim(p \vee \sim s)$
$(\sim c \to k) \to d$
$d \to (a \wedge \sim b)$
$(s \vee r) \leftrightarrow (\sim k \to c)$
___
$\sim b$

In questions 29–36, use the indirect approach or the conditional proof approach (when appropriate) to test the validity of each argument. If the argument is invalid, produce a counterexample chart.

29. $s \vee p$
$w \vee q$
$s \vee \sim w$
___
$p \vee q$

30. $a \vee s$
$w \to (p \to a)$
$\sim s \vee (p \wedge w)$
___
a

31. $p \wedge q$
$s \to (\sim q \vee r)$
$p \vee s$
___
$\sim s$

32. $(\sim c \vee q) \to a$
$\sim a \vee p$
$p \to r$
___
$\sim c \to r$

33. $(m \vee p) \to s$
$s \to (q \wedge w)$
$\sim a \to p$
$m \leftrightarrow a$
___
q

34. $p \wedge \sim s$
$\sim r \to \sim q$
$\sim a \leftrightarrow (c \vee q)$
$p \to (a \to s)$
___
$c \vee r$

35.
$$\begin{array}{l} p \to (\sim q \to r) \\ \sim r \\ r \vee s \\ \underline{(s \vee a) \leftrightarrow (p \vee w)} \\ \sim w \to q \end{array}$$

36.
$$\begin{array}{l} (a \wedge b) \wedge (c \to \sim b) \\ (p \leftrightarrow r) \to w \\ \sim c \to (p \vee q) \\ \underline{w \to (\sim a \vee c)} \\ p \vee w \end{array}$$

In questions 37–49, represent each argument symbolically. Then determine if it is valid or invalid. If it is invalid, produce a counterexample chart.

37. We will go to Disneyworld or to Las Vegas. If we go to Disneyworld we will stand on long lines. If we go to Las Vegas we stay up late. We don't stay up late. Therefore, we stand on long lines.

38. Callie is a good worker and she is a wiz at the computer. If Callie is a wiz at the computer then she can either diagnose and fix any problem or reprogram the computer. If she can fix any problem then her reports are not late. Therefore, if Callie can reprogram the computer, her reports are not late.

39. If your alarm clock rings you will not be late for work. If either your car does not start or you lose your electricity, you will be late for work. It is not true that both your car starts and you do not lose electricity. Therefore, your alarm clock does not ring.

40. It is not the case that if I received a paycheck then I spend it all. If I go to the bank during my lunch hour then I received a paycheck. I go to the bank during my lunch hour and I put money in my saving account. Therefore, if I put money into my savings account, I did not spend it all.

41. Either you stop at a service station or your car runs properly. If you either need gas or the gas gauge is broken, you have to stop at a service station. If you do not need gas then your gas gauge is broken. Therefore, your car does not run properly.

42. If Noah takes the day off, then he either goes out with Averi or goes out on his boat. If Noah works overtime, he does not go out with Averi. Either Noah works overtime, or he takes the day off but makes less money. Therefore, if Noah goes out on his boat, he makes less money.

43. If we can get two stars to be in the movie, we will produce it. DeNiro is a great actor, but we can't get two stars to be in the movie. Therefore, we will produce the movie or DeNiro is not a great actor.

44. Either Scott did not go to work today or it is not the case that if Scott goes to work early then the job was done on time. If Scott's entire department is out sick, then he did go to work today. Therefore, if Scott did not go to work early his entire department is not out sick.

45. If the neighborhood market closes then I drive further and shop in the next town. If I drive further then I spend more money on gas and cannot buy all the groceries I need. Either the neighborhood market closes or I get home early. I did not get home early. Therefore, I spend more money on gas.

46. If Gary puts his white clothes in the washing machine then he uses bleach. If he puts white clothes and red clothes in the washing machine, then some of his clothes turn pink. If some of his clothes turn pink then he uses bleach. Either Gary puts his red clothes in the washing machine or he does not use bleach. Therefore, Gary does not put his white clothes in the washing machine.

47. If the President comes to town, then there will be gridlock and the National Guard will be called. If the National Guard is called it will take twice as long to get home. If it takes twice as long to get home then either you will be very hungry or your dinner will get cold. Therefore, if the President comes to town you will be very hungry.

48. The computer program either does a payroll or an inventory update. If the program does a payroll it calculates federal tax and state tax. If the program does an inventory update, we cannot use the computer for three hours. We can't use the computer for three hours and the program did not compute federal tax. Therefore, the program computes state tax.

49. She is a Democrat or a Republican. If she is a Republican, she wants less government control and a revised tax plan. She does not want less government control. Therefore, she is a Democrat.

Sample Exam: Chapter 2

For questions 1–5, answer true, false or can't be determined.

1. If $\sim(\sim r \rightarrow w)$ is a premise, then what is the truth value of $r \rightarrow (s \vee w)$?
2. If $\sim(r \vee \sim w)$ is a premise, then what is the truth value of $\sim(\sim r \leftrightarrow w)$?
3. If $\sim r \wedge w$ is a premise, then what is the truth value of $\sim w \vee p$?
4. If $(w \rightarrow \sim r) \vee p$ is false, what is the truth value of $w \rightarrow p$?
5. If $\sim[(r \wedge w) \leftrightarrow s]$ is a premise and r is false, then what is the truth value of $s \vee p$?

For questions 6–9, construct a counterexample chart for each invalid argument.

6. $(a \wedge b) \vee c$
   $\sim c$
   $(a \vee d) \rightarrow (e \vee g)$
   ———
   $d \rightarrow g$

7. $(p \vee q) \rightarrow r$
   $\sim r \wedge s$
   $\sim p \rightarrow (a \vee b)$
   ———
   $s \leftrightarrow b$

8. p
   $p \rightarrow (q \vee r)$
   $q \rightarrow s$
   ———
   $s \vee a$

9. $p \rightarrow q$
   $(p \rightarrow q) \rightarrow (r \vee s)$
   $r \rightarrow w$
   ———
   $q \rightarrow w$

For questions 10–21, use the *TF* method to determine if each argument is valid or invalid. If it is invalid, produce a counterexample chart.

10. $a \vee (b \rightarrow c)$
    $c \rightarrow \sim w$
    $\sim a \wedge b$
    ———
    $w \rightarrow r$

11. $(m \rightarrow n) \rightarrow q$
    $\sim q$
    $(\sim n \vee w) \rightarrow s$
    ———
    $a \rightarrow s$

12. $(z \vee m) \rightarrow \sim n$
    $\sim z \rightarrow \sim(p \rightarrow q)$
    $n \wedge y$
    ———
    $\sim p$

13. $p \wedge q$
    $(q \vee r) \rightarrow \sim s$
    $s \rightarrow w$
    ———
    w

14. $p \wedge q$
$(q \vee r) \rightarrow \sim s$
$s \rightarrow w$
———
w

15. $\sim p \vee w$
$(p \rightarrow w) \rightarrow \sim r$
$r \vee s$
———
$s \rightarrow w$

16. $(p \rightarrow q) \rightarrow \sim a$
$q \vee s$
$a \wedge b$
$(s \vee w) \rightarrow r$
———
$r \leftrightarrow b$

17. $\sim a \vee b$
$\sim c \vee d$
$(a \rightarrow b) \rightarrow c$
$b \rightarrow d$
———
$\sim b$

18. $p \rightarrow (s \wedge r)$
$a \leftrightarrow b$
$(a \wedge c) \vee p$
———
$s \vee b$

19. $(\sim p \vee \sim s) \wedge r$
$w \vee q$
$q \rightarrow (p \wedge s)$
$(c \wedge d) \rightarrow \sim w$
———
$\sim c$

20. a
$(b \vee d) \rightarrow (p \rightarrow \sim q)$
$\sim a \leftrightarrow (b \rightarrow c)$
$(q \rightarrow \sim p) \rightarrow m$
———
m

21. $s \rightarrow \sim p$
$r \rightarrow (s \vee q)$
$(r \rightarrow s) \leftrightarrow \sim c$
$(\sim s \vee \sim p) \rightarrow c$
———
q

In questions 22–26, express each argument symbolically. Then use the *TF* method to determine if it is valid or invalid. If it is invalid, produce a counterexample chart.

22. If the butler was working the night of the crime, then he was a suspect. If he was not working the night of the crime, then the victim was not poisoned. The detective found his girlfriend locked in a closet and ruled the butler was not a suspect. Therefore, the victim was not poisoned.

23. If the Yankees win the baseball game then either the pitcher had nine strikeouts or the defense was outstanding. If the defense was not outstanding the crowd did not cheer. The crowd cheered. Therefore, the Yankees won.
24. If I go to college I will get a job. If I get a job then the company will pay for my master's degree. If the company pays for my master's degree then I can't work overtime. I worked overtime. Therefore, I didn't go to college.
25. It is not true that if you live in Idaho then you like potatoes. Either you like potatoes or you do not live on a farm. If you like horseback riding then you live on a farm. Therefore, you either don't like horseback riding or you like milking cows.
26. If you live in a glass house you shouldn't throw stones. If you throw stones then either you will break a window or the neighbors will call the police. If the neighbors call the police you will have to do community service. Therefore, if you live in a glass house you have to do community service.

CHAPTER THREE

FORMAL PROOFS

In Chapter Two, we explored arguments in an intuitive way. We now turn our attention to a more formal presentation that is used to prove an argument's validity.

3.1 Valid Argument Forms–Part I

Introduction

To formally prove that arguments are valid, we use tautologies extensively. While there are a rather large number of tautologies at our disposal, it turns out that only a few of them are frequently used when constructing formal proofs.

We now turn our attention to a group of tautologies that are implications. These conditional tautologies are often called *valid argument forms.* A concise listing of all the valid argument forms we study in this chapter can be found on page 190.

Modus Ponens

The first valid argument form we will consider is called *modus ponens*. This tautology can be represented symbolically as $[(p \to q) \wedge p] \to q$.

We may also express modus ponens in its argument form. If we think of the two components of the antecedent, $p \to q$ and p, as premises of an argument, and q as the conclusion, we can express the tautology as:

$$\begin{array}{l} p \to q \\ \underline{p \qquad} \\ \quad q \end{array}$$

Examining this tautology, we see that if one premise of an argument is an implication and the other premise is the LHS of this implication, we can conclude that the RHS of the implication must be true. It is critical to recognize the pattern for modus ponens. Modus ponens deals only with an argument whose premises are an implication and its LHS. When this occurs, we are guaranteed that the RHS of the implication must be true.

To verify that modus ponens is indeed a tautology, we will construct a truth table.

p	q	$p \to q$	$(p \to q) \wedge p$	$[(p \to q) \wedge p] \to q$
T	T	T	T	T
T	F	F	F	T
F	T	T	F	T
F	F	T	F	T

Example 1

Is the argument an illustration of modus ponens?

$$\begin{array}{l} q \to p \\ \underline{q \qquad} \\ \quad p \end{array}$$

Solution

The premises of the argument are an implication and its LHS. The conclusion is the RHS of the implication. This is exactly the pattern for modus ponens. ■

Example 2

Is the given argument an illustration of modus ponens?

$$\begin{array}{l} \sim p \to q \\ \underline{\sim p \qquad} \\ \quad q \end{array}$$

Solution

Do not let the appearance of the negation symbol lead you astray. The premises of the argument are an implication and its LHS. The conclusion is the RHS of the implication. This is exactly the pattern for modus ponens. ■

Example 3

Is the following argument an illustration of modus ponens?

$$\begin{array}{l} p \to q \\ \underline{q \qquad} \\ \quad p \end{array}$$

Solution

Since the premises of the argument are an implication and its *RHS*, this argument is not an illustration of modus ponens. ■

Example 4

Is the given argument an illustration of modus ponens?

$$\begin{array}{l} (\sim r \vee s) \to (w \to b) \\ \underline{\sim r \vee s \qquad\qquad\qquad} \\ \qquad w \to b \end{array}$$

Solution

Do not let the appearance of compound statements confuse you. The premises of the argument are an implication and its LHS. The conclusion is the RHS of the implication. This is exactly the pattern for modus ponens.

■

Example 5

Is the given argument an illustration of modus ponens?

$$\begin{array}{l} (\sim r \vee s) \rightarrow (w \rightarrow b) \\ \underline{r \vee \sim s \qquad\qquad} \\ \qquad w \rightarrow b \end{array}$$

Solution

In this case, one of the premises is an implication. However, the second premise is not its LHS. Therefore, this is not an illustration of modus ponens.

■

Modus Tollens

The *modus tollens* tautology can be represented symbolically as $[(p \rightarrow q) \wedge \sim q] \rightarrow \sim p$. We may also express modus tollens in its argument form. If we think of the two components of the antecedent, $p \rightarrow q$ and $\sim q$, as premises of an argument, and $\sim p$ as the conclusion, we can express the tautology as:

$$\begin{array}{l} p \rightarrow q \\ \underline{\sim q \qquad} \\ \sim p \end{array}$$

Examining this tautology, we see that if one premise of an argument is an implication and the other premise is the negation of its RHS, we can conclude that the negation of the LHS of the implication must be true. Again, recognition of the modus tollens pattern is crucial.

As with modus ponens, we will use a truth table to verify that modus tollens is a tautology.

p	q	$p \rightarrow q$	$(p \rightarrow q) \wedge \sim q$	$[(p \rightarrow q) \wedge \sim q] \rightarrow \sim p$
T	T	T	F	T
T	F	F	F	T
F	T	T	F	T
F	F	T	T	T

Example 6

Is the argument below an illustration of modus tollens?

$$\begin{array}{l} \sim p \rightarrow q \\ \underline{\sim q \quad\quad} \\ \quad p \end{array}$$

Solution

The premises of the argument are an implication and the negation of its RHS. The conclusion of the argument is the negation of the LHS of the implication. Therefore, the argument is an illustration of modus tollens.

■

Example 7

Is the argument below an illustration of modus tollens?

$$\begin{array}{l} p \rightarrow (q \wedge r) \\ \underline{\sim (q \wedge r) \quad} \\ \quad \sim p \end{array}$$

Solution

The premises of the argument are an implication and the negation of its RHS. We are not concerned that the RHS is a conjunction. The conclusion of the argument is the negation of the LHS of the implication. This is exactly the pattern required for us to conclude that the argument is valid because it is an illustration of the modus tollens tautology.

■

Example 8

Is the argument shown below an illustration of modus ponens, modus tollens or neither of these tautologies?

$$\begin{array}{l} \sim p \rightarrow \sim q \\ \underline{\sim q \quad\quad\quad} \\ \quad \sim p \end{array}$$

Solution

For the argument to have the modus ponens form, we must have an implication and its LHS as premises. This is not the case here. If the tautology is of the modus tollens form, it must have an implication and the negation of its RHS as premises. This is not the case either. This argument form is neither modus ponens nor modus tollens. In fact, the argument presented is not a tautology, and thus, is invalid. ■

To show a formal proof for an argument's validity, we construct a sequence of *true* statements, as well as reasons for their truth, in table form. A formal proof for the argument in Example 7 would be:

Statement	Reason
1. $p \to (q \wedge r)$	Premise
2. $\sim (q \wedge r)$	Premise
3. $\sim p$	Modus Tollens (1)(2)

The numbers after the words modus tollens refer to the statements that were used to deduce $\sim p$.

Example 9

Construct a formal proof for the following valid argument.

$$\begin{array}{l} r \to s \\ \sim s \\ \underline{\sim r \to p} \\ \quad p \end{array}$$

Solution

We record only true statements, beginning with the simplest premise.

Statement	Reason
1. $\sim s$	Premise

Since the first premise is an implication containing an s, we try to use either modus ponens or modus tollens in combination with $\sim s$. To use modus tollens, we need an implication and the negation of its RHS to be true. We know that $\sim s$ is true, so we can continue our formal proof and show that $\sim r$ is true.

Statement	Reason
1. $\sim s$	Premise
2. $r \to s$	Premise
3. $\sim r$	Modus Tollens (1)(2)

Now that we know $\sim r$ is true, we combine it with the third premise $\sim r \to p$ and use modus ponens to conclude that p is true.

Statement	Reason
1. $\sim s$	Premise
2. $r \to s$	Premise
3. $\sim r$	Modus Tollens (1)(2)
4. $\sim r \to p$	Premise
5. p	Modus Ponens (3)(4)

This completes the proof. ■

In-Class Exercises and Problems for Section 3.1

In-Class Exercises

In Exercises 1–8, state the valid argument form shown.

1. $\dfrac{\begin{array}{l}\sim s \to \sim k\\ k\end{array}}{s}$

2. $\dfrac{\begin{array}{l}\sim r \to w\\ \sim r\end{array}}{w}$

3. $\dfrac{\begin{array}{l}d \to (w \vee b)\\ \sim (w \vee b)\end{array}}{\sim d}$

4. $\dfrac{\begin{array}{l}(c \wedge h) \to \sim l\\ l\end{array}}{\sim (c \wedge h)}$

5. $\dfrac{\begin{array}{l}(d \to c) \to \sim p\\ d \to c\end{array}}{\sim p}$

6. $\dfrac{\begin{array}{l}\sim j \to \sim (s \vee \sim p)\\ \sim j\end{array}}{\sim (s \vee \sim p)}$

7. $\dfrac{\begin{array}{l}(m \vee n) \to \sim (r \to s)\\ r \to s\end{array}}{\sim (m \vee n)}$

8. $\dfrac{\begin{array}{l}\sim (w \to c) \to (a \wedge b)\\ \sim (a \wedge b)\end{array}}{w \to c}$

In Exercises 9–18, use the *TF* method to test each argument for validity. If it is valid, show a formal proof. If invalid, provide a counterexample chart.

9.
$$\begin{array}{l} \sim a \\ \sim a \rightarrow b \\ c \rightarrow \sim b \\ \hline \sim c \end{array}$$

10.
$$\begin{array}{l} r \rightarrow p \\ s \rightarrow w \\ \sim r \rightarrow s \\ \sim w \\ \hline p \end{array}$$

11.
$$\begin{array}{l} \sim r \rightarrow (a \vee b) \\ \sim (a \vee b) \rightarrow w \\ \sim w \\ \hline r \end{array}$$

12.
$$\begin{array}{l} \sim q \rightarrow s \\ s \rightarrow (p \vee r) \\ \sim (p \vee r) \\ \hline q \end{array}$$

13.
$$\begin{array}{l} (p \rightarrow q) \rightarrow (r \rightarrow s) \\ (r \rightarrow s) \rightarrow (m \rightarrow b) \\ p \rightarrow q \\ \hline m \rightarrow b \end{array}$$

14.
$$\begin{array}{l} \sim (w \wedge s) \\ k \rightarrow (w \wedge s) \\ p \rightarrow \sim k \\ \hline \sim p \end{array}$$

15.
$$\begin{array}{l} \sim w \rightarrow (c \vee s) \\ \sim r \rightarrow \sim p \\ p \rightarrow \sim w \\ \sim (c \vee s) \\ \hline r \end{array}$$

16.
$$\begin{array}{l} (r \vee \sim s) \rightarrow p \\ c \rightarrow (r \vee \sim s) \\ \sim p \\ \sim c \rightarrow d \\ \hline d \end{array}$$

17.
$$\begin{array}{l} \sim (p \vee r) \rightarrow s \\ \sim (a \rightarrow c) \\ s \rightarrow (k \wedge g) \\ (p \vee r) \rightarrow (a \rightarrow c) \\ \hline k \wedge g \end{array}$$

18.
$$\begin{array}{l} k \rightarrow (l \vee p) \\ (r \rightarrow w) \rightarrow q \\ \sim (l \vee p) \\ (r \rightarrow w) \rightarrow k \\ \hline \sim q \end{array}$$

Problems

For each formal proof in Problems 19–23, supply the reason that justifies each statement.

19. $g \to \sim p$

g

$\sim p \to n$

---

n

Statement	Reason
1. $g \to \sim p$	1.
2. g	2.
3. $\sim p$	3.
4. $\sim p \to n$	4.
5. n	5.

20. $r \to (p \vee \sim s)$

$q \to r$

$\sim (p \vee \sim s)$

$\sim q$

Statement	Reason
1. $\sim (p \vee \sim s)$	1.
2. $r \to (p \vee \sim s)$	2.
3. $\sim r$	3.
4. $q \to r$	4.
5. $\sim q$	5.

21. $(s \wedge b) \to k$

$\sim (s \wedge b) \to c$

$\sim k$

---

c

Statement	Reason
1. $(s \wedge b) \to k$	1.
2. $\sim k$	2.
3. $\sim (s \wedge b)$	3.
4. $\sim (s \wedge b) \to c$	4.
5. c	5.

22. $(a \to b) \to \sim w$

$p \to \sim (r \vee s)$

$a \to b$

$\sim w \to (r \vee s)$

$\sim p$

Statement	Reason
1. $a \to b$	1.
2. $(a \to b) \to \sim w$	2.
3. $\sim w$	3.
4. $\sim w \to (r \vee s)$	4.
5. $r \vee s$	5.
6. $p \to \sim (r \vee s)$	6.
7. $\sim p$	7.

23. $(s \vee w) \rightarrow \sim d$

$g \rightarrow (c \rightarrow h)$

d

$\sim (s \vee w) \rightarrow g$

$c \rightarrow h$

Statement	Reason
1. d	1.
2. $(s \vee w) \rightarrow \sim d$	2.
3. $\sim (s \vee w)$	3.
4. $\sim (s \vee w) \rightarrow g$	4.
5. g	5.
6. $g \rightarrow (c \rightarrow h)$	6.
7. $c \rightarrow h$	7.

In Problems 24–29, construct a formal proof for each valid argument.

24. $r \rightarrow \sim s$

r

$\sim s$

25. $\sim r \rightarrow s$

$\sim s$

---

r

26. $p \rightarrow (q \vee r)$

p

$q \vee r$

27. $s \rightarrow (p \rightarrow r)$

$\sim (p \rightarrow r)$

$\sim s$

28. $p \rightarrow r$

$r \rightarrow q$

p

---

q

29. $p \rightarrow r$

$r \rightarrow q$

$\sim q$

$\sim p$

In Problems 30–39, use the *TF* method to test each argument for validity. If it is valid, construct a formal proof. If invalid, produce a counterexample chart.

30. $p \rightarrow r$

$q \rightarrow \sim r$

p

$\sim q$

31. $p \rightarrow q$

$\sim q$

$p \rightarrow r$

---

r

32. $(p \to q) \to r$
$\sim s \to \sim r$
$p \to q$
$\overline{}$
s

33. $w \to (p \wedge a)$
$\sim (p \wedge a)$
$\sim w \to k$
$\overline{}$
k

34. $(p \wedge q) \to (r \vee s)$
$\sim (r \vee s)$
$\sim (p \wedge q) \to m$
$\overline{}$
m

35. $a \to (d \vee g)$
$(d \vee g) \to \sim m$
a
$\sim r \to \sim m$
$\overline{}$
r

36. $\sim a$
$\sim s \to (p \leftrightarrow q)$
$s \to (r \vee w)$
$(r \vee w) \to a$
$\overline{}$
$p \leftrightarrow q$

37. $(c \vee d) \to q$
$q \to \sim r$
$a \to r$
$b \to a$
$c \vee d$
$\overline{}$
$\sim b$

38. $r \to w$
$p \to \sim (r \to w)$
$(s \vee q) \to b$
$\sim p \to (s \vee q)$
$\overline{}$
b

39. $(m \vee n) \to w$
$\sim (a \to b) \to (p \wedge q)$
$\sim (m \vee n) \to (p \wedge q)$
$\sim w$
$\overline{}$
$a \to b$

In Problems 40–43, express each argument symbolically. If the argument is valid, construct a formal proof. If not, produce a counterexample chart.

40. If there is a south wind, there will not be rain. If there is no increase in barometric pressure, there will be rain. There is a south wind. Therefore, there is an increase in barometric pressure.

41. If the butler or the cook is awake then the kitchen light is on. If the kitchen light is on, then the dog does not bark. If there is a prowler, the dog barks. If there is no prowler, the maid is in the hall and she stole the silver. The butler or the cook is awake. Therefore, the maid is in the hall and she stole the silver.

42. If my alarm doesn't ring in the morning I will have to take a later train to work. If my alarm does ring then I will get up early and not oversleep. If I get up early and do not oversleep then I will have time for my morning coffee. I didn't have time for my morning coffee. Therefore, I took the later train to work.

43. If Harvey is your friend then your friend is a rabbit. If Harvey is your friend, you watch too much television. If your friend is a rabbit, then you have an active imagination. You don't have an active imagination. Therefore, you don't watch too much television.

3.2 Valid Argument Forms–Part II

Disjunctive Syllogism

The third valid argument form we will consider is called *disjunctive syllogism*. This tautology can be represented symbolically as $[(p \vee q) \wedge \sim p] \rightarrow q$ or $[(p \vee q) \wedge \sim q] \rightarrow p$.

In its argument form, we can present the disjunctive syllogism tautology as:

$$\begin{array}{c} p \vee q \\ \sim p \\ \hline q \end{array} \quad \text{or} \quad \begin{array}{c} p \vee q \\ \sim q \\ \hline p \end{array}$$

Examining this tautology, we see that if one of the premises of an argument is a disjunction and the other premise is the negation of one of its parts, we can conclude that the remaining part of the disjunction must be true. To verify that disjunctive syllogism is a tautology, we will construct a truth table.

p	q	$p \vee q$	$(p \vee q) \wedge \sim p$	$[(p \vee q) \wedge \sim p] \rightarrow q$
T	T	T	F	T
T	F	T	F	T
F	T	T	T	T
F	F	F	F	T

Again, recognition of the disjunctive syllogism pattern is crucial. The name of the tautology should provide a reminder that this tautology involves a disjunction, not an implication, as one of its premises.

Example 1

Is the argument below an illustration of disjunctive syllogism?

$$\begin{array}{l} r \vee \sim s \\ \underline{s \qquad} \\ \quad r \end{array}$$

Solution

The premises are a disjunction and the negation of one of its parts. Therefore, we are guaranteed that the remaining part of the disjunction must be true. This is exactly the disjunctive syllogism pattern. ■

Example 2

Is the following argument an illustration of disjunctive syllogism?

$$\begin{array}{l} r \rightarrow \sim s \\ \underline{s \qquad} \\ \quad r \end{array}$$

Solution

This argument cannot be an illustration of disjunctive syllogism since a disjunction does not appear as one of the premises. ■

Example 3

Is the following argument an illustration of disjunctive syllogism?

$$\begin{array}{l} p \vee q \\ \underline{q \qquad} \\ \quad p \end{array}$$

Solution

Although one of the premises is a disjunction, the other is not the negation of one of its parts. Therefore, this is not an illustration of disjunctive syllogism. ■

Example 4

Is the argument below an illustration of disjunctive syllogism?

$$\begin{array}{c} (p \to r) \vee q \\ \underline{\sim (p \to r)} \\ q \end{array}$$

Solution

The premises consist of a disjunction and the negation of one of its parts. Therefore, we are guaranteed that the remaining part of the disjunction must be true. This is the disjunctive syllogism pattern. ■

Example 5

Express the given valid argument symbolically and construct a formal proof.

If it is Halloween I will buy candy.
It is Halloween or Christmas.
I will not buy candy.
Therefore, it is Christmas.

Solution

The argument can be represented symbolically as:

$$\begin{array}{c} h \to b \\ h \vee c \\ \underline{\sim b} \\ c \end{array}$$

We begin the formal proof with the simplest premise, $\sim b$.

Statement	Reason
1. $\sim b$	Premise

Since the statement b appears in the first premise, we proceed to:

Statement	Reason
1. $\sim b$	Premise
2. $h \to b$	Premise

Which tautology should we use to combine the two premises? Since we have an implication and the negation of its RHS as premises, modus tollens is applicable.

Statement	Reason
1. $\sim b$	Premise
2. $h \to b$	Premise
3. $\sim h$	Modus Tollens (1) (2)

We now combine this result with the second premise of the argument.

Statement	Reason
1. $\sim b$	Premise
2. $h \to b$	Premise
3. $\sim h$	Modus Tollens (1) (2)
4. $h \vee c$	Premise

Statements three and four tell us that the disjunction is true and the negation of one of its parts is also true. Therefore, disjunctive syllogism guarantees that the remaining part of the disjunction is true.

Statement	Reason
1. $\sim b$	Premise
2. $h \to b$	Premise
3. $\sim h$	Modus Tollens (1) (2)
4. $h \vee c$	Premise
5. c	Disjunctive Syllogism (3)(4)

Thus, we have proven that the argument is valid. ■

Example 6

Use the *TF* method to determine whether the given argument is valid. If so, construct a formal proof of its validity. If it is not valid, produce a counterexample chart.

$$\begin{array}{l} q \vee \sim r \\ \sim p \\ \underline{p \vee \sim q} \\ \sim r \end{array}$$

Solution

Since $\sim p$ is true, p is false. We know that $p \vee \sim q$ is true and p is false, so $\sim q$ must be true. This means that q is false. But $q \vee \sim r$ is true, hence $\sim r$ must be true. Therefore, the argument is valid. Now we proceed with a formal proof, following the same line of reasoning. Begin by combining the second and third premises.

Statement	Reason
1. $\sim p$	Premise
2. $p \vee \sim q$	Premise
3. $\sim q$	Disjunctive Syllogism (1) (2)

Now use the first premise.

Statement	Reason
1. $\sim p$	Premise
2. $p \vee \sim q$	Premise
3. $\sim q$	Disjunctive Syllogism (1) (2)
4. $q \vee \sim r$	Premise
5. $\sim r$	Disjunctive Syllogism (3) (4)

■

Example 7

Use the *TF* method to determine whether the given argument is valid. If so, provide a formal proof of the validity. If not, produce a counterexample chart.

$$b \vee \sim c$$
$$c$$
$$\sim a \rightarrow b$$
$$\overline{\sim a}$$

Solution

Since c is true, $\sim c$ is false. We know $b \vee \sim c$ is true and $\sim c$ is false, so b must be true. But $\sim a \rightarrow b$ is true, so $\sim a$ may be true or false. Since the conclusion cannot be forced to be true, the argument is invalid and a counterexample must be produced. The counterexample chart is constructed using truth values for a, b, and c, as per our discussion above.

a	b	c
T	T	T

That is, when a is true, b is true, and c is true, every premise is true, but the conclusion is false. Thus, we have shown that the argument is invalid.

■

Disjunctive Addition

Another argument form that involves disjunctions is called *disjunctive addition.* This tautology can be represented symbolically as $p \to (p \lor z)$. The tautology tells us that if we know that a statement, p, is true, we can join any other statement to p using the disjunction connective, and the resulting disjunction must be true.

In its argument form, the disjunctive addition tautology may be expressed as:

$$\frac{p}{p \lor z}$$

Notice that in each of the three previous valid argument forms we have studied, there were two premises from which we drew a conclusion. The disjunctive addition tautology has only one premise.

Example 8

Use the *TF* method to determine whether the given argument is valid. If so, construct a formal proof. If not, produce a counterexample chart.

$$\begin{array}{c} g \\ (g \lor h) \to w \\ \hline w \end{array}$$

Solution

Since g is true, $g \lor h$ must be true. We know $(g \lor h) \to w$ is also true, so w must be true. Since the conclusion is forced to be true, the argument is valid. We now supply a formal proof. The first premise is the simple statement g. The statement g also appears as part of the antecedent of the second premise. Therefore, we want to establish the fact that the antecedent is true.

Statement	Reason
1. g	Premise
2. $g \vee h$	Disjunctive Addition (1)

The disjunction $g \vee h$ is true because we have joined h to a known true statement, g. Now that we know that the LHS of the second premise is true, the proof can be completed.

Statement	Reason
1. g	Premise
2. $g \vee h$	Disjunctive Addition (1)
3. $(g \vee h) \rightarrow w$	Premise
4. w	Modus Ponens (2) (3)

Notice that in step three it was necessary to state that $(g \vee h) \rightarrow w$ was true in order to apply the modus ponens tautology. ■

Example 9

Construct a formal proof for the following valid argument.

$$\begin{array}{l} p \\ s \rightarrow \sim r \\ r \vee \sim p \\ \underline{(\sim s \vee q) \rightarrow z} \\ \quad z \end{array}$$

Solution

It appears that the simple statement p can be combined with the disjunction $r \vee \sim p$. Here we are combining a disjunctive statement and the negation of one of its parts to deduce a conclusion.

Statement	Reason
1. p	Premise
2. $r \vee \sim p$	Premise
3. r	Disjunctive Syllogism (1) (2)

The true statement r can be combined with $s \rightarrow \sim r$.

Statement	Reason
1. p	Premise
2. $r \vee \sim p$	Premise
3. r	Disjunctive Syllogism (1) (2)
4. $s \rightarrow \sim r$	Premise
5. $\sim s$	Modus Tollens (3) (4)

If we could show that the LHS of the last premise, $(\sim s \vee q) \rightarrow z$, is true, we could use modus ponens to deduce z. We just showed that $\sim s$ is true. Since disjunctive addition allows us to join any statement to a true statement using the disjunctive connective, we proceed by stating that $\sim s \vee q$ is true. Finally, we apply modus ponens.

Statement	Reason
1. p	Premise
2. $r \vee \sim p$	Premise
3. r	Disjunctive Syllogism (1) (2)
4. $s \rightarrow \sim r$	Premise
5. $\sim s$	Modus Tollens (3) (4)
6. $\sim s \vee q$	Disjunctive Addition (5)
7. $(\sim s \vee q) \rightarrow z$	Premise
8. z	Modus Ponens (6) (7)

■

Example 10

Use the *TF* method to determine whether the given argument is valid. If so, construct a formal proof. If not, produce a counterexample chart.

$$\begin{array}{l} p \rightarrow q \\ p \\ \underline{q \rightarrow (r \vee s)} \\ \quad s \end{array}$$

Solution

We know p and $p \to q$ are true. Therefore, q is true. Since $q \to (r \vee s)$ is true, $r \vee s$ must be true. However, we cannot discern which component of this disjunction is true. In particular, we cannot force s to be true. Thus, the argument is invalid. The counterexample will show that when r is true and s is false, all the premises will be true, but the conclusion will be false.

p	q	r	s
T	T	T	F

■

In-Class Exercises and Problems for Section 3.2

In-Class Exercises

In Exercises 1–10, state the illustrated valid argument form.

1. $\begin{array}{l} \sim a \vee h \\ \underline{\sim h} \\ \sim a \end{array}$

2. $\begin{array}{l} (p \to q) \vee \sim b \\ \underline{b} \\ p \to q \end{array}$

3. $\begin{array}{l} \underline{\sim c} \\ \sim c \vee s \end{array}$

4. $\begin{array}{l} \underline{m} \\ m \vee (a \to c) \end{array}$

5. $\begin{array}{l} (k \wedge p) \vee \sim d \\ \underline{\sim (k \wedge p)} \\ \sim d \end{array}$

6. $\begin{array}{l} \underline{(n \leftrightarrow s)} \\ (n \leftrightarrow s) \vee \sim r \end{array}$

7. $\begin{array}{l} (a \vee \sim b) \to \sim d \\ \underline{d} \\ \sim (a \vee \sim b) \end{array}$

8. $\begin{array}{l} r \vee \sim (g \to \sim k) \\ \underline{\sim r} \\ \sim (g \to \sim k) \end{array}$

9. $\begin{array}{l} w \to (s \wedge \sim p) \\ \underline{w} \\ s \wedge \sim p \end{array}$

10. $\begin{array}{l} \underline{(s \to p) \wedge k} \\ [(s \to p) \wedge k] \vee (m \wedge n) \end{array}$

In Exercises 11–28, test each argument for validity. If it is valid, provide a formal proof. If it is invalid, provide a counterexample chart.

11. $$\begin{array}{l} a \\ b \rightarrow c \\ \sim a \vee b \\ \hline c \end{array}$$

12. $$\begin{array}{l} r \\ r \rightarrow (s \vee q) \\ \sim s \\ \hline q \end{array}$$

13. $$\begin{array}{l} p \rightarrow \sim q \\ q \\ \sim p \vee \sim r \\ \hline r \end{array}$$

14. $$\begin{array}{l} w \rightarrow c \\ w \\ (c \vee p) \rightarrow b \\ \hline b \end{array}$$

15. $$\begin{array}{l} p \vee (q \wedge r) \\ \sim (q \wedge r) \\ p \rightarrow s \\ \hline s \end{array}$$

16. $$\begin{array}{l} p \rightarrow w \\ \sim w \\ (\sim p \vee s) \rightarrow k \\ \hline k \end{array}$$

17. $$\begin{array}{l} (a \rightarrow b) \vee c \\ \sim c \\ \sim b \\ \hline \sim a \vee p \end{array}$$

18. $$\begin{array}{l} m \vee p \\ \sim p \\ w \rightarrow (m \vee l) \\ \hline \sim w \end{array}$$

19. $$\begin{array}{l} a \rightarrow \sim b \\ b \\ c \rightarrow a \\ c \vee d \\ \hline d \end{array}$$

20. $$\begin{array}{l} p \\ s \rightarrow \sim r \\ \sim p \vee r \\ \sim s \rightarrow w \\ \hline w \end{array}$$

21. $\sim p \vee d$
$\sim a$
$p \rightarrow \sim s$
$\sim a \rightarrow s$
———
d

22. $(w \vee z) \rightarrow k$
$p \rightarrow q$
$\sim q$
$\sim p \rightarrow w$
———
k

23. $\sim c \rightarrow \sim d$
$d \vee q$
$\sim q$
———
$c \vee (p \rightarrow s)$

24. $(r \rightarrow w) \rightarrow (s \vee g)$
$r \rightarrow w$
$\sim s$
———
$g \vee p$

25. $a \vee (b \rightarrow c)$
$\sim a \rightarrow \sim d$
$\sim (b \rightarrow c)$
———
d

26. $m \vee \sim s$
s
$n \rightarrow \sim (m \vee q)$
———
$\sim n$

27. $r \vee \sim s$
s
$(r \vee w) \rightarrow p$
$q \rightarrow \sim p$
———
$\sim q$

28. $\sim p \rightarrow q$
$\sim (s \vee c)$
$(\sim r \vee w) \rightarrow d$
$\sim (\sim p \rightarrow q) \vee [r \rightarrow (s \vee c)]$
———
d

For Exercises 29–31, symbolize each argument. If the argument is valid, provide a formal proof. If not, provide a counterexample chart.

29. If I travel to the southwest United States, I will see the Indian ruins. If I see either the Indian ruins or the Painted Desert, then I will not go to San Francisco. Either I go to San Francisco or my trip was cancelled. I travel to the southwest United States. Therefore, my trip was cancelled.

30. If it is very cold then I can ski and ice skate. If I can ski and ice skate then it is not fall. Either it is fall or spring. It is very cold. Therefore, it is spring.

31. If your major is engineering then you must take calculus. If you take calculus, you will not be able to work after school. You either work after school or take out a loan. You do not take out a loan. Therefore, your major is not engineering.

Problems

For each formal proof in Problems 32–36, supply the reason that justifies each statement.

32. $c \vee \sim d$
d
$p \to \sim c$
$\overline{\qquad}$
$\sim p$

Statement	Reason
1. d	1.
2. $c \vee \sim d$	2.
3. c	3.
4. $p \to \sim c$	4.
5. $\sim p$	5.

33. $w \to s$
w
$(s \vee m) \to \sim n$
$\overline{\qquad}$
$\sim n$

Statement	Reason
1. w	1.
2. $w \to s$	2.
3. s	3.
4. $s \vee m$	4.
5. $(s \vee m) \to \sim n$	5.
6. $\sim n$	6.

34. $a \vee b$

$(g \vee b) \to c$

$\sim q \to \sim c$

$\sim (a \vee b) \vee g$

---

q

Statement	Reason
1. $a \vee b$	1.
2. $\sim (a \vee b) \vee g$	2.
3. g	3.
4. $g \vee b$	4.
5. $(g \vee b) \to c$	5.
6. c	6.
7. $\sim q \to \sim c$	7.
8. q	8.

35. $\sim (p \to r)$

$w \vee q$

$\sim (p \to r) \to h$

$q \to \sim h$

$w \vee a$

Statement	Reason
1. $\sim (p \to r)$	1.
2. $\sim (p \to r) \to h$	2.
3. h	3.
4. $q \to \sim h$	4.
5. $\sim q$	5.
6. $w \vee q$	6.
7. w	7.
8. $w \vee a$	8.

36. p
$p \to (a \vee \sim c)$
$(w \to b) \to \sim s$
$(a \vee g) \to s$
c
$\overline{\sim (w \to b) \vee z}$

Statement	Reason
1. p	1.
2. $p \to (a \vee \sim c)$	2.
3. $a \vee \sim c$	3.
4. c	4.
5. a	5.
6. $a \vee g$	6.
7. $(a \vee g) \to s$	7.
8. s	8.
9. $(w \to b) \to \sim s$	9.
10. $\sim (w \to b)$	10.
11. $\sim (w \to b) \vee z$	11.

In Problems 37–42, construct a formal proof for each valid argument.

37. $q \vee \sim p$
$\sim q$
$\sim r \to p$
$\overline{r}$

38. $d \to \sim e$
e
$d \vee g$
$\overline{g}$

39. $(m \vee p) \to s$
$c \to m$
c
$\overline{s}$

40. $(c \to k) \vee (k \to p)$
$\sim (c \to k)$
$\sim p$
$\overline{\sim k}$

41. s
$(a \vee b) \to c$
$s \to a$
$\overline{c}$

42. $r \to (s \vee g)$
r
$\sim s$
$\overline{g \vee p}$

In Problems 43–56, test each argument for validity. If it is valid, construct a formal proof. If it is invalid, produce a counterexample chart.

43. w
$w \to (p \to q)$
p
———
$q \vee r$

44. $r \to q$
$\sim q$
$\sim w \vee \sim r$
———
w

45. $m \vee \sim n$
n
$m \to (p \vee r)$
$\sim r$
———
p

46. $s \vee p$
$b \vee a$
$p \to b$
$\sim s$
———
a

47. a
$(s \vee q) \to w$
$(a \vee p) \to k$
$k \to s$
———
w

48. $a \vee c$
$p \to r$
$\sim r$
$\sim p \to \sim a$
———
$c \vee q$

49. $(b \to c) \vee a$
$\sim a$
$(c \vee w) \to p$
b
———
p

50. $\sim r \to w$
$\sim s \vee p$
$\sim w$
$(r \vee q) \to p$
———
s

51. $s \vee w$
$w \to a$
$(a \vee b) \to p$
$\sim s$
———
$p \vee q$

52. $k \leftrightarrow l$
$\sim (k \leftrightarrow l) \vee p$
$\sim s \vee (a \to c)$
$(p \vee w) \to s$
———
$a \to c$

53. $\sim p \rightarrow (a \vee b)$
$\sim s$
$a \rightarrow q$
$\sim p$
$b \rightarrow s$
———
q

54. $(a \rightarrow b) \rightarrow (c \vee d)$
$\sim c$
$a \rightarrow b$
$(d \vee k) \rightarrow \sim b$
———
$\sim a$

55. $(m \leftrightarrow n) \vee r$
a
$\sim (m \leftrightarrow n)$
$r \rightarrow p$
$\sim (a \rightarrow m) \rightarrow \sim p$
———
m

56. w
$(p \vee k) \rightarrow c$
$w \rightarrow (p \vee s)$
$\sim z \vee c$
$\sim s$
———
z

For Problems 57–61, express each argument symbolically. If the argument is valid, construct a formal proof. If not, produce a counterexample chart.

57. If you like baseball, then you watch the World Series at home. If you have season tickets, then you do not watch the World Series at home. Either you have season tickets or you invite friends to your home. You like baseball. Therefore, you invite friends to your home.

58. If Emma invites Pat or Jean to her party then she invites Max. Emma does not invite Shawn to her party if she invites Max. Either Shawn is invited to her party or Emma's party is cancelled. Emma invites Pat to her party. Therefore, Emma's party is cancelled.

59. Either you do your homework or you do not pass your logic class. If you pass your logic class then you studied two hours a day. You did not study two hours a day. Therefore, you did not do your homework.

60. It will snow if the temperature is below freezing. Either it doesn't snow or I shovel my driveway. If the temperature is not below freezing or it rains, the roads become slippery. I don't shovel my driveway. Therefore, the roads become slippery.

61. If a murder is committed then either NCIS or the Criminal Minds Team is called in to investigate. If NCIS is called in to investigate then Gibbs is in charge of the case. Gibbs is not in charge of the case. A murder is committed. Therefore, either the Criminal Minds Team is called in to investigate or CSI Miami solved the case.

3.3 Valid Argument Forms–Part III

Conjunctive Addition

There are two commonly used argument forms that involve conjunctions. The first one that we will consider is *conjunctive addition*. This is expressed symbolically as:

$$\begin{array}{l} p \\ \underline{q \quad\;} \\ p \wedge q \end{array}$$

which states that if we know that two statements are true, we can join them with the conjunction connective and the resulting conjunction will be true.

This argument form is similar to disjunctive addition. However, with disjunctive addition we only needed to know that one statement was true in order to produce a true disjunctive statement. Since the only way a conjunction can be true is if both of its components are true, we need to know that two statements are true in order to state that their conjunction is true.

Example 1

Construct a formal proof for the following valid argument.

$$\begin{array}{l} (\sim r \wedge p) \rightarrow s \\ p \\ \underline{r \rightarrow \sim p \quad\;} \\ \quad s \vee b \end{array}$$

Solution

We begin with the second premise, p, and the third premise, $r \to \sim p$.

Statement	Reason
1. p	Premise
2. $r \to \sim p$	Premise
3. $\sim r$	Modus Tollens (1) (2)

Now that we know $\sim r$ is true, we can apply the conjunctive addition argument form to produce $\sim r \wedge p$, the LHS of the first premise.

Statement	Reason
1. p	Premise
2. $r \to \sim p$	Premise
3. $\sim r$	Modus Tollens (1) (2)
4. $\sim r \wedge p$	Conjunctive Addition (1) (3)

Since $\sim r \wedge p$ and $(\sim r \wedge p) \to s$ are both true, we apply modus ponens to deduce s.

Statement	Reason
1. p	Premise
2. $r \to \sim p$	Premise
3. $\sim r$	Modus Tollens (1) (2)
4. $\sim r \wedge p$	Conjunctive Addition (1) (3)
5. $(\sim r \wedge p) \to s$	Premise
6. s	Modus Ponens (4) (5)

Since we know s is true, disjunctive addition allows us to join any statement to it and the resulting disjunction will be true as well.

Statement	Reason
1. p	Premise
2. $r \to \sim p$	Premise
3. $\sim r$	Modus Tollens (1) (2)
4. $\sim r \wedge p$	Conjunctive Addition (1)(3)
5. $(\sim r \wedge p) \to s$	Premise
6. s	Modus Ponens (4) (5)
7. $s \vee b$	Disjunctive Addition (6)

Thus, we have shown that the argument is valid. ■

Example 2

Write a formal proof for the following valid argument.

$$\begin{array}{l} \sim s \\ \sim s \to q \\ (p \vee q) \to r \\ \hline q \wedge r \end{array}$$

Solution

Statement	Reason
1. $\sim s$	Premise
2. $\sim s \to q$	Premise
3. q	Modus Ponens (1) (2)
4. $p \vee q$	Disjunctive Addition (3)
5. $(p \vee q) \to r$	Premise
6. r	Modus Ponens (4) (5)
7. $q \wedge r$	Conjunctive Addition (3) (6)

Observe the use of both disjunctive addition and conjunctive addition in this proof. When disjunctive addition was applied in step four, it was only necessary to know one statement was true, namely, q. When conjunctive addition was applied in step seven, we had to know that both r and q were true.

■

Conjunctive Simplification

The second argument form that involves a conjunction is *conjunctive simplification.* This argument form can be expressed symbolically as:

$$\frac{p \wedge q}{p} \quad \text{or} \quad \frac{p \wedge q}{q}$$

which means that if a conjunction is true, each of its components must be true. Conjunctive simplification may be thought of as conjunctive addition "in reverse".

Notice that there is no analogous argument for the disjunction. That is, knowing that a disjunction is true does not provide us with information as to which specific component of the disjunction is true.

Example 3

Construct a formal proof for the following valid argument.

$$\begin{array}{c} p \wedge q \\ \sim r \rightarrow \sim p \\ \hline r \wedge q \end{array}$$

Solution

We begin by applying conjunctive simplification to $p \wedge q$.

Statement	Reason
1. $p \wedge q$	Premise
2. p	Conjunctive Simplification (1)
3. q	Conjunctive Simplification (1)

Since p is the negation of the RHS of $\sim r \rightarrow \sim p$, we can use modus tollens to deduce that r is true.

Statement	Reason
1. $p \wedge q$	Premise
2. p	Conjunctive Simplification (1)
3. q	Conjunctive Simplification (1)
4. $\sim r \to \sim p$	Premise
5. r	Modus Tollens (2) (4)

We now know that both r and q are true. Therefore, $r \wedge q$ is true by conjunctive addition.

Statement	Reason
1. $p \wedge q$	Premise
2. p	Conjunctive Simplification (1)
3. q	Conjunctive Simplification (1)
4. $\sim r \to \sim p$	Premise
5. r	Modus Tollens (2) (4)
6. $r \wedge q$	Conjunctive Addition (3) (5)

Thus, we have shown that the argument is valid. ■

Example 4

Use the *TF* method to determine whether the argument below is valid. If so, construct a formal proof. If not, produce a counterexample chart.

$$\begin{array}{c} (p \vee q) \to r \\ (w \wedge r) \to m \\ p \vee q \\ (r \vee s) \to w \\ \sim m \vee b \\ \hline b \end{array}$$

Solution

Since $p \vee q$ and $(p \vee q) \to r$ are premises and thus both true, r is true by modus ponens. By disjunctive addition, $r \vee s$ is true. Since $(r \vee s) \to w$ is a premise and its LHS is true, w, its RHS is true by modus ponens. Now, $w \wedge r$ is true by conjunctive addition. Since this is the LHS of the premise $(w \wedge r) \rightarrow m$, m is true by modus ponens. The disjunction $\sim m \vee b$ is a premise and is therefore true. But m is true, so b must be true by disjunctive syllogism. These steps are shown in the following formal proof.

Statement	Reason
1. $p \vee q$	Premise
2. $(p \vee q) \rightarrow r$	Premise
3. r	Modus Ponens (1) (2)
4. $r \vee s$	Disjunctive Addition (3)
5. $(r \vee s) \rightarrow w$	Premise
6. w	Modus Ponens (4) (5)
7. $w \wedge r$	Conjunctive Addition (3) (6)
8. $(w \wedge r) \rightarrow m$	Premise
9. m	Modus Ponens (7) (8)
10. $\sim m \vee b$	Premise
11. b	Disjunctive Syllogism (9) (10)

Therefore, we have shown that the argument is valid. ■

Example 5

Use the *TF* method to determine whether the argument below is valid. If so, construct a formal proof. If not, produce a counterexample chart.

$$\begin{array}{l} (r \vee s) \rightarrow (b \vee a) \\ \sim b \wedge r \\ (a \wedge c) \rightarrow d \\ \hline \qquad d \end{array}$$

Solution

Since the conjunction $\sim b \wedge r$ is true, each of its components, r and $\sim b$, is true. Since r is true, $r \vee s$ is true. Therefore, $b \vee a$ is true. We already know $\sim b$ is true, so a must be true. We are unable to proceed further, since we know nothing about the truth values of c and d in the third premise. This suggests that the argument may be invalid. If so, d would have to be false. If c were false as well, the third premise would be true, and yet the conclusion, d, would be false. We now have enough information to produce a counterexample chart. Notice that since the truth value of s is irrelevant, there are two, equally correct counterexample charts, one for the case when s is true, the other, when s is false. All the other truth values have been deduced from the above line of reasoning.

a	b	c	d	r	s
T	F	F	F	T	T
T	F	F	F	T	F

■

In-Class Exercises and Problems for Section 3.3

In-Class Exercises

In Exercises 1–12, state the illustrated valid argument form.

1. $$\frac{s \wedge \sim p}{s}$$

2. $$\frac{\begin{array}{c} k \\ \sim d \end{array}}{k \wedge \sim d}$$

3. $$\frac{\begin{array}{c} p \vee q \\ r \end{array}}{(p \vee q) \wedge r}$$

4. $$\frac{(\sim c \to a) \wedge \sim w}{\sim c \to a}$$

5. $$\frac{q \wedge \sim s}{(q \wedge \sim s) \vee m}$$

6. $$\frac{\begin{array}{c} j \\ p \to s \end{array}}{j \wedge (p \to s)}$$

7. $$\frac{\begin{array}{c} h \vee (s \wedge \sim n) \\ \sim (s \wedge \sim n) \end{array}}{h}$$

8. $$\frac{\begin{array}{c} p \to (a \wedge \sim b) \\ \sim (a \wedge \sim b) \end{array}}{\sim p}$$

9. $$\frac{(r \to s) \wedge (a \vee b)}{r \to s}$$

10. $$\frac{w \wedge k}{(w \wedge k) \vee (d \to a)}$$

11. $(s \to d) \to \sim(g \to h)$
$s \to d$
$\overline{}$
$\sim(g \to h)$

12. $\sim(\sim a \vee b)$
$c \to (\sim a \vee b)$
$\overline{}$
$\sim c$

In Exercises 13–28, test each argument for validity. If it is valid, provide a formal proof. If it is invalid, provide a counterexample chart.

13. $\sim p \vee s$
p
$(p \wedge s) \to q$
$\overline{}$
q

14. $p \to \sim q$
$q \to r$
$p \wedge b$
$\overline{}$
$\sim r$

15. $m \wedge n$
$p \to w$
$\sim m \vee p$
$\overline{}$
w

16. $s \to (a \vee w)$
$s \wedge g$
$d \to \sim(a \vee w)$
$\overline{}$
$\sim d$

17. $\sim p \vee (\sim s \to \sim w)$
$s \to \sim p$
p
$\overline{}$
$\sim s \wedge \sim w$

18. $(b \wedge p) \to s$
$a \to p$
$a \wedge b$
$\overline{}$
s

19. $r \to w$
$p \to (w \wedge s)$
$r \wedge s$
$\overline{}$
$\sim p$

20. $g \vee \sim h$
$k \to \sim g$
$a \wedge k$
$\overline{}$
$a \wedge \sim h$

21. $\sim a \wedge b$
$\sim d \to e$
$a \vee c$
$d \to \sim c$
$\overline{}$
e

22. $s \to b$
$s \vee c$
$(c \wedge \sim s) \to k$
$\sim b$
$\overline{}$
k

23. $r \vee \sim s$
$\sim w \rightarrow \sim p$
s
$r \rightarrow (p \wedge q)$
$\overline{\quad w \vee c \quad}$

24. $p \rightarrow s$
$c \wedge g$
$j \vee p$
$c \rightarrow \sim j$
$\overline{\quad s \wedge g \quad}$

25. k
$c \rightarrow (a \wedge b)$
$(b \vee z) \rightarrow m$
$c \vee \sim k$
$\overline{\quad m \quad}$

26. $(s \vee w) \rightarrow m$
$p \wedge b$
$r \rightarrow \sim q$
$s \vee r$
$p \rightarrow q$
$\overline{\quad m \quad}$

27. $(j \vee c) \rightarrow d$
$d \rightarrow k$
$\sim a \rightarrow j$
$(\sim a \wedge s) \vee p$
$\sim p$
$\overline{\quad k \wedge s \quad}$

28. $a \wedge \sim b$
$a \rightarrow e$
$\sim d \vee (c \vee \sim a)$
$(\sim b \wedge e) \rightarrow d$
$p \rightarrow c$
$\overline{\quad \sim p \quad}$

In Exercises 29–31, symbolize each argument. If the argument is valid, provide a formal proof. If not, provide a counterexample chart.

29. Every winter I go cross-country skiing or snowshoeing. If I go snowshoeing I have to buy new boots. If I go cross-country skiing but I don't go snowshoeing then I can save money. I didn't buy new boots. Therefore, I saved money.

30. If you like to take pictures then you should be a photographer. If you want to be a photographer then you own a camera and can develop your own film. If you can develop your own film then either you have a darkroom in your house or you have a

rich uncle. You like to take pictures but don't have a rich uncle. Therefore, you own a camera and have a darkroom in your house.

31. Jay likes to play his guitar or ride his motorcycle, but he doesn't like to get up early. If Jay doesn't wear his helmet he doesn't ride his motorcycle. If Jay wants to be considered a safe rider he wears his helmet. Jay does not play his guitar. Therefore, Jay is considered to be a safe driver.

Problems

In Problems 32–36, supply the reason that justifies each statement in the given valid argument.

32. $a \wedge \sim c$

$\sim s \rightarrow p$

$\underline{c \vee \sim p}$

s

Statement	Reason
1. $a \wedge \sim c$	1.
2. $\sim c$	2.
3. $c \vee \sim p$	3.
4. $\sim p$	4.
5. $\sim s \rightarrow p$	5.
6. s	6.

33. $\sim s$

$\sim s \rightarrow (q \wedge b)$

$\underline{\sim b \vee p}$

$p \wedge q$

Statement	Reason
1. $\sim s$	1.
2. $\sim s \rightarrow (q \wedge b)$	2.
3. $q \wedge b$	3.
4. b	4.
5. q	5.
6. $\sim b \vee p$	6.
7. p	7.
8. $p \wedge q$	8.

34. $(k \vee p) \rightarrow s$

$d \rightarrow \sim (s \wedge \sim m)$

$k \wedge \sim m$

$\sim d$

Statement	Reason
1. $k \wedge \sim m$	1.
2. k	2.
3. $\sim m$	3.
4. $k \vee p$	4.
5. $(k \vee p) \rightarrow s$	5.
6. s	6.
7. $s \wedge \sim m$	7.
8. $d \rightarrow \sim (s \wedge \sim m)$	8.
9. $\sim d$	9.

35. $(d \wedge p) \vee k$

$\sim a \rightarrow \sim p$

$(a \wedge d) \rightarrow m$

$\sim k$

$d \wedge m$

Statement	Reason
1. $\sim k$	1.
2. $(d \wedge p) \vee k$	2.
3. $d \wedge p$	3.
4. d	4.
5. p	5.
6. $\sim a \rightarrow \sim p$	6.
7. a	7.
8. $a \wedge d$	8.
9. $(a \wedge d) \rightarrow m$	9.
10. m	10.
11. $d \wedge m$	11.

36. $p \vee r$

$\sim(s \wedge w) \rightarrow \sim(p \vee r)$

$(w \vee a) \rightarrow g$

$\sim(g \wedge s) \vee k$

---

k

Statement	Reason
1. $p \vee r$	1.
2. $\sim(s \wedge w) \rightarrow \sim(p \vee r)$	2.
3. $s \wedge w$	3.
4. s	4.
5. w	5.
6. $w \vee a$	6.
7. $(w \vee a) \rightarrow g$	7.
8. g	8.
9. $g \wedge s$	9.
10. $\sim(g \wedge s) \vee k$	10.
11. k	11.

In Problems 37–42, construct a formal proof for each valid argument.

37. m

n

$\sim(m \wedge n) \vee q$

---

q

38. $b \wedge \sim s$

$\sim b \vee q$

$(q \wedge \sim s) \rightarrow l$

---

l

39. $q \rightarrow s$

$p \wedge q$

$w \rightarrow \sim s$

$\sim w$

40. $\sim a \rightarrow b$

$\sim b$

$(a \wedge \sim b) \rightarrow r$

---

r

41. w

$\sim(a \wedge c) \rightarrow \sim r$

$w \rightarrow r$

---

a

42. $\sim c \vee g$

$c \wedge d$

$(d \wedge g) \rightarrow p$

---

p

In Problems 43–56, test each argument for validity. If it is valid, construct a formal proof. If it is invalid, produce a counterexample chart.

43. s
$(s \vee q) \to p$
$\underline{\sim(p \wedge s) \vee a}$
a

44. $r \wedge \sim w$
$\sim(r \vee s) \vee q$
$\underline{q \to \sim d}$
$\sim d$

45. q
$(\sim p \to \sim q) \wedge (\sim r \vee s)$
$\underline{r}$
$p \wedge s$

46. $a \to g$
$\sim(d \wedge h) \to \sim(a \to g)$
$\underline{(d \vee n) \to s}$
$s \wedge h$

47. $\sim(s \wedge q) \to a$
$\sim p \vee s$
$\underline{p \wedge q}$
$\sim a$

48. $(p \wedge s) \to b$
$(r \leftrightarrow q) \to p$
s
$\underline{r \leftrightarrow q}$
b

49. $(p \vee q) \wedge w$
$\sim p$
$r \to \sim q$
$\underline{\sim r \to z}$
z

50. $(p \wedge q) \to r$
$\sim w \wedge \sim s$
$\sim q \to w$
$\underline{s \vee p}$
r

51. $\sim p \to \sim r$
$q \wedge r$
$\underline{(p \vee s) \to w}$
w

52. $k \to p$
$r \to s$
$\sim s \vee \sim c$
$\underline{r \wedge (p \vee c)}$
$\sim k$

53. $\sim w \vee (a \wedge b)$
$(b \wedge s) \rightarrow q$
$w \wedge s$
$(q \vee z) \rightarrow c$
―――――
c

54. $\sim s$
$r \vee q$
$\sim w \vee (a \wedge b)$
$r \rightarrow s$
$q \rightarrow w$
―――――
$b \vee j$

55. $\sim p \rightarrow q$
$(w \vee b) \rightarrow c$
$\sim p \wedge r$
$(s \wedge r) \rightarrow w$
$\sim q \vee s$
―――――
c

56. $(c \vee d) \wedge (p \rightarrow q)$
$d \rightarrow p$
$(d \wedge q) \rightarrow z$
a
$c \rightarrow \sim a$
―――――
z

In Problems 57–60, express each argument symbolically. If the argument is valid, construct a formal proof. If not, produce a counterexample chart.

57. If you join the circus you want to be a clown but not a lion tamer. Either you join the circus or become an acrobat. If you want to be a clown then you like to make people laugh. You don't become an acrobat. Therefore, you like to make people laugh or you enjoy standing on your head.

58. If Phil can do many card tricks and make quarters disappear, then Phil is a magician. If Phil can't make quarters disappear he needs more practice. Phil can either pull rabbits out of a hat or do many card tricks. Phil doesn't need more practice and he can't pull rabbits out of a hat. Therefore, Phil is a magician.

59. If you like downhill skiing then you like cold weather. Either you like downhill skiing or ice hockey. If you like ice skating then you don't like ice hockey. You like both ice skating and ice dancing. Therefore, you like ice dancing and cold weather.

60. If the Giants do not make the playoffs, then the Jets will. The Giants will make the playoffs if the Steelers don't, and either the Dolphins win or the Giants don't make the playoffs. If the Jets and Steelers make the playoffs then so will the Lions. The Dolphins do not win. Therefore, the Lions make the playoffs.

3.4 Using Equivalences in Proofs

Introduction

In this section, we will use the conditional equivalence, the contrapositive equivalence, the conditional negation equivalence, the DeMorgan's equivalences and the distributive equivalences in formal proofs, in much the same way we did in Section 2.4.

Using the Conditional Equivalence in Proofs

Recall that the conditional equivalence allows us to express a conditional statement as a disjunction to which it is logically equivalent. This equivalence is expressed symbolically as $(p \to q) \Leftrightarrow (\sim p \vee q)$, that is, $p \to q$ and $\sim p \vee q$ are logically equivalent. Notice that the left side of the disjunction is the negation of the LHS of the original conditional statement.

Example 1

Construct a formal proof for the following valid argument.

$$\begin{array}{l} p \to (r \to s) \\ p \\ (\sim r \vee s) \to w \\ \hline p \wedge w \end{array}$$

Solution

We begin with the premises p and $p \to (r \to s)$ to conclude $r \to s$ by modus ponens.

Statement	Reason
1. p	Premise
2. $p \to (r \to s)$	Premise
3. $r \to s$	Modus Ponens (1) (2)

We continue by recognizing that $r \to s$ is equivalent to $\sim r \vee s$ by the conditional equivalence.
This is useful since $\sim r \vee s$ is the LHS of the premise $(\sim r \vee s) \to w$, and thus w is true by modus ponens.

Statement	Reason
1. p	Premise
2. $p \to (r \to s)$	Premise
3. $r \to s$	Modus Ponens (1) (2)
4. $\sim r \vee s$	Conditional Equivalence (3)
5. $(\sim r \vee s) \to w$	Premise
6. w	Modus Ponens (4) (5)

Now, since both p and w are true, their conjunction is true by conjunctive addition.

Statement	Reason
1. p	Premise
2. $p \to (r \to s)$	Premise
3. $r \to s$	Modus Ponens (1) (2)
4. $\sim r \vee s$	Conditional Equivalence (3)
5. $(\sim r \vee s) \to w$	Premise
6. w	Modus Ponens (4) (5)
7. $p \wedge w$	Conjunctive Addition (1) (6)

Thus, we have shown that the argument is valid. ■

Example 2

Construct a formal proof for the following valid argument.

$$\begin{array}{c} (p \to q) \to (r \vee w) \\ \underline{(\sim p \vee q) \wedge \sim r} \\ w \vee z \end{array}$$

Solution

Begin with the conjunctive statement.

Statement	Reason
1. $(\sim p \vee q) \wedge \sim r$	Premise
2. $\sim p \vee q$	Conjunctive Simplification (1)
3. $\sim r$	Conjunctive Simplification (1)

Since $\sim p \vee q$ is equivalent to $p \to q$ by the conditional equivalence and $p \to q$ is the LHS of $(p \to q) \to (r \vee w)$, we can deduce $r \vee w$ by modus ponens. Notice that we must state both the conditional equivalence and modus ponens to deduce $r \vee w$.

Statement	Reason
1. $(\sim p \vee q) \wedge \sim r$	Premise
2. $\sim p \vee q$	Conjunctive Simplification (1)
3. $\sim r$	Conjunctive Simplification (1)
4. $p \to q$	Conditional Equivalence (2)
5. $(p \to q) \to (r \vee w)$	Premise
6. $r \vee w$	Modus Ponens (4) (5)

We know $\sim r$ is true from statement three. Therefore, w is true by disjunctive syllogism.
Finally, since w is true, we can join any statement to it using disjunctive addition, and the resulting disjunction will also be true. We can then conclude that $w \vee z$ is true.
Therefore, the argument is valid. The completed proof follows. Be sure to understand that statement four is true not because it is a premise, but rather because it is equivalent to statement two.

Statement	Reason
1. $(\sim p \vee q) \wedge \sim r$	Premise
2. $\sim p \vee q$	Conjunctive Simplification (1)
3. $\sim r$	Conjunctive Simplification (1)
4. $p \to q$	Conditional Equivalence (2)
5. $(p \to q) \to (r \vee w)$	Premise
6. $r \vee w$	Modus Ponens (4) (5)
7. w	Disjunctive Syllogism (3) (6)
8. $w \vee z$	Disjunctive Addition (7)

■

Using the Contrapositive Equivalence in Proofs

The contrapositive of a conditional statement is formed by interchanging its LHS with its RHS, and negating both sides of the new implication. In Section 1.6, we learned that any conditional statement is logically equivalent to its contrapositive. Symbolically, we write $(p \to q) \Leftrightarrow (\sim q \to \sim p)$.

Example 3

Construct a formal proof for the following valid argument.

$$\begin{array}{l} p \to (\sim q \to r) \\ (\sim p \vee s) \to w \\ \underline{\sim (\sim r \to q)} \\ \quad w \end{array}$$

Solution

The conditional statement within the parentheses in the third premise is the contrapositive of the RHS of the first premise. Therefore, the third premise is the negation of the RHS of the first premise. We can then deduce $\sim p$ by modus tollens.

Statement	Reason
1. $\sim (\sim r \to q)$	Premise
2. $\sim (\sim q \to r)$	Contrapositive Equivalence (1)
3. $p \to (\sim q \to r)$	Premise
4. $\sim p$	Modus Tollens (2) (3)

Since $\sim p$ is true, $\sim p \vee s$ is also true by disjunctive addition, and $\sim p \vee s$ is the LHS of $(\sim p \vee s) \to w$. Hence, we deduce w by modus ponens.

Statement	Reason
1. $\sim (\sim r \to q)$	Premise
2. $\sim (\sim q \to r)$	Contrapositive Equivalence (1)
3. $p \to (\sim q \to r)$	Premise
4. $\sim p$	Modus Tollens (2) (3)
5. $\sim p \vee s$	Disjunctive Addition (4)
6. $(\sim p \vee s) \to w$	Premise
7. w	Modus Ponens (5) (6)

This completes the proof. ■

Using the Conditional Negation Equivalence in Proofs

The conditional negation equivalence is expressed symbolically as $\sim(p \to q) \Leftrightarrow (p \wedge \sim q)$. That is, the negation of a conditional statement is logically equivalent to the conjunction of its LHS with the negation of its RHS.

Example 4

Produce a formal proof for the following valid argument.

$$\begin{array}{l} p \wedge (q \vee \sim z) \\ (r \wedge \sim s) \to z \\ \underline{(r \to s) \to \sim p} \\ \quad q \end{array}$$

Solution

We begin with the conjunctive statement.

Statement	Reason
1. $p \wedge (q \vee \sim z)$	Premise
2. p	Conjunctive Simplification (1)
3. $q \vee \sim z$	Conjunctive Simplification (1)

Now combine p with the third premise, $(r \to s) \to \sim p$ to obtain $\sim(r \to s)$ by modus tollens.

Statement	Reason
1. $p \wedge (q \vee \sim z)$	Premise
2. p	Conjunctive Simplification (1)
3. $q \vee \sim z$	Conjunctive Simplification (1)
4. $(r \to s) \to \sim p$	Premise
5. $\sim(r \to s)$	Modus Tollens (2) (4)

Now use the conditional negation equivalence to express $\sim(r \to s)$ as $r \wedge \sim s$.

Statement	Reason
1. $p \wedge (q \vee \sim z)$	Premise
2. p	Conjunctive Simplification (1)
3. $q \vee \sim z$	Conjunctive Simplification (1)
4. $(r \to s) \to \sim p$	Premise
5. $\sim (r \to s)$	Modus Tollens (2) (4)
6. $r \wedge \sim s$	Conditional Negation (5)

Since $r \wedge \sim s$ is the LHS of the premise $(r \wedge \sim s) \to z$, we deduce z by modus ponens.
Notice that in step three we showed $q \vee \sim z$ is true. Thus, q is true by disjunctive syllogism.

Statement	Reason
1. $p \wedge (q \vee \sim z)$	Premise
2. p	Conjunctive Simplification (1)
3. $q \vee \sim z$	Conjunctive Simplification (1)
4. $(r \to s) \to \sim p$	Premise
5. $\sim (r \to s)$	Modus Tollens (2) (4)
6. $r \wedge \sim s$	Conditional Negation (5)
7. $(r \wedge \sim s) \to z$	Premise
8. z	Modus Ponens (6) (7)
9. q	Disjunctive Syllogism (3) (8)

This completes the proof. ■

Example 5

Construct a formal proof for the following valid argument.

$$\begin{array}{c} p \wedge \sim (q \rightarrow r) \\ r \vee \sim k \\ (b \rightarrow w) \rightarrow k \\ \underline{(\sim w \vee z) \rightarrow s} \\ p \wedge s \end{array}$$

Solution

We begin with the first premise since it is a conjunction.

Statement	Reason
1. $p \wedge \sim (q \rightarrow r)$	Premise
2. p	Conjunctive Simplication (1)
3. $\sim (q \rightarrow r)$	Conjunctive Simplication (1)

Next, we negate the conditional statement to produce a conjunctive statement, and then apply conjunctive simplification.

Statement	Reason
1. $p \wedge \sim (q \rightarrow r)$	Premise
2. p	Conjunctive Simplication (1)
3. $\sim (q \rightarrow r)$	Conjunctive Simplication (1)
4. $q \wedge \sim r$	Conditional Negation (3)
5. q	Conjunctive Simplication (4)
6. $\sim r$	Conjunctive Simplication (4)

Now use the results of statement six with premise two.

Statement	Reason
1. $p \wedge \sim(q \to r)$	Premise
2. p	Conjunctive Simplication (1)
3. $\sim(q \to r)$	Conjunctive Simplication (1)
4. $q \wedge \sim r$	Conditional Negation (3)
5. q	Conjunctive Simplication (4)
6. $\sim r$	Conjunctive Simplication (4)
7. $r \vee \sim k$	Premise
8. $\sim k$	Disjunctive Syllogism (6) (7)

Realizing that $\sim k$ is the negation of the RHS of premise three, we apply modus tollens to obtain $\sim(b \to w)$.

Statement	Reason
1. $p \wedge \sim(q \to r)$	Premise
2. p	Conjunctive Simplication (1)
3. $\sim(q \to r)$	Conjunctive Simplication (1)
4. $q \wedge \sim r$	Conditional Negation (3)
5. q	Conjunctive Simplication (4)
6. $\sim r$	Conjunctive Simplication (4)
7. $r \vee \sim k$	Premise
8. $\sim k$	Disjunctive Syllogism (6) (7)
9. $(b \to w) \to k$	Premise
10. $\sim(b \to w)$	Modus Tollens (8) (9)

We now negate the conditional statement and simplify the resulting conjunction. Once we do this we can use disjunctive addition to produce the LHS of premise four, and then apply modus ponens to deduce s is true. Then, since s and p are both true, their conjunction is true.

Statement	Reason
1. $p \wedge \sim (q \to r)$	Premise
2. p	Conjunctive Simplication (1)
3. $\sim (q \to r)$	Conjunctive Simplication (1)
4. $q \wedge \sim r$	Conditional Negation (3)
5. q	Conjunctive Simplication (4)
6. $\sim r$	Conjunctive Simplication (4)
7. $r \vee \sim k$	Premise
8. $\sim k$	Disjunctive Syllogism (6) (7)
9. $(b \to w) \to k$	Premise
10. $\sim (b \to w)$	Modus Tollens (8) (9)
11. $b \wedge \sim w$	Conditional Negation (10)
12. b	Conjunctive Simplication (11)
13. $\sim w$	Conjunctive Simplication (11)
14. $\sim w \vee z$	Disjunctive Addition (13)
15. $(\sim w \vee z) \to s$	Premise
16. s	Modus Ponens (14) (15)
17. $p \wedge s$	Conjunctive Addition (2)(16)

■

Using DeMorgan's Equivalences in Proofs

DeMorgan's equivalences, expressed as $\sim (p \wedge q) \Leftrightarrow (\sim p \vee \sim q)$ and $\sim (p \vee q) \Leftrightarrow (\sim p \wedge \sim q)$, allow us to express the negation of a conjunction as a disjunction and the negation of a disjunction as a conjunction.

Example 6

Let's show a formal proof for the valid argument first encountered as Example 5 of Section 2.4.

$$\begin{array}{l} \sim p \to (q \wedge \sim r) \\ \sim q \vee r \\ \underline{p \to (s \wedge h)} \\ \quad h \end{array}$$

Solution

Note the relationship between the second premise and the RHS of the first premise. Using DeMorgan's equivalence, the second premise may be expressed as $\sim(q \wedge \sim r)$. Thus, the second premise is the negation of the RHS of premise one.

Statement	Reason
1. $\sim q \vee r$	Premise
2. $\sim(q \wedge \sim r)$	DeMorgan's Equivalence (1)
3. $\sim p \rightarrow (q \wedge \sim r)$	Premise

Now apply modus tollens to obtain p, which is the LHS of premise three.

Statement	Reason
1. $\sim q \vee r$	Premise
2. $\sim(q \wedge \sim r)$	DeMorgan's Equivalence (1)
3. $\sim p \rightarrow (q \wedge \sim r)$	Premise
4. p	Modus Tollens (2) (3)
5. $p \rightarrow (s \wedge h)$	Premise

Finally, apply modus ponens to statements four and five to deduce $s \wedge h$, and then use conjunctive simplification to infer the truth of h.

Statement	Reason
1. $\sim q \vee r$	Premise
2. $\sim(q \wedge \sim r)$	DeMorgan's Equivalence (1)
3. $\sim p \rightarrow (q \wedge \sim r)$	Premise
4. p	Modus Tollens (2) (3)
5. $p \rightarrow (s \wedge h)$	Premise
6. $s \wedge h$	Modus Ponens (4) (5)
7. h	Conjunctive Simplification (6)

■

Example 7

Use the *TF* method to test the following argument for validity. If it is valid, construct a formal proof. If not, produce a counterexample chart.

$$\begin{array}{l} (q \vee r) \rightarrow s \\ \sim q \rightarrow w \\ \sim s \\ \hline \quad w \end{array}$$

Solution

The premise $\sim s$ is true, so s is false. Since the premise $(q \vee r) \rightarrow s$ is true and s is false, $q \vee r$ is false. That means both q and r are false. Since q is false, $\sim q$, the LHS of the second premise, is true. Therefore, w must be true.

We proceed with the formal proof, following the same strategy outlined above. Begin by using the first and third premises to deduce $\sim (q \vee r)$.

Statement	Reason
1. $\sim s$	Premise
2. $(q \vee r) \rightarrow s$	Premise
3. $\sim (q \vee r)$	Modus Tollens (1) (2)

Now, apply DeMorgan's equivalence.

Statement	Reason
1. $\sim s$	Premise
2. $(q \vee r) \rightarrow s$	Premise
3. $\sim (q \vee r)$	Modus Tollens (1) (2)
4. $\sim q \wedge \sim r$	DeMorgan's Equivalence (3)

Since the conjunction is true, each of its components is true.

Statement	Reason
1. $\sim s$	Premise
2. $(q \vee r) \rightarrow s$	Premise
3. $\sim (q \vee r)$	Modus Tollens (1) (2)
4. $\sim q \wedge \sim r$	DeMorgan's Equivalence (3)
5. $\sim q$	Conjunctive Simplification (4)

Since $\sim q$ is the LHS of the second premise, we deduce w by modus ponens.

Statement	Reason
1. $\sim s$	Premise
2. $(q \vee r) \rightarrow s$	Premise
3. $\sim (q \vee r)$	Modus Tollens (1) (2)
4. $\sim q \wedge \sim r$	DeMorgan's Equivalence (3)
5. $\sim q$	Conjunctive Simplification (4)
6. $\sim q \rightarrow w$	Premise
7. w	Modus Ponens (5) (6)

■

Using the Distributive Equivalences in Proofs

Recall that the distributive equivalence involves two different connectives, and it has two forms. One form tells us that a conjunction distributes over a disjunction. This form can be expressed symbolically as $[p \wedge (q \vee r)] \Leftrightarrow [(p \wedge q) \vee (p \wedge r)]$

The other form of the distributive equivalence tells us that a disjunction distributes over a conjunction. It is expressed as $[p \vee (q \wedge r)] \Leftrightarrow [(p \vee q) \wedge (p \vee r)]$

Example 8

Show a formal proof for the following valid argument.

$$\begin{array}{c} \sim r \vee \sim s \\ w \rightarrow p \\ \underline{r \wedge (s \vee w)} \\ p \end{array}$$

Solution

Note that the distributive equivalence allows us to express the third premise as $(r \wedge s) \vee (r \wedge w)$, while DeMorgan's equivalence tells us that the first premise is logically equivalent to $\sim (r \wedge s)$. We begin our proof using these observations.

Statement	Reason
1. $r \wedge (s \vee w)$	Premise
2. $(r \wedge s) \vee (r \wedge w)$	Distributive Equivalence (1)
3. $\sim r \vee \sim s$	Premise
4. $\sim (r \wedge s)$	DeMorgan's Equivalence (3)

We now apply disjunctive syllogism to statements two and four, and then apply conjunctive simplification to the resulting statement.

Statement	Reason
1. $r \wedge (s \vee w)$	Premise
2. $(r \wedge s) \vee (r \wedge w)$	Distributive Equivalence (1)
3. $\sim r \vee \sim s$	Premise
4. $\sim (r \wedge s)$	DeMorgan's Equivalence (3)
5. $r \wedge w$	Disjunctive Syllogism (2) (4)
6. w	Conjunctive Simplification (5)

Finally, we can apply modus ponens after we state $w \rightarrow p$ is true. This will produce the desired conclusion.

Statement	Reason
1. $r \wedge (s \vee w)$	Premise
2. $(r \wedge s) \vee (r \wedge w)$	Distributive Equivalence (1)
3. $\sim r \vee \sim s$	Premise
4. $\sim (r \wedge s)$	DeMorgan's Equivalence (3)
5. $r \wedge w$	Disjunctive Syllogism (2) (4)
6. w	Conjunctive Simplification (5)
7. $w \rightarrow p$	Premise
8. p	Modus Ponens (6) (7)

■

In-Class Exercises and Problems for Section 3.4

In-Class Exercises

In Exercises 1–14, state the illustrated valid argument form or equivalence.

1. $\dfrac{\sim(a \to \sim q)}{a \wedge q}$

2. $\dfrac{\sim b \vee \sim c}{b \to \sim c}$

3. $\dfrac{\sim(n \wedge \sim p)}{\sim n \vee p}$

4. $\dfrac{\sim(k \to p)}{\sim(k \to p) \vee q}$

5. $\dfrac{r \to \sim d}{d \to \sim r}$

6. $\sim g$
   $\underline{(\sim h \vee a) \to g}$
   $\sim(\sim h \vee a)$

7. $\dfrac{q \vee (s \wedge \sim r)}{(q \vee s) \wedge (q \vee \sim r)}$

8. a
   $\underline{(s \to c) \vee \sim a}$
   $s \to c$

9. $\dfrac{\sim g \wedge a}{\sim(g \vee \sim a)}$

10. $q \wedge c$
    $\underline{(q \wedge c) \to \sim s}$
    $\sim s$

11. $\dfrac{(d \to s) \wedge a}{d \to s}$

12. $\dfrac{g \vee \sim m}{\sim(\sim g \wedge m)}$

13. $\dfrac{(a \wedge b) \to \sim c}{\sim(a \wedge b) \vee \sim c}$

14. $\dfrac{\sim[(p \to w) \to \sim k]}{(p \to w) \wedge k}$

In Exercises 15–22, construct a formal proof for each valid argument.

15. $\sim p \to s$
    $\underline{\sim(p \vee q)}$
    s

16. $b \to (a \to c)$
    b
    $\underline{\sim(\sim a \vee c) \vee n}$
    n

17. $\sim(s \to w)$
$w \vee p$
$a \to \sim p$
$\overline{\sim a}$

18. $s \to (\sim a \wedge b)$
$\sim s \to c$
$a \vee \sim b$
$\overline{c}$

19. $(\sim m \to p) \to q$
$a \vee \sim q$
$\sim p \to m$
$\overline{a \vee b}$

20. $(\sim s \vee p) \to k$
$d \vee c$
$s \to \sim(\sim d \to c)$
$\overline{k}$

21. $s \vee p$
$p \to (a \to \sim b)$
$\sim(w \to s)$
$r \to \sim(b \to \sim a)$
$\overline{\sim r}$

22. $w \to s$
$m \vee (q \to d)$
$\sim(p \vee \sim w)$
$m \to \sim(s \vee b)$
$\overline{\sim q \vee d}$

In Exercises 23–34, test each argument for validity. If valid, show a formal proof. If not, produce a counterexample chart.

23. $(p \to q) \to \sim r$
$r \vee c$
$(\sim p \vee q) \wedge s$
$\overline{c}$

24. $(a \wedge b) \to p$
$\sim w \to p$
$\sim(a \to \sim b)$
$\overline{w}$

25. $\sim p \vee a$
$\sim r \to q$
$r \to (p \wedge \sim a)$
$\overline{q}$

26. $\sim(\sim k \to q) \vee p$
$(p \vee w) \to n$
$r \wedge (\sim q \to k)$
$\overline{r \wedge n}$

27. $r \vee (m \to n)$
$\sim(\sim s \to r)$
$\sim(d \vee n)$
$\overline{\sim m}$

28. $(p \to w) \to s$
$c \wedge (\sim p \vee w)$
$\sim r \to \sim(s \wedge c)$
$\overline{r}$

29. $g \to \sim k$
$(\sim g \vee \sim k) \to p$
$(w \vee \sim s) \to \sim p$
$\overline{\quad s \quad}$

30. $\sim a \vee (p \wedge r)$
$(a \to p) \to (w \wedge \sim r)$
$w \vee \sim b$
$\overline{\quad b \quad}$

31. $(\sim p \vee q) \to w$
$\sim r \vee c$
$p \to q$
$(s \vee \sim r) \to \sim w$
$\overline{\quad c \quad}$

32. $\sim (w \to \sim s)$
$(p \wedge s) \to q$
$q \to (\sim a \vee b)$
$\sim w \vee p$
$\overline{\quad a \to b \quad}$

33. $h \vee a$
$p \to (a \to b)$
$s \to (b \vee c)$
$\sim (h \vee \sim p)$
$\overline{\quad \sim s \quad}$

34. $(b \to c) \to q$
$(s \wedge a) \to p$
$(a \wedge \sim b) \vee (a \wedge c)$
$(w \vee \sim s) \to \sim q$
$\overline{\quad p \quad}$

In Exercises 35–37, symbolize each argument. If the argument is valid, provide a formal proof. If not, provide a counterexample chart.

35. It is not true that if you work hard you will not be successful. If you work hard and are successful then you will be able to afford the car of your dreams. Either you cannot afford the car of your dreams or you buy an iPod. Therefore, you buy an iPod.

36. If I read a book I will read a novel but not a biography. If I do not read a book then I will either go to the opera or go to a museum. It is not the case that if I do not go to a museum I will go to the opera. Therefore, I will read a novel or go out to dinner.

37. If you like the outdoors then you should train for the New York Marathon and join a soccer team. It is not true that if you like to run you should become a basketball player. If you like to run then you should train for the New York Marathon. Either you want to become a basketball player or join a soccer team. Therefore, you like the outdoors.

Problems

For each formal proof in Problems 38–43, supply the reason that justifies each statement.

38. $\sim k \vee p$
$\sim(h \vee \sim k)$
$p \rightarrow a$
———
$a \wedge \sim h$

Statement	Reason
1. $\sim(h \vee \sim k)$	1.
2. $\sim h \wedge k$	2.
3. $\sim h$	3.
4. k	4.
5. $\sim k \vee p$	5.
6. p	6.
7. $p \rightarrow a$	7.
8. a	8.
9. $a \wedge \sim h$	9.

39. $\sim q \vee b$
$(a \wedge b) \rightarrow s$
$\sim(a \rightarrow \sim q)$
———
s

Statement	Reason
1. $\sim(a \rightarrow \sim q)$	1.
2. $a \wedge q$	2.
3. a	3.
4. q	4.
5. $\sim q \vee b$	5.
6. b	6.
7. $a \wedge b$	7.
8. $(a \wedge b) \rightarrow s$	8.
9. s	9.

40. $(s \to c) \to d$
$h \to \sim(d \vee p)$
$\sim s \vee c$
———
$\sim h$

Statement	Reason
1. $\sim s \vee c$	1.
2. $s \to c$	2.
3. $(s \to c) \to d$	3.
4. d	4.
5. $d \vee p$	5.
6. $h \to \sim(d \vee p)$	6.
7. $\sim h$	7.

41. $s \to \sim(\sim m \to p)$
$s \vee (n \to c)$
$\sim p \to m$
———
$\sim n \vee c$

Statement	Reason
1. $\sim p \to m$	1.
2. $\sim m \to p$	2.
3. $s \to \sim(\sim m \to p)$	3.
4. $\sim s$	4.
5. $s \vee (n \to c)$	5.
6. $n \to c$	6.
7. $\sim n \vee c$	7.

42. $p \wedge (a \vee b)$
$(c \vee \sim w) \to \sim b$
$\sim(p \wedge a)$
———
w

Statement	Reason
1. $p \wedge (a \vee b)$	1.
2. $(p \wedge a) \vee (p \wedge b)$	2.
3. $\sim(p \wedge a)$	3.
4. $p \wedge b$	4.
5. b	5.
6. $(c \vee \sim w) \to \sim b$	6.
7. $\sim(c \vee \sim w)$	7.
8. $\sim c \wedge w$	8.
9. w	9.

43. $\sim g \vee \sim c$

$(c \to \sim g) \to \sim(a \to \sim b)$

$s \vee \sim(a \vee p)$

$s \wedge b$

Statement	Reason
1. $\sim g \vee \sim c$	1.
2. $g \to \sim c$	2.
3. $c \to \sim g$	3.
4. $(c \to \sim g) \to \sim(a \to \sim b)$	4.
5. $\sim(a \to \sim b)$	5.
6. $a \wedge b$	6.
7. a	7.
8. b	8.
9. $a \vee p$	9.
10. $s \vee \sim(a \vee p)$	10.
11. s	11.
12. $s \wedge b$	12.

In Problems 44–51, construct a formal proof for each valid argument.

44. $\sim(a \to c)$

$q \vee c$

---

q

45. $(\sim m \to q) \to p$

$n \to \sim p$

$m \vee q$

$\sim n$

46. $s \to n$

$\sim n \vee p$

$\sim(\sim s \vee c)$

$p \wedge \sim c$

47. $\sim(p \to \sim q) \vee w$

$\sim r \to \sim w$

$q \to \sim p$

---

r

48. $(p \to w) \to s$

$\sim s$

$(\sim p \vee w) \vee a$

---

a

49. $r \to (p \wedge w)$

$\sim p \vee \sim w$

$r \vee s$

---

s

50. $s \vee a$
$\sim c \to r$
$\sim(\sim r \to s)$
$(\sim s \to a) \to b$
———
$b \wedge c$

51. $(p \vee \sim r) \to \sim(a \to \sim g)$
$r \to s$
$(s \vee w) \to m$
$g \to \sim a$
———
m

In Problems 52–65, test each argument for validity. If valid, construct a formal proof. If not, produce a counterexample chart.

52. $\sim s \vee q$
$q \to \sim b$
$\sim(s \to \sim p)$
———
$\sim b \wedge p$

53. $\sim(a \to b)$
$w \vee (\sim a \vee b)$
$(w \vee p) \to c$
———
c

54. p
$p \to \sim(c \vee \sim n)$
$(w \to \sim r) \to c$
———
r

55. $m \to (p \wedge \sim s)$
$(\sim m \vee a) \to w$
$\sim p \vee s$
———
w

56. $p \vee \sim s$
$\sim g \vee k$
$s \to \sim(g \to k)$
———
$\sim p$

57. $d \to \sim n$
$w \to (a \wedge b)$
$\sim(n \to \sim d) \vee w$
———
$b \vee c$

58. $\sim(a \wedge \sim b)$
$\sim r \vee p$
$(a \to b) \to (q \wedge r)$
———
p

59. $r \wedge (\sim a \to q)$
$\sim r \vee s$
$w \to (a \vee q)$
———
$s \wedge w$

60.
$$\begin{array}{l} p \vee r \\ (d \vee h) \rightarrow \sim p \\ \sim h \rightarrow q \\ \sim (s \rightarrow r) \\ \hline q \end{array}$$

61.
$$\begin{array}{l} \sim (c \wedge p) \rightarrow (s \rightarrow w) \\ s \wedge \sim w \\ (c \wedge s) \rightarrow n \\ \hline n \end{array}$$

62.
$$\begin{array}{l} (\sim r \wedge s) \rightarrow \sim (\sim p \vee q) \\ \sim (w \vee b) \vee c \\ p \rightarrow q \\ (r \vee \sim s) \rightarrow w \\ \hline c \end{array}$$

63.
$$\begin{array}{l} p \wedge q \\ (r \vee s) \rightarrow (w \rightarrow \sim a) \\ (a \rightarrow \sim w) \rightarrow \sim b \\ \sim r \rightarrow s \\ \hline p \wedge \sim b \end{array}$$

64.
$$\begin{array}{l} (\sim a \vee c) \vee (p \rightarrow s) \\ \sim w \rightarrow (a \wedge \sim c) \\ \sim (\sim p \vee s) \vee k \\ (w \wedge k) \rightarrow q \\ \hline q \end{array}$$

65.
$$\begin{array}{l} (a \wedge \sim b) \rightarrow c \\ (c \vee q) \rightarrow (w \rightarrow \sim s) \\ \sim (a \rightarrow b) \\ s \\ \hline a \wedge \sim w \end{array}$$

In Problems 66–69, express each argument symbolically. If the argument is valid, construct a formal proof. If not, produce a counterexample chart.

66. If Sam looks for a new job in another state then he wants to work part-time and learn to play tennis. Sam will either look for a new job in another state or not sell his house. If he decides to move to Arizona or Florida, he must sell his house. If he does not move to Arizona then he will definitely move to Florida. Therefore, Sam learns to play tennis.

67. Harry relaxes or he is hyperactive. It's not true that if Harry goes to the movies then he does not like videos. If Harry goes to the movies or eats popcorn, he relaxes. Therefore, Harry is hyperactive.

68. Kenny goes biking and follows the stock market. If Kenny goes biking it is not true that he either rides when the temperature is

above 60° or he doesn't ride alone. Either he rides when the temperature is above 60° or, if he follows the stock market then his computer is on. Therefore, Kenny's computer is on and he rides his bike alone.

69. Oleg either boots up his computer and plays games, or boots up his computer and does homework. If he boots up his computer, then he reads his e-mail if he accesses the internet. Oleg accesses the internet and does not do homework. Therefore, Oleg reads his e-mail and plays games.

3.5 Conditional Proofs

In Section 2.5, we saw that if the conclusion of an argument was a conditional statement, we could often employ a subtle line of reasoning to construct a proof. Recall that if the conclusion of an argument is a conditional statement and its LHS is false, the argument is valid, regardless of the truth value of the RHS. However, if the LHS of the conclusion is true, then in order to show that the argument is valid, we must prove that the RHS is true as well.

Therefore, if we *assume* the LHS of the conclusion is true and we are able to show that under this assumption, its RHS is true, we will have shown that the argument is valid. This sort of strategy is called a *conditional proof.* It is important to realize that this strategy may only be used if the conclusion is a conditional statement.

When we assume that the LHS of a conclusion is true, the reason we supply in our formal proof is called the *assumption for a conditional proof*, abbreviated *ACP*.

Example 1

Write a conditional proof for the following valid argument.

$$\begin{array}{c} p \rightarrow q \\ q \rightarrow \sim r \\ \hline p \rightarrow \sim r \end{array}$$

Solution

Since the conclusion of the argument is a conditional statement, we can use a conditional proof and assume that its LHS is true. We then proceed as we did previously, employing all the tools at our disposal.

Statement	Reason
1. p	ACP
2. $p \to q$	Premise
3. q	Modus Ponens (1) (2)
4. $q \to \sim r$	Premise
5. $\sim r$	Modus Ponens (3) (4)
6. $p \to \sim r$	Conditional Proof (1) (5)

Notice that in statement six, the reason we use to show that the conclusion, $p \to \sim r$, is true is "Conditional Proof." ■

When constructing a formal proof of an argument whose conclusion is a conditional statement, the assumption for a conditional proof need not be the first step in our proof. We may assume the LHS of the conclusion is true at any step in a proof, as our next example illustrates.

Example 2

Write a conditional proof for the following valid argument.

$$\begin{array}{c} \sim a \vee b \\ \sim c \to d \\ \sim b \\ \underline{\sim a \to (d \to \sim e)} \\ e \to c \end{array}$$

Solution

Since the conclusion of this argument is a conditional statement, we will attempt to use a conditional proof. However, if we inspect the premises, we note that the assumption that e is true will not allow us to proceed. While e appears in the premise $\sim a \to (d \to \sim e)$, we cannot reason further, since the truth values of $\sim a$ and d are as yet unknown.

We choose to begin with the first and third premises.

Statement	Reason
1. $\sim a \vee b$	Premise
2. $\sim b$	Premise
3. $\sim a$	Disjunctive Syllogism (1) (2)

Now we proceed to the fourth premise, knowing its LHS is true.

Statement	Reason
1. $\sim a \vee b$	Premise
2. $\sim b$	Premise
3. $\sim a$	Disjunctive Syllogism (1) (2)
4. $\sim a \to (d \to \sim e)$	Premise
5. $d \to \sim e$	Modus Ponens (3) (4)

Now it seems reasonable to assume the LHS of the conclusion, *e,* is true, since it is the negation of the RHS of $d \to \sim e$. We then proceed in the usual way.

Statement	Reason
1. $\sim a \vee b$	Premise
2. $\sim b$	Premise
3. $\sim a$	Disjunctive Syllogism (1) (2)
4. $\sim a \to (d \to \sim e)$	Premise
5. $d \to \sim e$	Modus Ponens (3) (4)
6. e	ACP
7. $\sim d$	Modus Tollens (5) (6)
8. $\sim \mathrm{c} \to d$	Premise
9. c	Modus Tollens (7) (8)
10. $e \to c$	Conditional Proof (6) (9)

Thus, we have shown that the argument is valid. ■

In-Class Exercises and Problems for Section 3.5

In-Class Exercises

In Exercises 1–12, test each argument for validity. If valid, construct a formal proof. If not, produce a counterexample chart.

1. $\dfrac{c \to (d \wedge g)}{c \to g}$

2. $\dfrac{\begin{array}{c}\sim p \to \sim r \\ \sim p \vee q\end{array}}{r \to q}$

3.
$$\begin{array}{l} r \vee s \\ \sim a \vee \sim s \\ \hline \sim r \to \sim a \end{array}$$

4.
$$\begin{array}{l} s \to \sim w \\ h \to \sim s \\ \hline w \to h \end{array}$$

5.
$$\begin{array}{l} \sim k \to (p \wedge b) \\ \sim b \vee a \\ \hline \sim k \to a \end{array}$$

6.
$$\begin{array}{l} \sim (h \vee p) \vee l \\ s \to \sim l \\ \hline h \to \sim s \end{array}$$

7.
$$\begin{array}{l} p \\ (p \wedge a) \to n \\ \sim n \vee (c \wedge d) \\ \hline a \to c \end{array}$$

8.
$$\begin{array}{l} r \to \sim p \\ \sim r \vee (q \to s) \\ q \\ \hline p \to \sim s \end{array}$$

9.
$$\begin{array}{l} \sim p \to \sim q \\ (p \wedge r) \to w \\ (r \vee s) \to q \\ \hline r \to w \end{array}$$

10.
$$\begin{array}{l} a \\ \sim s \vee r \\ (p \vee q) \to w \\ a \to (w \to s) \\ \hline p \to r \end{array}$$

11.
$$\begin{array}{l} \sim d \vee p \\ p \to (\sim m \to n) \\ d \\ (m \wedge d) \to r \\ \hline \sim n \to (r \vee a) \end{array}$$

12.
$$\begin{array}{l} (a \wedge d) \to w \\ b \vee d \\ a \wedge (b \to p) \\ c \to (w \vee s) \\ \hline \sim p \to c \end{array}$$

Problems

In Problems 13–14, supply the reason that justifies each statement in the formal proof.

13. $\sim p \vee q$
$r \to \sim q$
$\sim r \to s$
―――――
$p \to s$

Statement	Reason
1. p	1.
2. $\sim p \vee q$	2.
3. q	3.
4. $r \to \sim q$	4.
5. $\sim r$	5.
6. $\sim r \to s$	6.
7. s	7.
8. $p \to s$	8.

14. $a \vee \sim b$
b
$c \to \sim d$
$a \to (\sim d \to q)$
―――――
$\sim q \to \sim c$

Statement	Reason
1. b	1.
2. $a \vee \sim b$	2.
3. a	3.
4. $a \to (\sim d \to q)$	4.
5. $\sim d \to q$	5.
6. $\sim q$	6.
7. d	7.
8. $c \to \sim d$	8.
9. $\sim c$	9.
10. $\sim q \to \sim c$	10.

In Problems 15–20, construct a formal proof for each valid argument.

15. $$\begin{array}{l} (c \vee p) \to s \\ \hline p \to (s \vee b) \end{array}$$

16. $$\begin{array}{l} a \vee b \\ a \to c \\ \hline \sim b \to c \end{array}$$

17. $$\begin{array}{l} \sim a \to \sim w \\ (a \vee d) \to \sim p \\ \hline w \to \sim p \end{array}$$

18. $$\begin{array}{l} (w \vee \sim s) \to p \\ \sim a \to w \\ \hline \sim p \to (a \vee q) \end{array}$$

19. $$\begin{array}{l} \sim p \\ (\sim p \wedge s) \to q \\ \sim q \vee \sim r \\ \hline s \to \sim r \end{array}$$

20. $$\begin{array}{l} \sim(\sim p \vee s) \\ p \to (q \to l) \\ q \vee m \\ \hline \sim l \to m \end{array}$$

In Problems 21–30, test each argument for validity. If valid, construct a formal proof. If not, produce a counterexample chart.

21. $$\begin{array}{l} d \to \sim a \\ r \to (a \wedge \sim b) \\ d \vee \sim q \\ \hline r \to \sim q \end{array}$$

22. $$\begin{array}{l} s \wedge k \\ k \to (\sim d \vee c) \\ b \to c \\ \hline d \to b \end{array}$$

23. $$\begin{array}{l} \sim s \to (\sim c \vee q) \\ \sim d \\ d \vee \sim s \\ q \to p \\ \hline c \to p \end{array}$$

24. $$\begin{array}{l} g \to \sim n \\ (p \vee q) \to r \\ r \to w \\ g \vee p \\ \hline n \to w \end{array}$$

25. $$\begin{array}{l} p \wedge \sim w \\ \sim w \to (b \vee c) \\ p \to (a \to c) \\ \hline a \to b \end{array}$$

26. $$\begin{array}{l} a \\ \sim a \vee b \\ (b \vee d) \to (p \vee s) \\ \sim k \to \sim s \\ \hline \sim p \to k \end{array}$$

27. $\sim w \to \sim a$
$\sim(w \wedge p) \vee s$
$s \to b$
$\sim(a \wedge \sim p)$
$\overline{\quad a \to b \quad}$

28. $\sim r$
$(b \vee c) \to r$
$\sim d \vee m$
$d \to \sim(c \vee p)$
$\overline{\quad p \to m \quad}$

29. $(a \wedge \sim b) \to (c \vee d)$
$g \to (w \wedge \sim d)$
$\sim(a \to b)$
$\sim c \vee z$
$\overline{\quad g \to z \quad}$

30. $(p \to a) \to (\sim m \vee c)$
$r \to \sim(c \vee q)$
$\sim r \to k$
$\sim p \vee a$
$\overline{\quad m \to k \quad}$

In Problems 31–33, symbolize each argument. If the argument is valid, construct a formal proof. If not, produce a counterexample chart.

31. Jim is having a surprise party for Theresa. If Lou is unable to come to the party then Jim will not have the party. If Emad is invited to the party then Jane won't come. If Lou comes to the party then either Jane or Ken will attend. Therefore, if Ken doesn't come to the party then either Emad won't come or Rochelle will stay home.

32. Maria goes swimming and skydiving. If she rides her bike and goes Rollerblading then she won't go swimming. If Maria doesn't ride her bike, then if she goes skydiving she has to buy a parachute. Therefore, if Maria goes Rollerblading she buys a parachute.

33. Tom will either work on his new computer program or he will build his daughter a dollhouse. If he works on his new computer program then he won't have time to go boating. If he stays home to finish his model airplane then, he won't go boating or he won't go horseback riding. Therefore, if Tom doesn't build his daughter a dollhouse then he will finish his model airplane.

3.6 Indirect Proofs

As we saw in Section 2.6, an alternate method of testing for validity is the *indirect approach* or *indirect method.* Recall that the strategy for an indirect approach is to assume that the argument is invalid, i.e., that the premises are all true but the conclusion is false. If, in truth, the argument is valid, a contradiction will arise.

In practical terms, this means that we assume the negation of the conclusion is true. Then, if we are able to show that any statement and its negation are both true, we will have arrived at a contradiction. This means that the assumption was incorrect and thus, the conclusion is true, making the argument valid.

This technique can be employed regardless of the form of the conclusion; that is, the conclusion of the argument may be a simple sentence, or any compound statement, including a conditional statement.

Example 1

Construct a formal proof for the following valid argument using the indirect method.

$$\begin{array}{l} p \vee r \\ r \rightarrow s \\ \underline{p \rightarrow q} \\ q \vee s \end{array}$$

Solution

Begin by assuming $\sim(q \vee s)$ is true. This is the assumption for the indirect proof, abbreviated AIP. We next apply DeMorgan's equivalence and then conjunctive simplification.

Statement	Reason
1. $\sim(q \vee s)$	AIP
2. $\sim q \wedge \sim s$	DeMorgan's Equivalence (1)
3. $\sim q$	Conjunctive Simplification (2)
4. $\sim s$	Conjunctive Simplification (2)

Now combine $\sim q$ with the third premise using modus tollens to conclude $\sim p$. (We could just as easily have used $\sim s$ with the second premise.)

Statement	Reason
1. $\sim(q \vee s)$	AIP
2. $\sim q \wedge \sim s$	DeMorgan's Equivalence (1)
3. $\sim q$	Conjunctive Simplification (2)
4. $\sim s$	Conjunctive Simplification (2)
5. $p \rightarrow q$	Premise
6. $\sim p$	Modus Tollens (3) (5)

Using the first premise, $p \vee r$, and the result from step six, we can reason r by disjunctive syllogism.

Statement	Reason
1. $\sim(q \vee s)$	AIP
2. $\sim q \wedge \sim s$	DeMorgan's Equivalence (1)
3. $\sim q$	Conjunctive Simplification (2)
4. $\sim s$	Conjunctive Simplification (2)
5. $p \rightarrow q$	Premise
6. $\sim p$	Modus Tollens (3) (5)
7. $p \vee r$	Premise
8. r	Disjunctive Syllogism (6) (7)

Combining r with the second premise produces s by modus ponens. However, in step four above, we showed that $\sim s$ was true. This is a contradiction. Therefore, the assumption in step one must have been incorrect. Hence, $q \vee s$ is true.

Statement	Reason
1. $\sim(q \vee s)$	AIP
2. $\sim q \wedge \sim s$	DeMorgan's Equivalence (1)
3. $\sim q$	Conjunctive Simplification (2)
4. $\sim s$	Conjunctive Simplification (2)
5. $p \to q$	Premise
6. $\sim p$	Modus Tollens (3) (5)
7. $p \vee r$	Premise
8. r	Disjunctive Syllogism (6) (7)
9. $r \to s$	Premise
10. s	Modus Ponens (8) (9)
11. $q \vee s$	Contradiction (4) (10)

We have shown that the argument is valid. ■

Example 2

Write a formal proof for the following valid argument using the indirect approach.

$$\begin{array}{l} m \to (\sim p \to q) \\ n \\ r \to \sim p \\ m \vee \sim n \\ \hline \quad q \vee \sim r \end{array}$$

Solution

As in a conditional proof, the assumption need not be stated in the first line of an indirect proof. We choose to begin this proof by combining the second and fourth premises, to deduce m by disjunctive syllogism.

Statement	Reason
1. $m \vee \sim n$	Premise
2. n	Premise
3. m	Disjunctive Syllogism (1) (2)

Combining this result with the first premise will produce $\sim p \to q$ by modus ponens.

Statement	Reason
1. $m \vee \sim n$	Premise
2. n	Premise
3. m	Disjunctive Syllogism (1) (2)
4. $m \to (\sim p \to q)$	Premise
5. $\sim p \to q$	Modus Ponens (3) (4)

It is at this point that we choose to state our assumption for the indirect proof, and assume that the negation of $q \vee \sim r$ is true. We next apply DeMorgan's equivalence followed by conjunctive simplification.

Statement	Reason
1. $m \vee \sim n$	Premise
2. n	Premise
3. m	Disjunctive Syllogism (1) (2)
4. $m \to (\sim p \to q)$	Premise
5. $\sim p \to q$	Modus Ponens (3) (4)
6. $\sim (q \vee \sim r)$	AIP
7. $\sim q \wedge r$	DeMorgan's Equivalence (6)
8. $\sim q$	Conjunctive Simplification (7)
9. r	Conjunctive Simplification (7)

We now combine r with the third premise to deduce $\sim p$. Then using $\sim p$ in combination with statement five we deduce q. However, in statement eight we showed $\sim q$ is true. Hence, we have a contradiction based on our incorrect assumption in step six.

Statement	Reason
1. $m \vee \sim n$	Premise
2. n	Premise
3. m	Disjunctive Syllogism (1) (2)
4. $m \to (\sim p \to q)$	Premise
5. $\sim p \to q$	Modus Ponens (3) (4)
6. $\sim (q \vee \sim r)$	AIP
7. $\sim q \wedge r$	DeMorgan's Equivalence (6)
8. $\sim q$	Conjunctive Simplification (7)
9. r	Conjunctive Simplification (7)
10. $r \to \sim p$	Premise
11. $\sim p$	Modus Ponens (9) (10)
12. q	Modus Ponens (5) (11)
13. $q \vee \sim r$	Contradiction (8) (12)

Therefore, the conclusion is true and the argument is valid. ■

Example 3

Construct an indirect proof for the following valid argument.

$$\begin{array}{l} p \to q \\ q \to \sim r \\ \hline p \to \sim r \end{array}$$

Solution

We already have shown that this argument is valid using a conditional proof in Example 1 of Section 3.5. Since the indirect method can be used for *any* valid argument, we should be able to construct a formal proof using an indirect approach as well.

First, assume the negation of the conclusion is true, and then apply the conditional negation equivalence followed by conjunctive simplification.

Statement	Reason
1. $\sim(p \to \sim r)$	AIP
2. $p \wedge r$	Conditional Negation (1)
3. p	Conjunctive Simplification (2)
4. r	Conjunctive Simplification (2)

Using p with premise one produces q by modus ponens. Then, using q with premise two produces $\sim r$, again by modus ponens.

Statement	Reason
1. $\sim(p \to \sim r)$	AIP
2. $p \wedge r$	Conditional Negation (1)
3. p	Conjunctive Simplification (2)
4. r	Conjunctive Simplification (2)
5. $p \to q$	Premise
6. q	Modus Ponens (3) (5)
7. $q \to \sim r$	Premise
8. $\sim r$	Modus Ponens (6) (7)

Steps four and eight are contradictory. Therefore, our assumption was incorrect, so $p \to \sim r$ is true.

Statement	Reason
1. $\sim(p \to \sim r)$	AIP
2. $p \wedge r$	Conditional Negation (1)
3. p	Conjunctive Simplification (2)
4. r	Conjunctive Simplification (2)
5. $p \to q$	Premise
6. q	Modus Ponens (3) (5)
7. $q \to \sim r$	Premise
8. $\sim r$	Modus Ponens (6) (7)
9. $p \to \sim r$	Contradiction (4) (8)

■

In-Class Exercises and Problems for Section 3.6

In-Class Exercises

In Exercises 1–12, test the each argument for validity. If valid, construct a formal proof. If not, produce a counterexample chart.

1. $m \to n$
   $\underline{m \vee n}$
   n

2. $\underline{a \to (c \to r)}$
   $(a \wedge c) \to r$

3. $c \to g$
   $l \to m$
   $\underline{c \vee l}$
   $g \vee m$

4. $p \to s$
   $(d \vee r) \to k$
   $\underline{p \vee \sim k}$
   $d \vee s$

5. $(p \vee n) \to a$
   $\sim a \vee (b \wedge c)$
   $\underline{b \to (c \wedge n)}$
   $g \vee n$

6. $r \to \sim p$
   $\sim (s \vee \sim n)$
   $\underline{\sim s \to (\sim p \to w)}$
   $\sim w \to \sim r$

7. $\sim a \vee w$
   $d \to s$
   $\underline{(a \to w) \to (p \vee d)}$
   $p \vee s$

8. $(r \vee p) \to m$
   $(p \vee s) \to w$
   $\underline{\sim m \vee (w \to \sim s)}$
   $\sim (r \wedge s)$

9. $\sim s \vee h$
   $d \to a$
   $\sim s \to (a \to \sim c)$
   $\underline{\sim h}$
   $c \to \sim d$

10. $\sim c$
    $p \vee q$
    $\sim b \to (q \to c)$
    $\underline{(\sim p \to q) \to (a \vee \sim b)}$
    a

11. $\sim z \vee (\sim s \rightarrow w)$

$a \rightarrow \sim s$

$\sim z \rightarrow \sim k$

k

$w \vee \sim a$

12. r

$\sim (p \wedge r) \vee s$

$(k \vee c) \rightarrow \sim w$

$s \rightarrow (c \rightarrow w)$

$c \rightarrow \sim p$

Problems

Each argument in Problems 13–16 is valid. In each problem, supply the reasons that justify the statements made.

13. $a \vee (s \wedge q)$

$q \rightarrow p$

$(s \vee w) \rightarrow \sim p$

---

a

Statement	Reason
1. $\sim a$	1.
2. $a \vee (s \wedge q)$	2.
3. $s \wedge q$	3.
4. s	4.
5. q	5.
6. $q \rightarrow p$	6.
7. p	7.
8. $(s \vee w) \rightarrow \sim p$	8.
9. $\sim (s \vee w)$	9.
10. $\sim s \wedge \sim w$	10.
11. $\sim s$	11.
12. a	12.

14. $(\sim c \vee r) \rightarrow a$

$\sim a \vee p$

$p \rightarrow q$

$\sim c \rightarrow q$

Statement	Reason
1. $\sim(\sim c \rightarrow q)$	1.
2. $\sim c \wedge \sim q$	2.
3. $\sim c$	3.
4. $\sim q$	4.
5. $\sim c \vee r$	5.
6. $(\sim c \vee r) \rightarrow a$	6.
7. a	7.
8. $\sim a \vee p$	8.
9. p	9.
10. $p \rightarrow q$	10.
11. q	11.
12. $\sim c \rightarrow q$	12.

15. $n \vee (\sim b \wedge s)$

$(c \vee p) \rightarrow (a \rightarrow b)$

$s \rightarrow a$

$n \vee \sim c$

Statement	Reason
1. $\sim(n \vee \sim c)$	1.
2. $\sim n \wedge c$	2.
3. $\sim n$	3.
4. c	4.
5. $n \vee (\sim b \wedge s)$	5.
6. $\sim b \wedge s$	6.
7. $\sim b$	7.
8. s	8.
9. $s \rightarrow a$	9.
10. a	10.
11. $c \vee p$	11.
12. $(c \vee p) \rightarrow (a \rightarrow b)$	12.
13. $a \rightarrow b$	13.
14. b	14.
15. $n \vee \sim c$	15.

16. $p \wedge \sim s$

$\sim s \rightarrow (w \rightarrow k)$

$\sim w \rightarrow \sim r$

$\overline{\quad k \vee \sim r \quad}$

Statement	Reason
1. $p \wedge \sim s$	1.
2. p	2.
3. $\sim s$	3.
4. $\sim s \rightarrow (w \rightarrow k)$	4.
5. $w \rightarrow k$	5.
6. $\sim (k \vee \sim r)$	6.
7. $\sim k \wedge r$	7.
8. $\sim k$	8.
9. r	9.
10. $\sim w$	10.
11. $\sim w \rightarrow \sim r$	11.
12. $\sim r$	12.
13. $k \vee \sim r$	13.

In Problems 17–22, construct a formal proof for each valid argument.

17. $\sim h \vee k$

$\sim k \rightarrow h$

$\overline{\quad k \quad}$

18. $a \rightarrow (p \wedge s)$

$\overline{\quad \sim a \vee p \quad}$

19. $(p \rightarrow q) \rightarrow s$

$\sim s \vee k$

$\sim k \vee r$

$\overline{\quad \sim p \rightarrow r \quad}$

20. b

$(p \rightarrow b) \rightarrow a$

$a \rightarrow c$

$\overline{\quad c \quad}$

21. $\sim n \rightarrow s$

$(n \vee b) \rightarrow w$

$\sim w \vee \sim q$

$\overline{\quad \sim q \vee s \quad}$

22. r

$r \rightarrow (g \vee c)$

$(c \vee p) \rightarrow k$

$\overline{\quad \sim g \rightarrow k \quad}$

In Problems 23–32, test each argument for validity. If valid, construct a formal proof. If not, produce a counterexample chart.

23. $a \vee (c \rightarrow \sim p)$
$\sim a \rightarrow (\sim p \vee c)$
$\overline{\sim p \vee a}$

24. $(\sim s \wedge p) \vee b$
$(m \vee p) \rightarrow s$
$\overline{b}$

25. $(\sim c \vee d) \rightarrow e$
$\sim e \vee d$
$d \rightarrow z$
$\overline{z}$

26. $p \vee h$
$w \rightarrow (k \rightarrow p)$
$\sim h \vee (k \wedge w)$
$\overline{p}$

27. $(s \vee a) \rightarrow q$
$(c \vee \sim d) \rightarrow \sim q$
$d \rightarrow p$
$\overline{s \rightarrow \sim p}$

28. $(a \rightarrow h) \vee m$
$(p \rightarrow m) \rightarrow \sim a$
$\sim h \rightarrow \sim m$
$\overline{h \vee p}$

29. $\sim (d \rightarrow l)$
$d \rightarrow (b \vee s)$
$s \rightarrow (m \wedge \sim c)$
$\overline{b \vee \sim c}$

30. $(p \rightarrow c) \rightarrow a$
$(w \vee p) \rightarrow \sim b$
$\sim a \vee s$
$\overline{w \rightarrow s}$

31. $\sim r$
$s \rightarrow r$
$\sim s \rightarrow (d \rightarrow l)$
$c \vee d$
$\overline{\sim l \rightarrow c}$

32. $(p \vee q) \rightarrow \sim a$
$a \vee \sim s$
$\sim w \rightarrow (p \wedge s)$
$w \rightarrow r$
$\overline{r}$

In Problems 33–35, symbolize each argument. If the argument is valid, construct a formal proof. If not, produce a counterexample chart.

33. You either fly with a charter airline or you can't go on a vacation. If you pay full price for your airline ticket or your flight departs between 9:00 a.m. and 5:00 p.m., then you didn't fly with a

charter airline. If you didn't pay full price for your airline ticket you can stay in a better hotel. Therefore, if you go on a vacation, then you can stay in a better hotel.

34. If Mike doesn't play nine holes of golf then he will play in his jazz band. If Mike goes to Virginia to visit his friend, then if the weather is nice he will play nine holes of golf. Either Mike will not play in his jazz band, or the weather is nice and he will go to Virginia to visit his friend. Therefore, Mike plays nine holes of golf.

35. Rich will either create a crossword puzzle or a cryptic crossword puzzle. If Rich goes to Connecticut then either he will not judge a crossword puzzle contest or he will create a crossword puzzle. Rich will either go to Connecticut and judge a crossword puzzle contest or not create a cryptic crossword puzzle. Therefore, Rich creates a crossword puzzle.

Valid Argument Forms

1. Modus Ponens

Given: An implication — $p \rightarrow q$

The LHS of the implication — p

Conclusion: The RHS of the implication — q

$$\begin{array}{l} p \rightarrow q \\ p \\ \hline q \end{array}$$

2. Modus Tollens

Given: An implication — $p \rightarrow q$

The negation of the RHS of the implication — $\sim q$

Conclusion: The negation of the LHS of the implication — $\sim p$

$$\begin{array}{l} p \rightarrow q \\ \sim q \\ \hline \sim p \end{array}$$

3. Disjunctive Syllogism

Given: A disjunction

The negation of one side of the disjunction

Conclusion: The other side of the disjunction

$$\begin{array}{l} p \vee q \\ \sim p \\ \hline q \end{array} \qquad \begin{array}{l} p \vee q \\ \sim q \\ \hline p \end{array}$$

4. Disjunctive Addition

Given: A statement

Conclusion: The disjunction of the given statement with any statement

$$\begin{array}{l} p \\ \hline p \vee z \end{array}$$

5. Conjunctive Addition

Given: A statement

Another statement

Conclusion: The conjunction of the two given statements

$$\begin{array}{l} p \\ q \\ \hline p \wedge q \end{array}$$

6. Conjunctive Simplification

Given: A conjunction

Conclusion: Either side of the conjunction

$$\begin{array}{l} p \wedge q \\ \hline p \end{array} \qquad \begin{array}{l} p \wedge q \\ \hline q \end{array}$$

Chapter 3 Review

In questions 1–10, match each valid argument to the equivalence or valid argument form it illustrates.

1. $\dfrac{\sim[r \to \sim(w \vee s)]}{r \wedge (w \vee s)}$

2. $\dfrac{\sim[r \wedge (w \to s)]}{\sim r \vee \sim(w \to s)}$

3. $\dfrac{\begin{array}{l} r \to \sim(w \vee s) \\ w \vee s \end{array}}{\sim r}$

4. $\dfrac{\begin{array}{l} r \vee \sim(w \vee s) \\ w \vee s \end{array}}{r}$

5. $\dfrac{r}{r \vee (w \vee s)}$

6. $\dfrac{r \wedge \sim(w \vee s)}{\sim(w \vee s)}$

7. $\dfrac{\begin{array}{l} r \to \sim(w \vee s) \\ r \end{array}}{\sim(w \vee s)}$

8. $\dfrac{\begin{array}{l} \sim(w \vee s) \\ r \end{array}}{r \wedge \sim(w \vee s)}$

9. $\dfrac{r \to \sim(w \vee s)}{(w \vee s) \to \sim r}$

10. $\dfrac{r \to \sim(w \vee s)}{\sim r \vee \sim(w \vee s)}$

a. Conjunctive Addition
b. Modus Tollens
c. Disjunctive Syllogism
d. Conditional Negation
e. Conditional Equivalence
f. Modus Ponens
g. DeMorgan's Equivalence
h. Conjunctive Simplification
i. Disjunctive Addition
j. Contrapositive Equivalence

Questions 11–13 each show the proof of a valid argument. Supply the reason that justifies each statement.

11. $b \wedge c$

$(r \vee s) \rightarrow \sim w$

$\sim r \rightarrow (p \vee \sim c)$

$(b \vee d) \rightarrow w$

---

p

Statement	Reason
1. $b \wedge c$	1.
2. b	2.
3. c	3.
4. $b \vee d$	4.
5. $(b \vee d) \rightarrow w$	5.
6. w	6.
7. $(r \vee s) \rightarrow \sim w$	7.
8. $\sim (r \vee s)$	8.
9. $\sim r \wedge \sim s$	9.
10. $\sim r$	10.
11. $\sim r \rightarrow (p \vee \sim c)$	11.
12. $p \vee \sim c$	12.
13. p	13.

12. $s \rightarrow p$

$\sim c \rightarrow (d \wedge r)$

$\sim p \vee b$

$c \rightarrow \sim b$

$s \rightarrow r$

Statement	Reason
1. s	1.
2. $s \rightarrow p$	2.
3. p	3.
4. $\sim p \vee b$	4.
5. b	5.
6. $c \rightarrow \sim b$	6.
7. $\sim c$	7.
8. $\sim c \rightarrow (d \wedge r)$	8.
9. $d \wedge r$	9.
10. r	10.
11. $s \rightarrow r$	11.

13. $r \to (a \to b)$

$(c \vee q) \to a$

$(c \to b) \to w$

$\sim r \vee w$

Statement	Reason
1. $\sim(\sim r \vee w)$	1.
2. $r \wedge \sim w$	2.
3. r	3.
4. $\sim w$	4.
5. $r \to (a \to b)$	5.
6. $a \to b$	6.
7. $(c \to b) \to w$	7.
8. $\sim(c \to b)$	8.
9. $c \wedge \sim b$	9.
10. c	10.
11. $\sim b$	11.
12. $c \vee q$	12.
13. $(c \vee q) \to a$	13.
14. a	14.
15. $\sim a$	15.
16. $\sim r \vee w$	16.

In questions 14–21, construct a formal proof for each valid argument. Where appropriate, you may use either a direct proof, conditional proof or indirect proof.

14. $(\sim p \vee q) \to r$

$\sim p$

$a \to \sim r$

$\sim a$

15. $(c \vee d) \wedge a$

$b \to \sim (c \vee d)$

$b \vee \sim s$

$\sim s$

16. $(\sim d \vee a) \to s$

$d \vee \sim s$

$\sim p \to \sim d$

---

p

17. $\sim (p \wedge s) \to q$

$a \wedge \sim q$

$a \to (\sim s \vee \sim c)$

$p \wedge \sim c$

18.
$$\begin{array}{l} b \to c \\ \sim c \\ (\sim b \vee \sim p) \to s \\ \hline s \vee q \end{array}$$

19.
$$\begin{array}{l} \sim d \to c \\ k \to p \\ (d \vee c) \to k \\ \hline p \end{array}$$

20.
$$\begin{array}{l} m \to \sim n \\ \sim b \vee s \\ m \wedge p \\ \sim n \to (a \wedge b) \\ \hline s \end{array}$$

21.
$$\begin{array}{l} a \to (p \to \sim s) \\ l \vee a \\ p \\ (\sim s \vee k) \to b \\ \hline \sim l \to b \end{array}$$

In questions 22–35, test each argument for validity. If valid, construct a formal proof. Where appropriate, you may use either a direct proof, a conditional proof or an indirect proof. If invalid, produce a counterexample chart.

22.
$$\begin{array}{l} \sim a \to g \\ a \to \sim g \\ \hline g \end{array}$$

23.
$$\begin{array}{l} p \vee (\sim r \to s) \\ \sim (p \vee a) \wedge \sim s \\ \hline r \end{array}$$

24.
$$\begin{array}{l} k \vee \sim (a \vee \sim b) \\ b \to c \\ \hline \sim k \to (c \vee n) \end{array}$$

25.
$$\begin{array}{l} \sim (p \vee \sim s) \wedge \sim (a \to \sim b) \\ \sim w \to (s \wedge a) \\ \hline w \end{array}$$

26.
$$\begin{array}{l} (a \vee g) \to l \\ \sim l \\ g \vee (p \vee q) \\ \hline \sim p \to q \end{array}$$

27.
$$\begin{array}{l} (a \vee g) \to l \\ a \\ (\sim m \vee w) \to \sim l \\ \hline m \to g \end{array}$$

28.
$$\begin{array}{l} (a \vee s) \to n \\ a \wedge \sim b \\ \sim n \vee \sim p \\ \hline \sim (p \vee b) \end{array}$$

29.
$$\begin{array}{l} \sim (s \to \sim q) \\ \sim q \vee p \\ (p \vee d) \to c \\ \hline c \end{array}$$

30. $\sim s \to q$
$a \vee \sim b$
$(\sim p \vee q) \to b$
$\overline{\sim a \to s}$

31. $(a \vee d) \to w$
$(\sim e \vee b) \to p$
$a \vee \sim e$
$\overline{w}$

32. a
$\sim r \vee w$
$(a \vee b) \to r$
$(m \vee n) \to \sim w$
$\overline{\sim m}$

33. $r \vee (s \wedge p)$
$\sim a \vee b$
$(\sim r \to s) \to a$
$(\sim w \vee q) \to \sim b$
$\overline{w}$

34. $w \to \sim a$
$(\sim w \wedge b) \to c$
$\sim c \vee p$
$\sim (a \wedge \sim b)$
$\overline{a \to p}$

35. $\sim (p \vee \sim r)$
$(r \wedge w) \to k$
$k \to (a \vee q)$
$a \to s$
$\overline{w \to q}$

For questions 36–37, each set of premises is followed by four possible conclusions. Select the conclusion(s) that makes the argument valid.

36. If the program is run, the computer prints headings and reads data. The withholding routine is performed if there is tax due. If data is read, then the computer performs the calculation routine. The program is run and there is not tax due. Therefore,
   a. the calculation routine is performed.
   b. the withholding routine is performed.
   c. the withholding routine is not performed.
   d. headings are printed.

37. If the stock price is high Frank will sell his stock but if the stock price is low then he buys stock. Frank gets a dividend if the stock price is high or if it is low. If Frank buys stock then the stock exchange is open. The stock price is high but the stock exchange is not open. Therefore,
   a. Frank sells his stock.
   b. Frank buys stock.
   c. Frank gets a dividend.
   d. the stock price is low.

Sample Exam: Chapter 3

In questions 1–8, state the illustrated equivalence or valid argument form.

1. $\sim(a \to d) \vee (r \wedge q)$
   $a \to d$
   ____
   $r \wedge q$

2. $(\sim n \vee c) \to (p \to \sim q)$
   $\sim(p \to \sim q)$
   ____
   $\sim(\sim n \vee c)$

3. $(l \vee \sim m) \wedge k$
   ____
   $l \vee \sim m$

4. $\sim(g \to \sim h)$
   ____
   $g \wedge h$

5. $\sim s \vee \sim p$
   ____
   $s \to \sim p$

6. $a \to \sim b$
   ____
   $(a \to \sim b) \vee c$

7. $\sim r \vee w$
   ____
   $\sim(r \wedge \sim w)$

8. $\sim(a \vee m) \to (d \to j)$
   $\sim(a \vee m)$
   ____
   $d \to j$

In questions 9–14, each set of premises is followed by four possible conclusions. Which of the four conclusions makes thc argumcnt valid?

9. $\sim(\sim p \vee q)$
   $(\sim w \vee r) \to s$
   $w \to \sim p$
   ____

   a. r b. s
   c. w d. $\sim p$

10. $(\sim m \vee c) \to l$
    $l \to (p \to \sim r)$
    $\sim m \wedge p$
    ____

    a. r b. $\sim l$
    c. m d. $\sim r$

11. $\sim(p \to \sim q)$
    $(q \vee r) \to \sim s$
    $p \to (w \vee s)$
    ____

    a. $p \to \sim q$ b. $\sim s \wedge \sim w$
    c. $w \vee a$ d. $\sim q \vee s$

12. $\sim w \to \sim p$
    $p \wedge \sim k$
    $\sim w \vee n$
    ____

    a. $p \to \sim w$ b. $w \to \sim p$
    c. $k \vee \sim n$ d. $n \wedge \sim k$

13. $\sim p \vee q$
$\sim(\sim q \to \sim p) \vee r$
$r \to (s \vee b)$

a. s b. b
c. $s \vee b$ d. $\sim r$

14. $(a \vee b) \wedge c$
$c \to \sim a$
$b \to d$

a. $\sim d$ b. $c \wedge a$
c. d d. $\sim a \to \sim d$

In question 15, supply the reason that justifies each statement.

15. $\sim w \to \sim a$
$e \to (d \to g)$
$\sim(e \to g)$
$d \vee (c \vee a)$

$c \vee w$

Statement	Reason
1. $\sim(e \to g)$	1.
2. $e \wedge \sim g$	2.
3. e	3.
4. $\sim g$	4.
5. $e \to (d \to g)$	5.
6. $d \to g$	6.
7. $\sim d$	7.
8. $d \vee (c \vee a)$	8.
9. $c \vee a$	9.
10. $\sim(c \vee w)$	10.
11. $\sim c \wedge \sim w$	11.
12. $\sim c$	12.
13. $\sim w$	13.
14. $\sim w \to \sim a$	14.
15. $\sim a$	15.
16. c	16.
17. $c \vee w$	17.

In questions 16–17, construct a formal proof for each valid argument.

16. $(\sim s \vee r) \rightarrow \sim d$

d

$r \vee \sim b$

$\sim b \vee n$

17. $\sim s \rightarrow \sim a$

$d \rightarrow (a \wedge \sim c)$

$(s \vee q) \rightarrow r$

$d \rightarrow r$

In questions 18–21, test the argument for validity. If valid, construct a formal proof. If not, produce a counterexample chart.

18. $(b \vee \sim c) \rightarrow \sim p$

$\sim a \rightarrow (k \rightarrow b)$

$\sim a \wedge p$

$\sim k$

19. $g \rightarrow k$

$(p \vee a) \rightarrow c$

$(\sim g \vee k) \rightarrow p$

$c \wedge p$

20. $(w \rightarrow r) \vee n$

$b \rightarrow s$

$w \rightarrow (a \wedge b)$

$w \rightarrow n$

21. $a \vee (p \wedge \sim s)$

$a \rightarrow k$

$(s \vee b) \rightarrow \sim (a \vee p)$

---

k

In questions 22 and 23, symbolize each argument. If the argument is valid, construct a formal proof. If not, produce a counterexample chart.

22. If you want to eat a cannoli then you should also order cappuccino. If you do not want to eat a cannoli then you should order gelato. If you order espresso then you should not order cappuccino. You ordered espresso. Therefore, you should order gelato.

23. If you go to college it's fun to be part of Greek life. If you are part of Greek life then you will either join a fraternity or a sorority. If you join a fraternity then you will go to many parties and meet a lot of girls. Therefore, if you go to college you will meet a lot of girls.

CHAPTER FOUR

APPLICATIONS OF LOGICAL REASONING

The skills we developed in the previous three chapters can be applied to a variety of situations in which reasoning power is paramount. In this chapter we will examine three such applications.

4.1 Fuzzy Statements and Truthfulness

Introduction

In modern computing, logic is used to create computer programs. A computer program is a series of commands that perform a task. The task may be as simple as computing a subtotal or as complicated as controlling the temperature in a room. More complicated tasks require a computer program to accept many inputs, process them, and then create an effective output. In this chapter we will study a branch of logic that was developed to perform these complicated tasks.

The statements that we have studied so far have been referred to as *TF* statements. A *TF* statement can have one of two truth values–true or false. Systems involving such statements are referred to as two-valued logic systems. However, some statements are neither completely true nor completely false.

Consider the statement "Felicia is stubborn." Is this true or false? You may have to know Felicia or someone who knows her. Ask her sister and she may say "Yes." Ask her teacher and he may say "No." Why could there be two different answers to this question? Investigate further and her sister may say "Felicia is stubborn because whenever I ask to borrow her clothes, she says 'No'." Ask her teacher and he may say "Felicia is not stubborn because when many other students refuse to rewrite an assignment, Felicia would never say 'No'."

Fuzzy Statements

How do we decide whether Felicia is stubborn or not? There is no simple answer to this question. There is a degree to Felicia's stubbornness. She's not the most stubborn person in the world, yet she's not going to cave in to every demand either. This kind of statement is called a *fuzzy statement*, since it is not totally true nor totally false. Its truth value lies somewhere between true and false.

Example 1

Determine which of the following statements are fuzzy statements.

a. The lizard is big.
b. The phone is ringing.
c. The entire house is painted white.
d. The sofa is heavy.
e. The car is old.

Solutions

a. This is a fuzzy statement. How big is big? If one lives in the Southwest around lizards, a lizard that is a foot long may not seem big at all while a Northerner might think a foot-long lizard is frighteningly big.
b. This is not a fuzzy statement. If there is no series of intermittent sounds coming from the phone, nor lights flashing, it is not ringing.
c. This is not a fuzzy statement. White is a specifically defined color and the house is either painted entirely white or not.
d. This is a fuzzy statement. Is a sofa considered heavy when it takes two people to lift it, or when three people are needed to lift it?
e. This is a fuzzy statement. When is a car old? After 5 years? After 10 years?

■

Example 2

Change each fuzzy statement above into a *TF* statement.

Solution

There are many ways to change each statement. Here is a reasonable solution for each of the fuzzy statements in Example 1.

a. The lizard is 2 feet long.
d. The sofa weighs 300 lbs.
e. The car is 8 years old.

■

Membership and Truthfulness

A statement is deemed true or false by determining whether or not its subject is a member of a particular set. (A set is simply a collection of objects.) For example, the statement "The entire house is painted white" would be true if you collected all objects that had the property of being entirely white, and found this particular house in the collection. Therefore, the house would have *total membership* in the collection of entirely white objects. Similarly, the statement "The phone is ringing" would be false if you collected all ringing phones and your phone was not part of this collection. In this case we say that your phone has *non-membership* in the collection of all ringing phones.

Let's again consider the statement "Felicia is stubborn." If you look up the word "stubborn" in the dictionary, you might find the following definition: "refusing to yield, obey or comply." Does this sound like Felicia? She seems not to comply with her sister's wishes to borrow clothes, yet obeys her teacher when he asks her to rewrite an assignment. Therefore, Felicia may be considered "a bit stubborn."

Now, if we were to collect all stubborn people, would Felicia be a member of our set? She would have *partial membership* in our set.

Example 3

Decide whether the following objects have total, partial or non-membership in the set of red objects.

a. A candy cane
b. A red ball
c. The American flag
d. A banana

Solutions

a. A candy cane has partial membership, since it is red and white.
b. A red ball has total membership, since it is entirely red.
c. The American flag has partial membership because it is red, white, and blue.
d. Since a banana is yellow, it has non-membership.

■

We would like to express the extent to which an object has membership in a set as a numerical value. This value will be called its *degree of truthfulness*. If a statement is definitively true (the subject has total membership in the defined set), its degree of truthfulness is one. If a statement is definitively false (the subject has no membership in the defined set), its degree of truthfulness is zero. If a statement is partly true (the subject has partial membership in the defined set), its degree of truthfulness is some value between zero and one.

A degree of truthfulness can be determined in many ways. For the example involving Felicia's stubbornness, we could take a poll and find out that 50% of the people asked thought she was stubborn. We could then assign a degree of truthfulness of 0.5 to the statement "Felicia is stubborn." For the example of the American flag, we may determine that 40% of its area is colored red, and consequently assign a degree of truthfulness of 0.4 to the statement "The American flag is red."

In Chapter One, we used a letter to represent a *TF* statement. For example, we could write p: The American flag is red. To express the degree of truthfulness of "The American flag is red" we will write

$\tau(p) = 0.4$. The symbol τ is the Greek letter tau. This notation tells us that the degree of truthfulness for the statement "The American flag is red" is 0.4.

Example 4

The following statements have degrees of truthfulness of 0, 0.3, 0.7, and 1. Match each statement with the degree of truthfulness that could reasonably describe it.

a. Winning the lottery is fun.
b. Going to the amusement park is fun.
c. Going to jail is fun.
d. Going to traffic court is fun.

Solution

a. τ(Winning the lottery is fun) = 1.
b. τ(Going to the amusement park is fun) = 0.7.
c. τ(Going to jail is fun) = 0.
d. τ(Going to traffic court is fun) = 0.3.

■

What makes each of the statements in the above example fuzzy? It's the fact that all of the conditions (winning the lottery, going to the amusement park, etc.) are described as being "fun." The four statements above can be written using one, general fuzzy statement, "X is fun" where X is the condition that is being considered fun. For example, in the statement, "Winning the lottery is fun," X would be "winning the lottery."

Earlier in this section, we discussed how the degree of truthfulness of a statement may be estimated by taking a survey. This is not always possible. Sometimes, the relationship between two quantities can be expressed using a formula. The following example will illustrate how one can obtain a formula to estimate the degree of truthfulness of a statement that has an associated numerical value.

Example 5

During tax season, Suzanne took 3 hours to do her taxes, Joan took 5 hours to do her taxes and Tony took $6\frac{1}{2}$ hours to do his taxes. Paula held the record, requiring only 2 hours while Steve took an outrageous amount of time, 12 hours. Estimate the degree of truthfulness for each of the following statements:

a. Joan took a long time to do her taxes.
b. Tony took a long time to do his taxes.

Solution

First, rank the numerical values (hours) in order: 2 (Paula), 3 (Suzanne), 5 (Joan), $6\frac{1}{2}$ (Tony) and 12 (Steve). Then assign a degree of truthfulness of zero to the lowest value and a degree of truthfulness of one to the highest value. Therefore,

$$\tau(\text{Paula took a long time to do her taxes}) = 0$$

$$\tau(\text{Steve took a long time to do his taxes}) = 1.$$

Next, find the difference in hours associated with Paula and Steve. Since Steve spent 12 hours and Paula spent 2 hours, the difference is $12 - 2 = 10$. This is called the *total change.*

To calculate τ(Joan took a long time to do her taxes), we calculate the difference between the number of hours required for Joan to do her taxes and the least number of hours required for anyone to do taxes. For the case of Joan, calculate $5 - 2 = 3$. For the case of Tony, calculate $6\frac{1}{2} - 2 = 4\frac{1}{2} = 4.5$. Finally, divide each of the above differences by the total change to find the respective degrees of truthfulness. Hence,

$$\tau(\text{Joan took a long time to do her taxes}) = \frac{3}{10} = 0.3, \text{ and}$$

$$\tau(\text{Tony took a long time to do his taxes}) = \frac{4.5}{10} = 0.45.$$

■

In summary, there are two ways to assign degrees of truthfulness to a statement. If data is gathered through a survey and the survey is being used as supporting evidence for the statement, then

$$\tau(x) = \frac{\text{number of positive responses}}{\text{total number of responses}}.$$

If the data describes numerical values (rather than survey results) that will be used as supporting evidence for the statement, then

$$\tau(x) = \frac{\text{value of the data point} - \text{minimum data value}}{\text{maximum data value} - \text{minimum data value}}.$$

In-Class Exercises and Problems for Section 4.1

In-Class Exercises

For Exercises 1–2, circle the fuzzy statements.

1. My car is old. It's a Ford and has 85,000 miles on it. If my car is old and has 85,000 miles on it, then I will buy a new one soon. If my car is a Ford, then I will buy a better model. Therefore, I will buy a better model.
2. She is lazy or uncooperative. If she is uncooperative, she will not volunteer to help. If she is lazy then she will not do a good job. If she does not do a good job, then we will not ask her again. Therefore, if she volunteers, we will not ask her again.
3. Rewrite the paragraph in exercise 1 so that all statements are *TF* statements and construct a formal proof for the argument.

For Exercises 4–6, read each paragraph and then answer the questions posed.

4. BuyUsedBooks.com is trying to decide on a new logo to include on its web page. Three logos have been presented to the advertising department. One is a picture resembling Ernest Hemmingway, another is a book that looks used but valuable, and the third is an abstract logo, in orange and purple, using the name of the website. The company decided to obtain the opinions of the websurfers. Of the 1,200 websurfers who responded to the question "Which is the best logo?", the following results were tallied:

Ernest Hemmingway	650
Valuable, Old Book	200
Orange & Purple Abstract	350

Find a reasonable degree of truthfulness for each of the following statements:

a. The "Ernest Hemmingway picture" is a great logo.
b. The "Valuable, Old Book" is a great logo.
c. The "Orange & Purple Abstract" is a great logo.

5. Fred feels he is always buying ink cartridges for his printer. The cartridge he purchased from a computer speciality store lasted for 3.5 months. The one he bought from an office supply store lasted 3 months. The cartridge he bought from the company that manufactured his printer lasted 4 months. In speaking to his friend, he found out that the best cartridge to buy is one that is sold on a certain internet site because that cartridge lasts 5 months. He also found out that a refillable cartridge will last only 2.5 months.

   Find a reasonable degree of truthfulness for the following statements.

   a. A cartridge purchased from the computer speciality store will last a long time.
   b. A cartridge purchased from the office supply store will last a long time.
   c. A cartridge purchased from the printer manufacturer will last a long time.
   d. A cartridge bought on the internet will last a long time.
   e. A refillable cartridge will last a long time.

6. I thought my car got great gas mileage. I get 22 mpg. I asked around and found out that Nick gets 20 mpg, Barbara gets 18 mpg and Jenine only gets 14 mpg. Then I spoke to Deanna and she gets 24 mpg.

   Find a reasonable degree of truthfulness for the following statements.

   a. I get great gas mileage.
   b. Nick gets great gas mileage.
   c. Barbara gets great gas mileage.
   d. Jenine gets great gas mileage.
   e. Deanna gets great gas mileage.

Problems

In Problems 7–16, determine which statements are fuzzy statements.

7. Carol is a great softball player.
8. Steve got an A in statistics.
9. The puppy is cute.

10. The kitten is soft and cuddly.
11. If you score greater than 1200 on the SAT then you will receive a scholarship.
12. You will be asked to play in the symphony if and only if you are one of the best violinists.
13. Dad went to play golf or went bowling.
14. The gas tank is full and I can't start the car.
15. She is blond but old.
16. I will do her the favor if she is nice to me.

In Problems 17–21, change each fuzzy statement to a *TF* statement.

17. Alfonse is a good student.
18. The movie I saw last night was horrible.
19. That dish you prepared was tasty.
20. That book you want me to read is too long.
21. California is a nice place to live.

In Problems 22–26, change each *TF* statement to a fuzzy statement.

22. It is 25°F outside today.
23. It is raining and the wind is blowing outside.
24. She did not comb her hair today.
25. The limit on my credit card is $25,000.
26. Richard is 5 feet 2 inches tall.

In Problems 27–29, match each statement to its reasonable degree of truthfulness.

27. The teacher gave a test on Chapter 1. Davis scored 100%, Joshua scored 75%, and Maria scored 40%.

Statements	Degrees of truthfulness
Joshua understood Chapter 1.	0.4
Maria understood Chapter 1.	0.75
Davis understood Chapter 1.	1.00

28. The Mr. Tasty Muffin Company polled their consumers as to whether or not they liked their new muffins. Of the 50 people polled, 45 liked the Strawberry Creme, 30 liked the Blueberry

Creme, 38 like the Chocolate Banana and 20 liked the Pumpkin Spice.

Statements	Degrees of truthfulness
Blueberry Creme is a tasty muffin.	0.4
Chocolate Banana is a tasty muffin.	0.6
Pumpkin Spice is a tasty muffin.	0.76
Strawberry Creme is a tasty muffin.	0.9

29. Mike's car needed a new set of tires. He priced them at four different places: Tire World wanted $350 for the set, Autoland wanted $500 for the set, Ray's Wholesale wanted $200 for the set and Custom Auto Design wanted $750 for the set.

Statements	Degrees of truthfulness
Autoland is expensive.	0.0
Custom Auto Design is expensive.	0.27
Ray's Wholesale is expensive.	0.55
Tire World is expensive.	1.00

In Problems 30–31, read each paragraph and, using the information, calculate a reasonable degree of truthfulness for each of the given statements. Explain your calculation.

30. One hundred New Yorkers were asked where they bring out-of-town visitors when they come to New York. The results were as follows: 76 go to the theater, 85 go to Rockefeller Center, 62 visit the museums and 36 attend a sporting event.

Statements

The NY museums are a great tourist attraction.
Rockefeller Center is a great tourist attraction.
NY sports are a great tourist attraction.
NY theaters are a great tourist attraction.

31. Jodi has been babysitting for several families in her area. The Clarkes give her $7.50 an hour, the Davidsons give her $6 an hour, the Martuccis give her $8 an hour and the Waxmans give her $6.50 an hour. She hears from her friends that the best deal going is babysitting for the Jordan family, who will pay $10 an hour. The worst family to babysit for are the Yeardleys, who pay only $5 an hour.

Statements
The Clarkes pay their babysitters well.
The Davidsons pay their babysitters well.
The Martuccis pay their babysitters well.
The Waxmans pay their babysitters well.
The Jordans pay their babysitters well.
The Yeardleys pay their babysitters well.

4.2 Fuzzy Logic

Introduction

For the two-valued logic system we studied in Chapters One, Two, and Three, we learned how to determine the truth value of a compound statement, given the truth values of the component simple statements. In fuzzy logic, a statement does not necessarily have a truth value of either true or false. Instead, it has a degree of truthfulness that is a numerical value between zero and one.

Negations

If statement p has a degree of truthfulness of $\tau(p)$, then $\sim p$ has a degree of truthfulness of $\tau(\sim p) = 1 - \tau(p)$.

Example 1

Suppose g: Gail is strong. Then, $\sim g$: Gail is not strong. If g has a degree of truthfulness of 0.85, that is, $\tau(g) = 0.85$, its negation, $\sim g$, has a degree of truthfulness of $\tau(\sim g) = 1 - 0.85 = 0.15$.

■

Conjunctions

If the statement p has a degree of truthfulness of $\tau(p)$, and statement q has a degree of truthfulness of $\tau(q)$, then the conjunction $p \wedge q$ has a degree of truthfulness equal to the smaller of the two values. Symbolically, this relationship is expressed as:

$$\tau(p \wedge q) = \min\{\tau(p), \tau(q)\}.$$

Example 2

Suppose the statement "Charlie is handsome" has a degree of truthfulness equal to 0.9, and the statement "Charlie is tall" has a degree of truthfulness equal to 0.65. Hence, the statement "Charlie is tall and handsome" has a degree of truthfulness that is equal to $\min\{0.9, 0.65\} = 0.65$.

■

Disjunctions

If the statement p has a degree of truthfulness of $\tau(p)$, and statement q has a degree of truthfulness of $\tau(q)$, then the disjunction $p \vee q$ has a degree of truthfulness equal to the larger of the two values. Symbolically, this relationship is expressed as:

$$\tau(p \vee q) = \max\{\tau(p), \tau(q)\}.$$

Example 3

Let r: The table is round, and let s: The table is square. Let $\tau(r) = 0.72$ and $\tau(s) = 0.57$. Therefore, the degree of truthfulness for the statement $r \vee s$ is $\max\{0.72, 0.57\} = 0.72$.

■

Conditionals

The way in which we compute the degree of truthfulness for conditional statements is based on the conditional equivalence, $(p \to q) \leftrightarrow (\sim p \vee q)$. If the statement p has a degree of truthfulness of $\tau(p)$, and statement q has a degree of truthfulness of $\tau(q)$, then the conditional statement $p \to q$ has a degree of truthfulness equal to the degree of truthfulness for $\sim p \vee q$. Since the degree of truthfulness for $\sim p$ is $1 - \tau(p)$, the degree of truthfulness for $\sim p \vee q$ is symbolically expressed as $\max\{1 - \tau(p), \tau(q)\}$. Therefore, we have $\tau(p \to q) = \max\{1 - \tau(p), \tau(q)\}$.

Example 4

Let a and b be two simple sentences. If $\tau(a) = 0.23$ and $\tau(b) = 0.49$, then the degree of truthfulness for the conditional statement $a \to b$ is $\tau(a \to b) = \max\{1 - 0.23, 0.49\} = \max\{0.77, 0.49\} = 0.77$.

■

Biconditionals

The way in which we compute the degree of truthfulness for a biconditional statement is based on the definition of a biconditional statement

$$(p \leftrightarrow q) \leftrightarrow [(p \to q) \wedge (q \to p)],$$

and on the conditional equivalence

$$[(p \to q) \wedge (q \to p)] \leftrightarrow [(\sim p \vee q) \wedge (\sim q \vee p)].$$

Therefore, the degree of truthfulness for $p \leftrightarrow q$ is the minimum of two maximum functions.

Symbolically, we express this degree of truthfulness as $\tau(p \leftrightarrow q) = \min\{\max[1 - \tau(p), \tau(q)], \max[1 - \tau(q), \tau(p)]\}$.

Example 5

Assume that the statement "The movie is bad" has a degree of truthfulness of 0.7 and the statement "The movie has horrible acting" has a degree of truthfulness of 0.45. The degree of truthfulness of the biconditional statement "The movie is bad if and only if it has horrible acting" is computed as

$$\min\{\max(1-0.7, 0.45),\ \max(1-0.45, 0.7)\}.$$

When simplified, we obtain

$$\min\{\max(0.3, 0.45),\ \max(0.55, 0.7)\} = \min\{0.45, 0.7\} = 0.45. \quad ■$$

In-Class Exercises and Problems for Section 4.2

In-Class Exercises

For Exercises 1–10, let $\tau(p) = 0.37$, $\tau(q) = 0.45$, $\tau(r) = 0.81$, and $\tau(s) = 0.12$. Find the degree of truthfulness for each statement.

1. $p \wedge \sim q$
2. $(r \vee s) \vee p$
3. $\sim (q \wedge \sim s)$
4. $\sim (p \vee q) \wedge (\sim p \vee q)$
5. $s \rightarrow r$
6. $\sim r \rightarrow p$
7. $\sim (r \rightarrow p)$
8. $s \leftrightarrow q$
9. $(s \wedge \sim q) \leftrightarrow \sim r$
10. $\sim (p \leftrightarrow q) \vee (p \rightarrow q)$
11. a. Find the degree of truthfulness for $\sim p \vee r$.

    b. Find the degree of truthfulness for $\sim (p \wedge \sim r)$.

    c. What do you notice about the answers to (a) and (b)?

    d. What does your answer to part (c) suggest about the statements $\sim p \vee r$ and $\sim (p \wedge \sim r)$?

12. a. Find the degree of truthfulness for $\sim(q \rightarrow s)$.
    b. Find the degree of truthfulness for $\sim q \vee s$.
    c. What do you notice about the answers to (a) and (b)?
    d. What does your answer to part (c) suggest about the statements $\sim(q \rightarrow s)$ and $\sim q \vee s$?

Problems

For problems 13–27, determine the degree of truthfulness for each statement, given that $\tau(p) = 0.4$, $\tau(q) = 0.35$, and $\tau(r) = 0.7$.

13. $\sim r$
14. $p \wedge q$
15. $r \vee p$
16. $p \vee \sim q$
17. $r \wedge \sim r$
18. $r \vee \sim r$
19. $q \rightarrow p$
20. $\sim(p \vee q)$
21. $p \leftrightarrow q$
22. $\sim r \rightarrow p$
23. $(p \wedge q) \vee r$
24. $\sim p \leftrightarrow r$
25. $(p \wedge r) \wedge (\sim q \vee r)$
26. $\sim(p \rightarrow \sim r) \wedge (q \vee \sim p)$
27. $\sim[\sim(p \rightarrow q) \wedge (q \rightarrow \sim p)]$

4.3 Actions for Fuzzy Inputs

Introduction

In Section 4.1, we mentioned some methods used to determine degrees of truthfulness. One such method is to take a poll, asking the question "Do you think *x* has characteristic *y*?" We then use the percentage obtained from the number of positive responses as a degree of truthfulness for the statement "*x* has characteristic *y*."

Sometimes, the degree of truthfulness for particular values are related by a function, which in many cases can be represented by a mathematical formula. Functions can also be represented using a table of values. These representations show the relationship between a statement and its degree of truthfulness. In this section, we will discuss how to read tables to obtain relationships between statements and degrees of truthfulness.

Reading Input Tables

A table that relates a statement to a degree of truthfulness is called an *input table.* Each statement contains an object x, whose membership in a given set is being considered. Recall that the extent to which x is a member of this set is referred to as the degree of truthfulness. Consider the following input table for the statement "The lecture on x was boring."

INPUT TABLE

Object x	Degree of Truthfulness τ(Lecture on x was boring)
Poetry	0.8
Accounting	0.7
Geology	0.5
Mathematics	0.3
Politics	0.2

Each object can be placed into the statement "The lecture on x was boring." For example, using the first object as an input, we would have the statement "The lecture on poetry was boring." Using the table, we see that assigned to this statement is a degree of truthfulness of 0.8.

Example 1

Using the above table, determine the degree of truthfulness of the statement "The geology lecture was boring."

Solution

Locate the object, geology, in the table. It is associated with a degree of truthfulness of 0.5.

■

Reading Action Tables

A table that relates a particular degree of truthfulness to a resulting action is called an *action table.* For example, the statement "Lecture x is boring" could have the action, "I won't stay long" associated with it. What do we mean by "staying long"? This itself is a fuzzy statement. The following table shows actions taken given the degrees of truthfulness of the statement "Lecture x was boring."

ACTION TABLE

Degree of Truthfulness	Action
1.0	Stay 20 minutes
0.9	Stay 25 minutes
0.8	Stay 30 minutes
0.7	Stay 35 minutes
0.6	Stay 40 minutes
0.5	Stay 50 minutes
0.4	Stay 55 minutes
0.3	Stay 60 minutes
0.2	Stay 65 minutes
0.1	Stay 70 minutes
0.0	Stay 75 minutes

Notice that the output of the previous input table became the input for the action table. To find the action for the statement "The math lecture was boring," we first reference the input table. The degree of truthfulness of this statement is 0.3. Note that the action table associates a degree of truthfulness of 0.3 with the action "Stay 60 minutes." Hence, for a math lecture that is boring with a degree of truthfulness of 0.3, you will stay for 60 minutes.

Example 2

Determine the action for the statement "The accounting lecture was boring."

Solution

Using the input table, the degree of truthfulness for the statement "The accounting lecture was boring," is 0.7. The action table tells us that a degree of truthfulness of 0.7 is associated with the action "Stay 35 minutes."

■

Multiple Inputs and Action Tables

Sometimes, a situation has more than one input and each input may have its own associated action. For example, consider the input statements "Eric is hungry" and "Eric is pressed for time" and their corresponding action statements, "Eric will eat a large meal" and "Eric will eat a small meal."

If Eric is hungry, he will probably want to eat a big meal. Therefore, the degree of truthfulness of the statement "Eric is hungry" will determine how big a meal Eric will eat. If Eric is pressed for time, he will probably eat a small meal since small meals take less time to eat. The degree of truthfulness of the statement "Eric is pressed for time" will determine the size of the meal Eric will eat.

Suppose Eric goes into a cafeteria and sees the menu of meals shown below. He must order by number. The size of the meal increases as the numbers increase.

Order by Number	Description
#1	Cup of Yogurt
#2	Order of French Fries
#3	Club Sandwich
#4	Hot Dog and French Fries
#5	3 Slices of Pizza
#6	Pasta and Garlic Bread

The action table that follows shows actions both for "being hungry" and "being pressed for time." Each action indicates the meal that would be chosen for a given degree of truthfulness.

Suppose the degree of truthfulness for the statement "Eric is hungry" is 0.8 and the degree of truthfulness of the statement "Eric is pressed for time" is 0.9. What course of action should Eric take?

ACTION TABLE

Degree of Truthfulness	Action for Being Pressed for Time	Action for Being Hungry
0.9	Order #1	Order #6
0.8	Order #2	Order #5
0.6	Order #3	Order #4
0.4	Order #3	Order #3
0.2	Order #4	Order #2

The degree of truthfulness for the statement "Eric is hungry" is 0.8. Therefore, we look at the column in the action table labeled

"Action for being hungry" and find the meal that is assigned to a degree of truthfulness of 0.8. This is meal #5.

The degree of truthfulness for the statement "Eric is pressed for time" is 0.9, and being pressed for time would cause someone to eat a small meal. Therefore, we look at the column in the action table labeled "Action for being pressed for time" and find the meal that is assigned to a degree of truthfulness of 0.9. This is meal #1.

However, since Eric is both hungry and pressed for time, we need to find one meal that best satisfies both conditions. Since the degree of truthfulness of the statement "Eric is pressed for time" is larger than the degree of truthfulness of the statement "Eric is hungry", it makes sense that the action resulting from "being pressed for time" should weigh more heavily in Eric's final decision than the action resulting from "being hungry". Therefore, we calculate the final action by computing a *weighted average.*

A weighted average calculation is unlike the usual average calculation. To compute an average, you add up all of the values and divide by the number of values that were added. To compute a weighted average, some values are counted more often in the sum than others.

The values we need to compute this weighted average are the following:

i. the degree of truthfulness for the first input, τ_1,
ii. a value for the action resulting from the degree of truthfulness of the first input, A_1,
iii. the degree of truthfulness for the second input, τ_2,
iv. a value for the action resulting from the degree of truthfulness of the second input, A_2.

The formula for computing the final action is given by

$$\text{Final action} = \frac{\tau_1 \times A_1 + \tau_2 \times A_2}{\tau_1 + \tau_2}.$$

In our example, the necessary values are: $\tau_1 = 0.8$, $A_1 = 5$, $\tau_2 = 0.9$, and $A_2 = 1$. Therefore, our final action is

$$\text{Final action} = \frac{0.8 \times 5 + 0.9 \times 1}{0.8 + 0.9} \approx 2.88.$$

It is impossible to have a meal with order #2.88. Therefore, we round 2.88 to the nearest acceptable action value, which is 3.

Therefore, Eric should order meal #3 for his current state of hunger and time constraint.

Example 3

Determine the action Eric should take if the degree of truthfulness of the statement "Eric is hungry" is 0.4 and the degree of truthfulness of the statement "Eric is pressed for time" is 0.2.

Solution

The action for Eric's being hungry is to eat a large meal. Using the action table, we see that if the statement "Eric is hungry" has a degree of truthfulness of 0.4, he should order meal #3. The action for Eric's being pressed for time is to order a small meal. If the statement "Eric is pressed for time" has a degree of truthfulness of 0.2, he should order meal #4. Hence,

$$\text{Final Action} = \frac{0.4 \times 3 + 0.2 \times 4}{0.4 + 0.2} \approx 3.33.$$

Rounded to the nearest acceptable action value, we find that Eric should opt for meal #3.

■

Compound Inputs and Action Tables

Sometimes, several actions must be combined to produce a final action. If we are given the input statement "The road is dry and even," its degree of truthfulness depends upon two characteristics, the road being dry and the road being even. Using techniques from Section 4.2, we can determine the degree of truthfulness of the entire compound statement. Let's assume that the degree of truthfulness of the statement "The road is dry" is 0.7 and the degree of truthfulness of the statement "The road is even" is 0.2. The degree of truthfulness of the compound statement "The road is dry and even" would be $\min\{0.7, 0.2\} = 0.2$.

The action associated with the above compound statement might be "Drive at a reasonable speed." The following action table shows possible values for the statement "Drive at a reasonable speed."

ACTION TABLE

Degree of Truthfulness	Action
1	Drive at 60 mph
0.7	Drive at 50 mph
0.5	Drive at 45 mph
0.2	Drive at 35 mph
0	Drive at 20 mph

Therefore, the associated action for the statement "The road is dry and even" with the degree of truthfulness of 0.2 is to drive at 35 mph.

Example 4

Find the action for the statement "The road is dry or even" if the degree of truthfulness of the statement "The road is dry" is 1 and the degree of truthfulness of the statement "The road is even" is 0.2.

Solution

The degree of truthfulness of the statement "The road is dry or even" is $\max\{1, 0.2\} = 1$. Therefore, according to the output table, the action taken should be to drive at 60 mph.

■

Example 5

Consider the input statements, "Gail feels cold and wet" and "Heating bills are high." Their corresponding action statement is "Change the setting on the thermostat." The following action table describes what setting the thermostat should be set to for various degrees of truthfulness of the input statements.

ACTION TABLE

Degree of Truthfulness	Action for Gail feels cold and wet	Action for the heating bills are high
0.9	Set thermostat to 6	Set thermostat to 1
0.7	Set thermostat to 5	Set thermostat to 2
0.6	Set thermostat to 4	Set thermostat to 3
0.5	Set thermostat to 3	Set thermostat to 4
0.3	Set thermostat to 2	Set thermostat to 5
0.2	Set thermostat to 1	Set thermostat to 6

Assume that the degree of truthfulness for the statement "Gail is cold" is 0.3, "Gail is wet" = 0.7, and "Heating bills are high" = 0.2. Using the action table above, find the appropriate final action.

Solution

The degree of truthfulness for the statement "Gail is cold and wet" is $\min\{0.3, 0.7\} = 0.3$. This statement corresponds to the action "Turn the thermostat to setting 2". The degree of truthfulness for the statement "Heating bills are high" is 0.2. This statement corresponds to the action "Turn the thermostat to setting 6". The final action is computed using the weighted average as follows:

$$\text{Final action} = \frac{0.3 \times 2 + 0.2 \times 6}{0.3 + 0.2} = 3.6$$

Rounded to the nearest acceptable action value, the thermostat should be set to setting 4. ■

Given a particular input situation, one of three techniques is used to compute the corresponding output. These techniques are summarized in the table below.

Type of Input	Technique Used to Compute Output
A single statememt	Read τ from the INPUT table. Find output from the ACTION table.
Two statements each with its own action	Read τ from the INPUT table. Find the corresponding action for each statement from the ACTION table. Use the final action formula to calculate the output.
A compound statement	Use fuzzy logic to determine τ for the compound statement. Find output from the ACTION table.

In-Class Exercises and Problems for Section 4.3

In-Class Exercises

1. Five students took a fuzzy logic test. Nancy got an A, Tom got a D, Ricky got a D-, Marianne got a B, and Lilia got a B-. The following tables describe the statement "x did well on the test", and an appropriate action "Judging by the last test grade, x should study y hours for the next test."

   Input: x did well on the exam.

   Action: x should study y hours for the next test.

INPUT TABLE

Object (x)	τ
Ricky	0.4
Tom	0.6
Lilia	0.7
Marianne	0.8
Nancy	0.9

ACTION TABLE

τ	y values
0.9	1
0.8	2
0.7	3
0.6	4
0.5	5
0.4	6

a. Find an appropriate action for the statement "Lilia did well on the test."
b. Find an appropriate action for the statement "Ricky did well on the test."
c. Find an appropriate action for the statement "Nancy did well on the test."
d. Which input has the action "Judging by the last test grade, x should study 4 hours for the next test"?
e. Which input has the action "Judging by the last test grade, x should study 5 hours for the next test"?

2. Headlights on some new cars automatically adjust to the external conditions. Consider the statements "It is dark outside at time x" and "It is foggy out at time x." Both have the action statement "Adjust the headlights to brightness y." Consider the input and action tables that follow.

INPUT TABLE

Object (x)	τ for "It is dark."	τ for "It is foggy."
6 am	0.7	0.9
11 am	0.4	0.5
2 pm	0.1	0.1
5 pm	0.5	0.1
10 pm	0.9	0.7
11 pm	0.9	0.6

ACTION TABLE

τ	y values for "It is dark"	y values for "It is foggy"
0.1	1	1
0.2	1	1
0.3	2	1
0.4	2	2
0.5	3	2
0.6	4	2
0.7	5	3
0.8	5	3
0.9	5	3

a. Find a final action for the input statement "It is dark at 6 am" with the statement "It is foggy at 6 am."
b. Find a final action for the input statement "It is dark at 2 pm" with the statement "It is foggy at 2 pm."
c. Find a final action for the input statement "It is dark at 5 pm" with the statement "It is foggy at 5 pm."
d. Find a final action for the input statement "It is dark at 11 pm" with the statement "It is foggy at 11 pm."

3. Cathy wants to buy a mattress. She must consider the quality of the mattress, the comfort of the mattress and the price of the mattress before she decides which to purchase. Obviously, an expensive mattress is extremely comfortable and of the highest quality, whereas a cheap mattress will not be as comfortable nor is it of the highest quality. She has a choice of 6 mattresses. Consider the action table below.

ACTION TABLE

$\tau(x)$	Buy mattress y
0.1	Relax-A-Mat
0.2	EcoRest
0.3	EcoRest
0.4	Restful Comfort
0.5	Restful Comfort
0.6	Quality Comfort
0.7	Quality Comfort
0.8	Exquisite Comfort
0.9	Royal Comfort

Let q: Cathy wants a quality mattress, c: Cathy wants a comfortable mattress, and e: Cathy wants an expensive mattress.

a. Consider the statement "Cathy wants a comfortable and quality mattress." If $\tau(q) = 0.7$ and $\tau(c) = 0.8$, which mattress will Cathy buy?
b. Consider the statement "Cathy wants a comfortable or quality mattress." If $\tau(q) = 0.9$ and $\tau(c) = 0.5$, which mattress will Cathy buy?

c. Consider the statement "Cathy wants either a comfortable and inexpensive mattress, or a quality mattress." Suppose $\tau(q) = 0.4$, $\tau(c) = 0.5$, and $\tau(e) = 0.3$. Which mattress will Cathy buy?

d. Consider the statement "Cathy does not want an expensive mattress but a comfortable mattress." If $\tau(c) = 0.6$ and $\tau(e) = 0.7$, which mattress will Cathy buy?

Problems

For Problems 4–5, determine the appropriate action.

4. Input statement: "A \$$x$ house is valuable."
   Action: "Buy a \$$y$ security system."

INPUT TABLE

Object x	τ(A \$$x$ house is valuable)
\$180,000	0.1
195,000	0.1
225,000	0.3
290,000	0.6
310,000	0.6
350,000	0.7
500,000	0.8

ACTION TABLE

τ	Purchase a security system for
0.8	\$3,500
0.7	3,000
0.6	2,500
0.3	1,000
0.1	850

a. What is the action for the statement "A \$225,000 house is valuable"?

b. What is the action for the statement "A \$350,000 house is valuable"?

c. What is the action for the statement "A \$180,000 house is valuable"?

5. Input statement: "x is tired."
   Action: Sleep y hours"

INPUT TABLE

Object (x)	$\tau(x \text{ is tired})$
Francis	0.1
Kenton	0.2
Elly	0.6
Jayson	0.6
Teri	0.7
Shelley	0.8
Philipe	0.9

ACTION TABLE

τ	Sleep y hours
0.9	12
0.8	10
0.7	8
0.6	7
0.4	6
0.2	5
0.1	4

a. What is the action for the statement "Elly is tired"?
b. What is the action for the statement "Francis is tired"?
c. What is the action for the statement "Jayson is tired"?
d. What is the action for the statement "Philipe is tired"?

For Problems 6–7, determine the final action given the degree of truthfulness of each input.

6. Input Statement 1: "The air inside the car at x°F is warm"

   Action: "Turn the air conditioning to setting y_1."

   Input statement 2: "The air outside the car at x°F is warm."

   Action: "Turn the air conditioning to setting y_2."

INPUT TABLE

Object (x)	τ
50 degrees	0.1
55 degrees	0.3
60 degrees	0.5
65 degrees	0.5
70 degrees	0.7
75 degrees	0.8
80 degrees	0.9

ACTION TABLE

τ	y_1 values	y_2 values
0.1	2	1
0.3	3	1
0.5	4	2
0.7	4	2
0.8	5	3
0.9	5	4

a. Determine the final action for the input statement "The air inside the car at 65 degrees is warm" with the statement "The air outside the car at 50 degrees is warm."
b. Determine the final action for the input statement "The air inside the car at 50 degrees is warm" with the statement "The air outside the car at 50 degrees is warm."
c. Determine the final action for the input statement "The air inside the car at 50 degrees is warm" with the statement "The air outside the car at 65 degrees is warm."
d. Determine the final action for the input statement "The air inside the car at 70 degrees is warm" with the statement "The air outside the car at 55 degrees is warm."
e. Determine the final action for the input statement "The air inside the car at 80 degrees is warm" with the statement "The air outside the car at 80 degrees is warm."

7. Input Statement: "A staff of x_1 people makes a short-handed staff."

Action: "Allow y_1 days to complete the order."
Input Statement: "A staff that takes on x_2 orders is busy"
Action: "Allow y_2 days to complete the order."

INPUT TABLE

Object (x_1)	Object (x_2)	τ
20	300	0.9
30	250	0.7
40	200	0.5
45	180	0.5
50	150	0.3
80	50	0.1

ACTION TABLE

τ	y_1 values	y_2 values
0.9	30	25
0.7	25	20
0.5	20	15
0.3	15	10
0.1	10	5

a. Determine the final action for the statement "A staff of 45 people makes a short-handed staff" with the statement "A staff that takes on 200 orders is busy."

b. Determine the final action for the statement "A staff of 20 people makes a short-handed staff" with the statement "A staff that takes on 300 orders is busy."

c. Determine the final action for the statement "A staff of 80 people makes a short-handed staff" with the statement "A staff that takes on 50 orders is busy."

d. Determine the final action for the statement "A staff of 20 people makes a short-handed staff" with the statement "A staff that takes on 50 orders is busy."

e. Determine the final action for the statement "A staff of 80 people makes a short-handed staff" with the statement "A staff that takes on 300 orders is busy."

For Problems 8–9, find the action for the given compound inputs.

8. Input statement: "A crowd of x_1 people is a large crowd and opening x_2 attractions makes for a busy day."
Action: "Hire y staff members."

INPUT TABLE

Object x_1	Object x_2	τ
1,000	8	0.4
2,000	10	0.5
3,000	11	0.6
4,000	12	0.7
5,000	13	0.8
6,000	14	0.9

ACTION TABLE

τ	y values
0.4	50
0.5	100
0.6	150
0.7	200
0.8	250
0.9	300

a. Determine the action for the statement "A crowd of 5,000 people is a large crowd and opening 10 attractions makes for a busy day."

b. Determine the action for the statement "A crowd of 3,000 people is a large crowd and opening 12 attractions makes for a busy day."

c. Determine the action for the statement "A crowd of 1,000 people is a large crowd and opening 10 attractions makes for a busy day."

d. Determine the action for the statement "A crowd of 6,000 people is a large crowd and opening 14 attractions makes for a busy day."

e. Determine the action for the statement "A crowd of 2,000 people is a large crowd and opening 12 attractions makes for a busy day.

9. Input Statement: "The weather is bad and the traffic is bad."
Action: "Allow y minutes to get to work."

ACTION TABLE

τ	y values
0.9	60
0.8	50
0.6	40
0.4	30
0.3	25
0.1	20

a. Determine the action for the statement "The weather is bad and the traffic is bad" if the degree of truthfulness for "The weather is bad" is 0.6 and the degree of truthfulness for the traffic is bad is 0.4.

b. Determine the action for the statement "The weather is bad and the traffic is bad" if the degree of truthfulness for "The weather is bad" is 0.1 and the degree of truthfulness for the traffic is bad is 0.4.

c. Determine the action for the statement "The weather is bad and the traffic is bad" if the degree of truthfulness for "The weather is bad" is 0.8 and the degree of truthfulness for the traffic is bad is 0.6.

d. Determine the action for the statement "The weather is bad or the traffic is bad" if the degree of truthfulness for "The weather is bad" is 0.6 and the degree of truthfulness for the traffic is bad is 0.4.

e. Determine the action for the statement "The weather is bad or the traffic is bad" if the degree of truthfulness for "The weather is bad" is 0.4 and the degree of truthfulness for the traffic is bad is 0.1.

4.4 Introduction to Flowcharts

Introduction

The system of logic we have been investigating is related to the logic used in modern computers. Computer programs, or software, are written to solve complex problems. To solve a problem, it must be fully understood. Often, questions are asked and then, based on the responses, actions are taken. These questions and their answers are expressed using simple statements, conjunctions, disjunctions, conditionals and negations. A pictorial way to display these questions and answers is with a *flowchart*, or outline. A flowchart is used to help design a program that will provide a solution to the problem.

A system of logic is also found in the computer's hardware, or physical components. Information passes through the internal workings of the computer through electronic circuits. Circuits are built to correspond to conjunction, disjunction, and conditional statements as well as their negations. In this chapter, we will explore the logic involved in creating flowcharts and take a closer look at some basic circuit theory.

Outlining the Solution to a Problem

Let's explore the logic used in solving a problem. Suppose you have been asked to design the logic that will be used in a vending machine. The machine sells candy that may cost 50¢ or 75¢, chips that may cost 65¢ or 90¢, and cakes that may cost 90¢ or $1.25. The machine accepts dollar bills and coins and works in the following way: The customer inserts money, selects an item and, if the amount deposited is equal to or greater than the cost of the item, receives the item and perhaps some change. The customer must deposit at least the cost of an item to receive it. If she does not do so, the machine will display a message asking for more money or will refund the money and not dispense the item.

Example 1

What information does the machine need to know?

Solution

The machine must know how much money was entered and what item was selected. This information is called *input*. The machine cannot proceed unless this information, or input, is provided.

■

Example 2

What decisions must the machine make?

Solution

The machine must decide if the desired item costs more, less, or the same as the amount deposited. Once this information is known, the machine can proceed.

■

Example 3

What are the various actions the machine can take?

Solution

If the cost of the item is the same as the amount of money deposited, the machine dispenses the item. If the amount of money deposited is more than the cost of the item, the machine must calculate the correct change and dispense the item and the change. If the cost of the item is greater than the amount of money deposited, the machine must display a message asking for more money.

If the person inserts more money, then the machine must decide if the total amount of money now received is exactly the same, less than, or greater than the cost of the item and repeat the process described above. If, after a given number of tries, the amount deposited is still insufficient, the money is returned.

■

The logic needed to solve the problem of the vending machine interacting with a customer had three major stages. The first stage required input, the second stage involved decisions and the third stage produced actions based on these decisions. These stages, in a flowchart, correspond to *input questions, Yes/No questions* and *commands*.

Questions

A question, unlike a *TF* statement, is an expression whose purpose is to elicit information. The answer to a question can be "yes", "no" or another response, called *input.* Any question whose answer is either yes or no is called a *Yes/No question*. Otherwise, it is called an *input question.*

Example 4

The sentence "Is today Tuesday?" is not a *TF* statement, but a question. This is a Yes/No question.

■

Example 5

The sentence "Where do you live?" is not a *TF* statement, but rather a question. This is an input question.

■

Commands

A *command* is a sentence that is neither a statement nor a question, but rather an instruction to complete a task and/or provide information. There are three classifications of such sentences. A *simple command* requires only the completion of a task. An *input command* requires the completion of a task followed by the passing on of the information learned. A *conditional command*, requires that we complete a task and then take any one of a number of actions, depending upon the result of having completed this task.

Example 6

The sentence "Answer the phone" is a simple command because it requires the completion of a task.

■

Example 7

The sentence "Find out what time it is" is an input command because it asks to complete a task (finding out what time it is) and then provide the information learned (the time).

■

Example 8

The sentence "If you find Joe, give him the money, otherwise keep it and return it to me" is a conditional command because it asks us to complete a task (look for Joe) and then to take an action depending on the outcome of the task.

■

In a flowchart, the only commands we use are simple commands. Any situation that warrants an input command would be represented by an input question in a flowchart. Similarly, any conditional command would be represented by a Yes/No question, followed by simple commands that are based on the answer to the Yes/No question.

Example 9

The input command "Tell me your name" can be represented as the input question "What is your name?"

■

Example 10

The conditional command "Catch the ball if it is hit to you" can be rewritten as the the Yes/No question, "Is the ball hit to you?" and the simple commands "Catch it" if the answer to the question is "yes" and "Stand there and pay attention to the game" if the answer to the question is "no."

■

In-Class Exercises and Problems for Section 4.4

In-Class Exercises

For Exercises 1–2, supply an appropriate Yes/No question.

1. Linda and Richie are going out tonight. Linda asks, "What time is it?" Richie responds, "7 pm." What is an appropriate Yes/No question that can be asked by Linda to get the response, "Rush to get ready!"?

2. Marjorie asks Dottie, "What are you doing?" Dottie responds, "I am busy!" What is an appropriate Yes/No question that can be asked by Marjorie to get the response, "Go use another computer!"?

For Exercises 3–4, supply an appropriate input question.

3. Samir asks himself, "Do I have enough to pay for the car?" If he does, he will pay in full. If not, he will borrow the money. Supply an appropriate input question that must be asked by Samir before he decides whether he has enough money.

4. Nick asks Pat "Do you want to play golf?" If Pat wants to play, they will go to the golf course. If not, they will go to the movies. Supply an appropriate input question that Nick can ask Pat before asking whether or not he wants to play golf.

For Exercises 5–7, decide which type of command is used for the "no" response. Then, supply an appropriate command of the same type for a "yes" response.

5. Joey asks Barbara, "What is the weather like in NY?" "Lousy," she responds. "Is it raining?" Joey asks. "No," replies Barbara. "Find out if it will rain later since I have a flight to NY in 3 hours," Joey commands. If Barbara had replied "Yes," write an appropriate command Joey could have issued instead.

6. "What's the score?" Andrea asks Patrick. "It's a blowout," he replies. "Is your team winning?" asks Andrea. "No" responds Patrick, "I have to cheer louder!" If Patrick had responded "Yes," write an appropriate command he could have given to Andrea.

7. "How do you feel?" Marisa asks her daughter. "Horrible!" whispers Elana. "Do you have a sore throat, dear?" Marisa asks. "No, Mom" replies Elana. "Then eat something," says Marisa. If Elana had responded "Yes," write an appropriate command her mother could have given her.

For Exercises 8–11, write a conditional command as indicated.

8. A cashier has the following note at her register: deluxe BBQ regularly $179.99, but with a club card, $159.99. A customer approaches her register with a deluxe BBQ. Issue a conditional command telling the cashier how much to charge for the BBQ.

9. A buffet-style restaurant charges adults $13.95, but children under 12 are charged only $6.95. The waiter forgot to charge one person in a group that just ate. He must add a price to the bill. Issue a conditional command telling the waiter how much to charge for this person.

10. The sign at the curb reads: No parking Monday & Thursday. A driver needs to park on this street. Issue a conditional command telling the driver where to park.

11. A shopper approaches the checkout counters at the supermarket and sees signs that read: Lane 1: 10 items or less; Lane 2: 20 items or less; Lane 3: credit cards accepted; Lanes 4 to 8: all transactions accepted. The shopper needs to choose a lane. Issue a conditional command that tells the shopper which lane to choose.

Problems

For Problems 12–31, classify each sentence as either a Yes/No question, an input question, a simple command, an input command or a conditional command.

12. Put the box down on the floor.

13. Are you my cousin's friend, John?

14. Ask her if she is ready to go.

15. How much is 81 minus 60?

16. If you drop it, please pick it up.
17. Does Shannon have a red car?
18. Have you seen Rob?
19. Calculate the area of the kitchen floor.
20. If you see him, tell him to call me.
21 Is Cathy coming to work today?
22. Do you have the memo?
23. Please give me directions to the restaurant.
24. Tell me which road to take.
25. Have a great day.
26. Leave a message.
27. If it's over $100, charge it. Otherwise pay cash.
28. What is her favorite team?
29. Find the rates for a 3-month CD.
30. If you have time, drop off my dry-cleaning.
31. Is this statement a *TF* statement?

For Problems 32–34, write a short outline describing the logic needed to solve the problem. Your outline should include the three stages of solving the problem: getting input, making decisions and creating actions for the decisions.

32. What is the logic used by a receptionist to assist a customer who is trying to contact a particular telephone extension in an office?
33. What is the logic used by a VCR in order to tape a show?
34. What is the logic used by the self-scanning checkout machine at a supermarket? Assume you scan your own items and pay by credit card.

4.5 Decisions Using Flowcharts

Introduction

When you attempt to solve a problem, it is often necessary to ask key questions, and then make a decision based on the answers. This same procedure is employed when a computer is used to solve a problem. This form of logic can be illustrated using flowcharts, which are pictorial representations of the reasoning process used to solve a problem.

In a flowchart, the shapes used have specific meanings.

The parallelogram  is used to enclose an input question.

The diamond 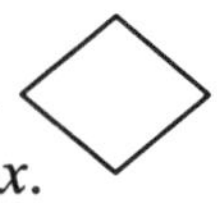 is used to enclose Yes/No questions. It is called a *decision box.*

The rectangle  is used to enclose a simple command or for the result of decision-making.

Example 1

Nick ordered a pizza a while ago. There's a knock at the door. Nick checks to see if it's the delivery boy. If it is, he will open the door and pay him. If it's not, he'll ignore the knock and wait for the person to leave. Write a flowchart that describes the actions taken.

Solution

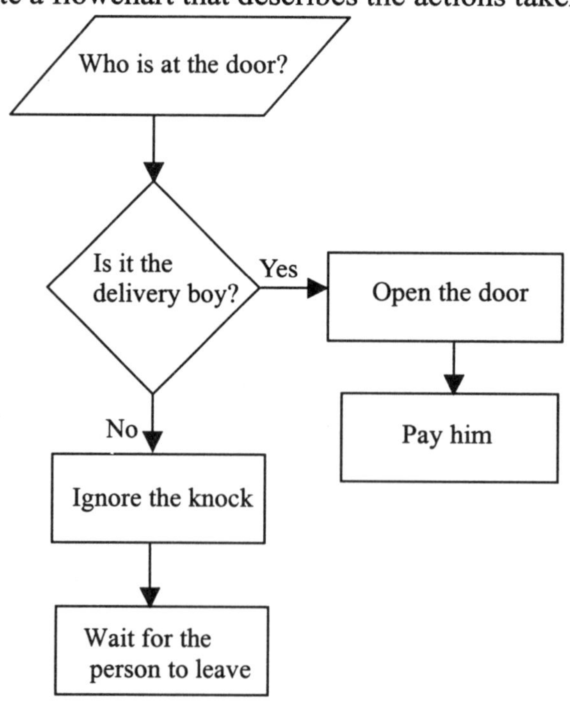

■

Notice that the sentence "Who is at the door?" is an input question. Therefore, it appears inside a parallelogram in the flowchart. The sentence "Is it the delivery boy?" is a Yes/No question. It appears inside a diamond. The sentences "Open the door," "Pay him," "Ignore the knock" and "Wait for person to leave" are simple commands, so these sentences appear inside rectangles in the flowchart.

Example 2

Consider the following flowchart for calculating net salary.

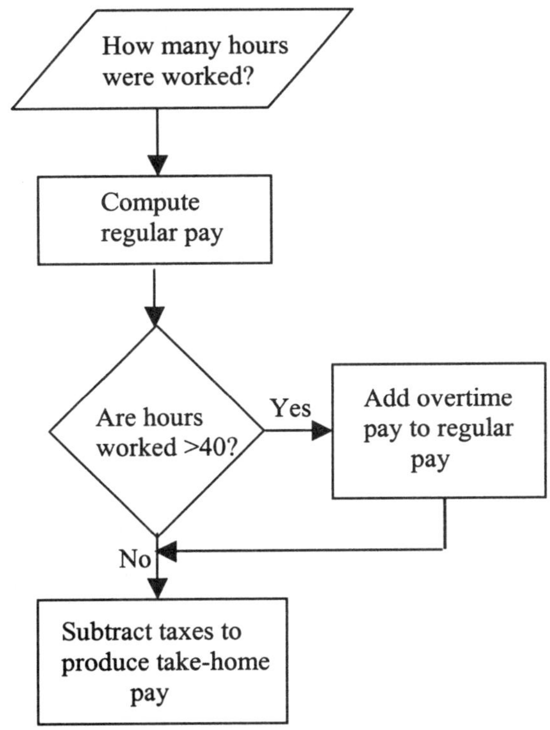

Answer each of the following questions.

a. What is the input question?
b. What is the Yes/No question?
c. What are the three simple commands?
d. What happens if the number of hours is greater than 40?
e. What happens if the number of hours is not greater than 40?

Solution

The input question is "How many hours were worked?" The Yes/No question is "Are hours worked greater than 40?" The three simple commands are "Compute regular pay," "Add overtime pay to regular pay," and "Subtract taxes to produce take-home pay."

The flowchart shows that if the number of hours is greater than 40, overtime pay is added to the regular pay and then taxes are subtracted to produce take-home pay. If the number of hours is not greater than 40, we just subtract the taxes to produce the take-home pay. ■

Example 3

The flowchart below describes how a company assigns a commission rate to its salespeople.

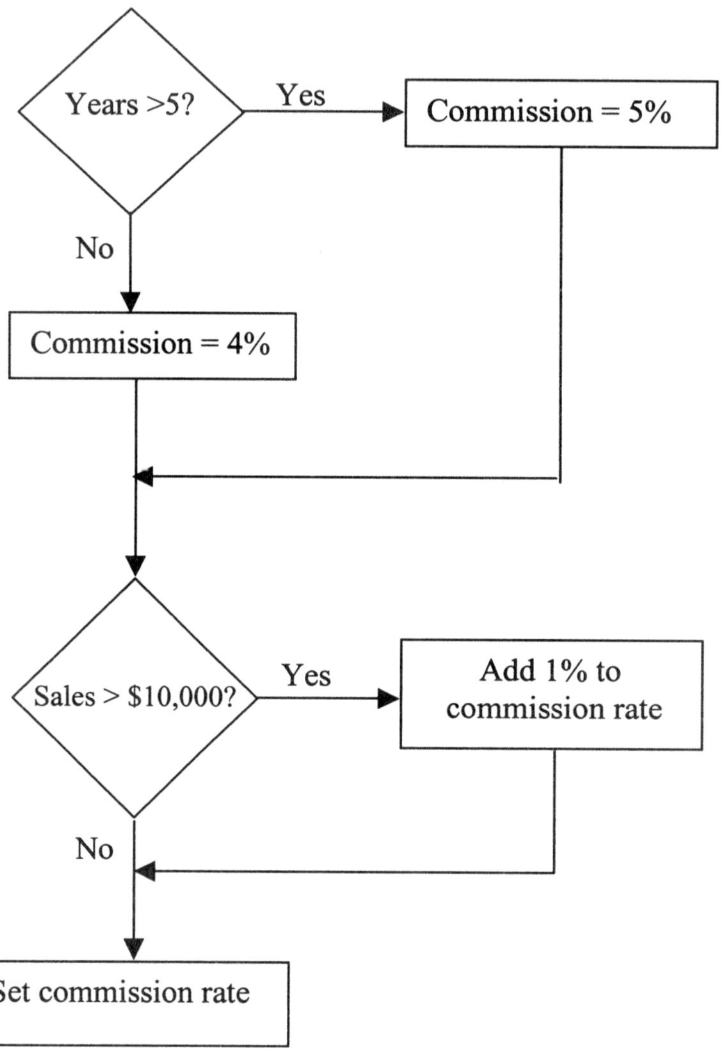

a. Suppose a salesperson has been with the company four years and had sales of $8,000. What is the commission rate for this salesperson?

b. If a salesperson has been with the company six years and had $12,000 in sales, what would be the commission rate?

Solution

For part (a) we come in through the first decision box and proceed straight down, since the length of service was less than five years. At the second decision box, we again proceed straight down since the sales were less than $10,000. Therefore, the commission rate is 4%. In part (b) we proceed to the right after entering the first decision box since the length of service was greater than five years. This means that the commission rate at this point is 5%. When the second decision box is reached, we proceed to the right, since sales were greater than $10,000. At this point, another 1% is added to the commission rate. Therefore, the commission rate for this salesperson will be 6%. ■

Example 4

The policy for withdrawal from a mathematics course at a college states that anyone can withdraw from a course for any reason before the tenth week of class. During and after the tenth week, if a student is passing, he or she may withdraw from the course provided that no more than three classes were missed. Provide a flowchart for this policy.

Solution

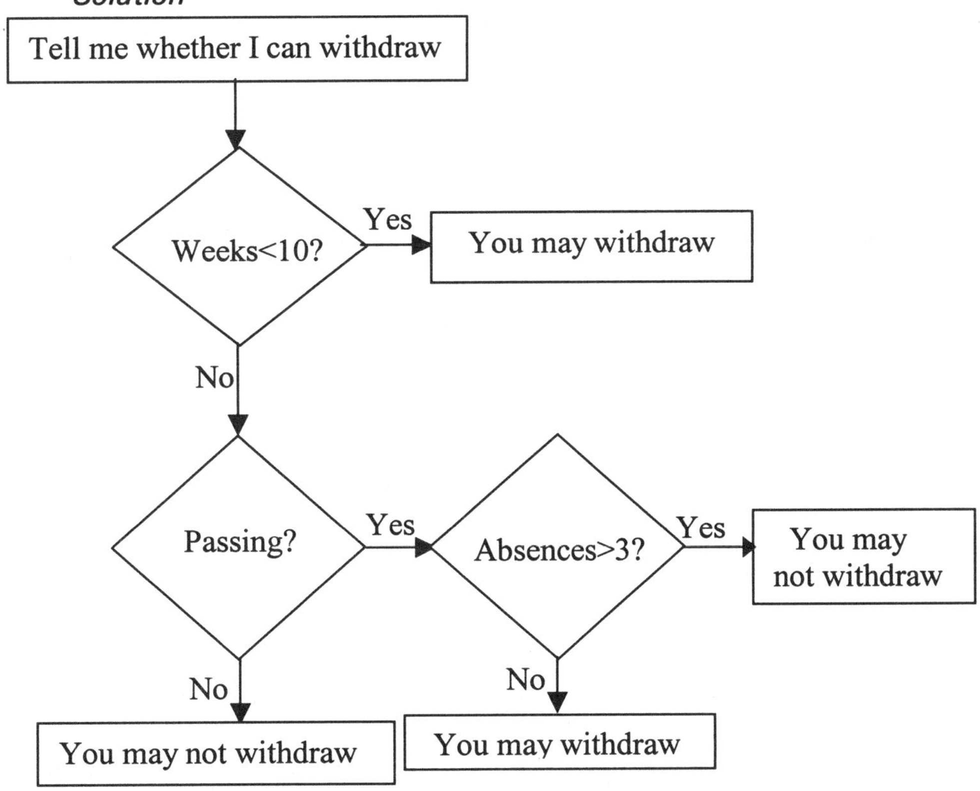

■

In-Class Exercises and Problems for Section 4.5

In-Class Exercises

1. In order to get to a classmate's house, you need to get to New Hyde Park Road. You must decide if you will take the Expressway or the Northern Parkway. If the Parkway is open, and if there is a New Hyde Park Road exit, you will take the Parkway. If there is not a New Hyde Park Road exit, you will ask for alternate directions. If the Parkway is closed and the Expressway is open and if there is a New Hyde Park Road exit, you will take the Expressway. If the Expressway does not have a New Hyde Park Road exit, you will ask for alternate directions. If the Expressway is closed, you will take the service road.
   a. Construct a flowchart that depicts your decision-making process.
   b. How will you get to New Hyde Park Road if the Northern Parkway is closed and the Expressway is open but does not have an exit to New Hyde Park Road?
   c. How will you get to New Hyde Park Road if the Northern Parkway is open and there is an exit to New Hyde Park Road?
   d. How will you get to New Hyde Park Road if both the Northern Parkway and the Expressway are closed?
2. If you are in good academic standing, have access to the internet, are self-motivated and have taken an online course before, you may register for at most two online courses. If you are not in good academic standing or if you do not have internet access, you are not permitted to take any online courses. If you are not self-motivated or if you have not taken an online course before, you must speak to a counselor before a final decision can be made about enrolling in an online course.
   a. Construct a flowchart for the above situation.
   b. Vijay is a student in good academic standing, has no access to the internet and would like to take an online course. Can she take the class?
   c. Euel is a self-motivated student. He is in good academic standing, has internet access and has taken two online courses in the past. Does he need to see a counselor in order to register for online courses?

Problems

3. Use the flowchart below to answer each question.
   a. When is pizza eaten?
   b. What happens for dinner on weekdays?

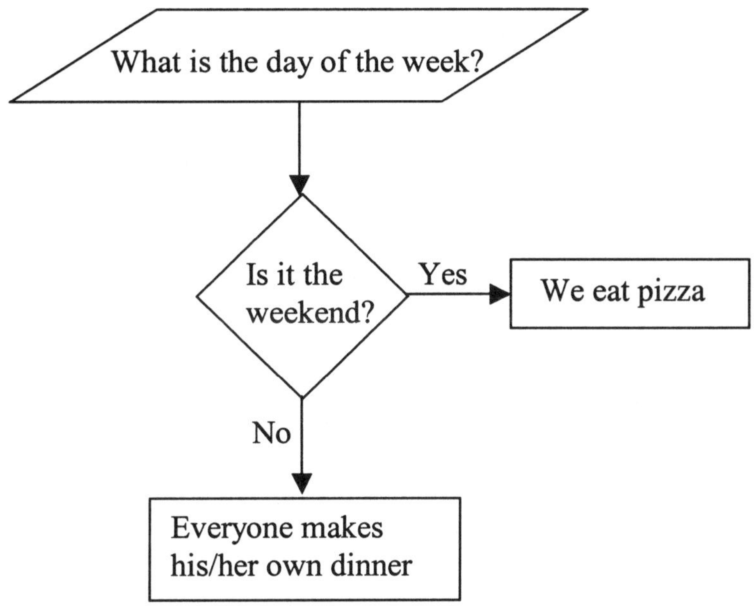

4. Use the following flowchart to answer each question.
   a. Lanna has seven absences. Can she drop her lowest grade?
   b. Oscar has perfect attendance. Can he drop his lowest grade?

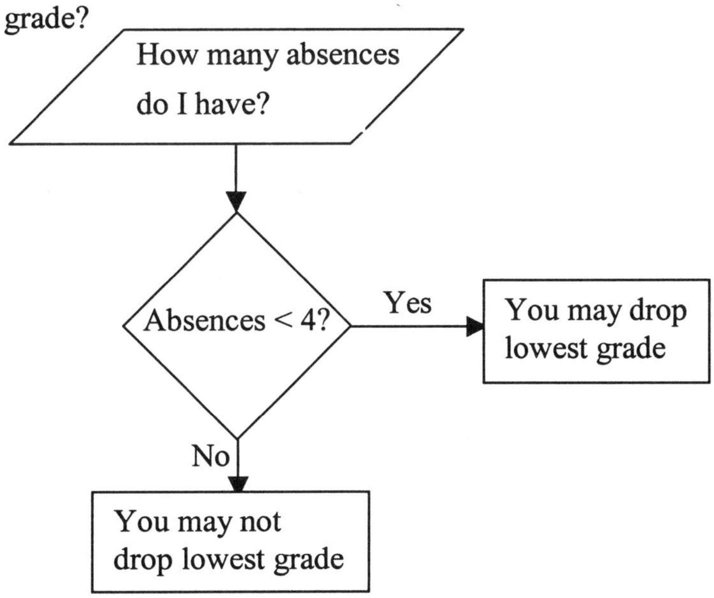

5. Using the flowchart below, determine the fine for each question asked.

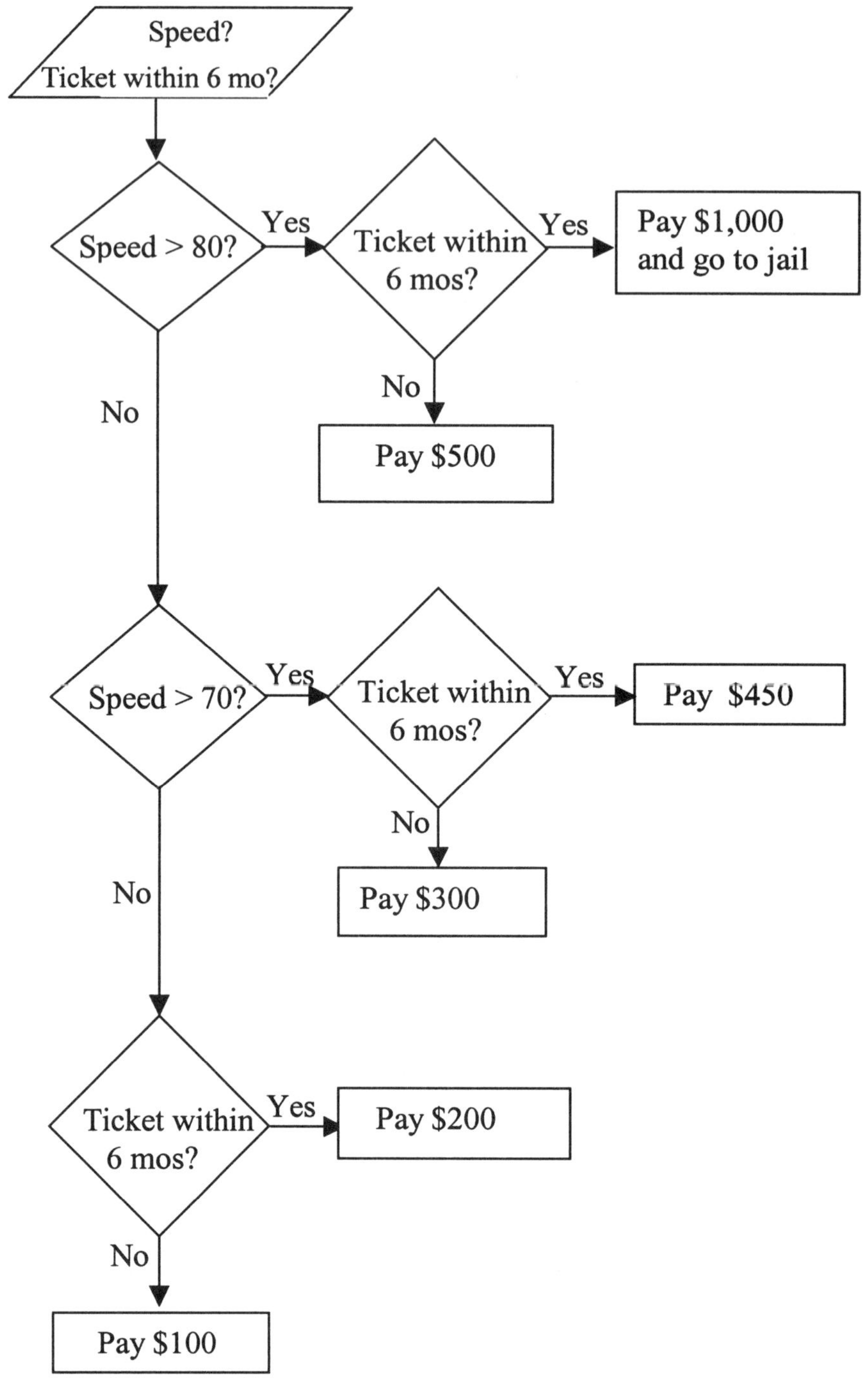

a. You were travelling at 65 mph and you received a speeding ticket 4 months ago.
b. You have never received a speeding ticket before, but you were travelling at 67 mph.
c. You were travelling at 73 mph and received a speeding ticket 18 months ago.
d. You were travelling at 83 mph and received a speeding ticket 5 months ago.
e. You were travelling at 74 mph and received a speeding ticket 1 month ago.

6. Using the flowchart shown in Example 4, determine if the following conditions allow a student to drop.
   a. The student has 4 absences and is passing in the seventh week.
   b. The student is passing in week 4 and has 5 absences.
   c. The student is failing in week 12 and has 2 absences.
   d. The student is passing in week 12 and has 3 absences.
   e. The student is failing in week 6 and has 6 absences.

7. Everything in Cheap Charlie's Shop sells for one dollar or less. Construct a flowchart that one can use to give change, *using the minimum number of coins*, to a customer who used a dollar bill to purchase one item.

4.6 Electronic Circuit Reduction

The ideas learned in logic can be used to simplify complicated *electronic circuits*. An electronic circuit consists of a wire, a source of electricity *S*, and a switch or switches that can be turned on and off. When a switch is on (i.e., the circuit is closed), current may flow through it, beginning at *S*, and terminating at a point *T*. If the switch is off (i.e., the circuit is open), current does not flow through it.

Consider the circuit below that begins at *S*, ends at *T*, and has two switches, *A* and *B*.

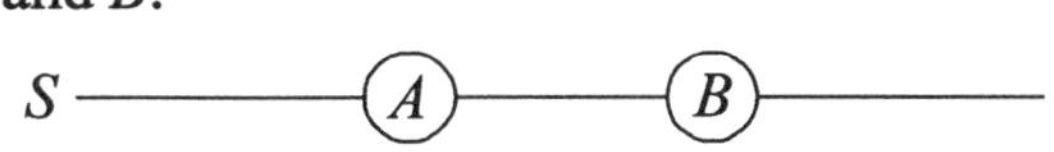

If both *A* and *B* are on, current will flow from *S* to *T*. If either or both switches are off, current will not flow from *S* to *T*. When both switches in a circuit must be on in order for current to flow, the switches are said to be arranged in *series*. We say A is in series with B, or there is a *series circuit* connecting A with B.

If we construct a table that shows us the conditions under which current flows in a series circuit containing two switches, A and B, it looks satisfyingly familiar.

A	B	Current flows from S to T
On	On	Yes
On	Off	No
Off	On	No
Off	Off	No

The table is analogous to the conjunction truth table in logic, with the "On" condition being equivalent to "true" and the "Off" condition being equivalent to "false". Therefore, we can symbolically represent the arrangement of two switches in series as $A \wedge B$. Light strings for Christmas trees are often wired in series. That's why when one bulb burns out, the entire string does not work.

If current can travel from S to T over two branches of a circuit as shown below, the circuit is said to be a *parallel circuit*, and switches A and B are said to be in *parallel*.

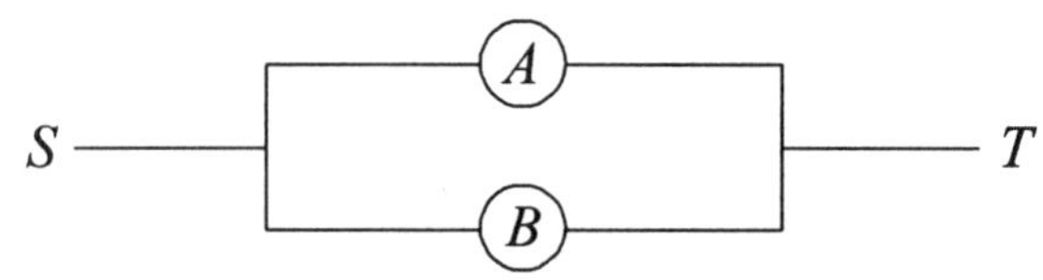

Notice that if a circuit is arranged in parallel, current will flow from S to T if either switch is on or if both switches are on.

If we construct a table that shows us the conditions under which current flows in a parallel circuit containing two switches, A and B, it too looks familiar.

A	B	Current flows from S to T
On	On	Yes
On	Off	Yes
Off	On	Yes
Off	Off	No

The table is analogous to the disjunction truth table in logic. Thus, we can symbolically represent the arrangement of two switches in parallel as $A \vee B$. Most lights in your home are wired in parallel. That's why when one bulb burns out, the others remain on.

If two switches are arranged such that whenever one switch is on the other is off, we say that the switches are *complementary* switches. If A represents a switch, then its complementary switch is denoted as $\sim A$. Often, these kinds of switches are found at the bottom and top of a staircase. This is what allows you to open and close the light over a staircase whether you are at the bottom or the top of the staircase.

If complementary switches are in series, current will not flow from S to T. In logic notation, this is expressed as $A \wedge \sim A =$ false. If a switch and its complement are connected in series, current *never* flows from S to T.

On the other hand, if complementary switches are in parallel, current will flow from S to T. In logic notation, this is expressed as $A \vee \sim A =$ true. That is, if a switch and its complement are connected in parallel, current *always* flows from S to T.

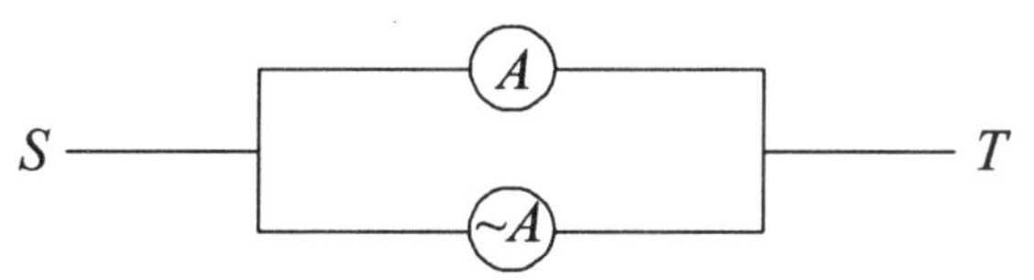

In most applications, we encounter more than two switches in a circuit.

Example 1

Consider the following circuit.

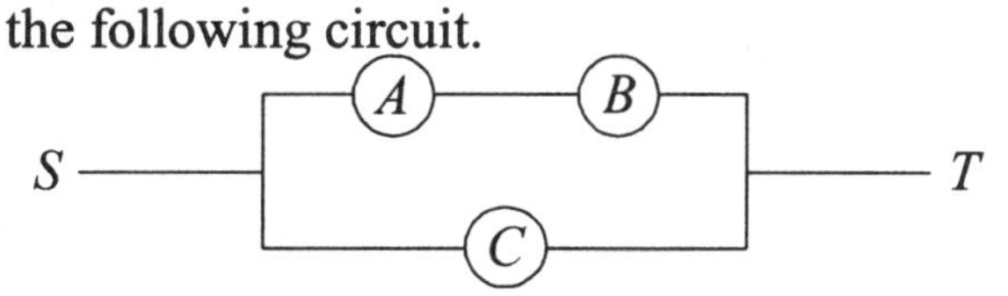

Express this circuit symbolically.

Solution

First, we notice that A and B are in series. We express this symbolically as $A \wedge B$. Switch C is in parallel with $A \wedge B$. Therefore, we can express the circuit as $(A \wedge B) \vee C$.

■

Example 2

For the circuit shown in Example 1, under what conditions will current flow from S to T?

Solution

We answer this question by examining the equivalent logic question, i.e., under what conditions is $(A \wedge B) \vee C$ on. To do this, we construct a truth table.

A	B	C	$A \wedge B$	$(A \wedge B) \vee C$
On	On	On	On	On
On	On	Off	On	On
On	Off	On	Off	On
On	Off	Off	Off	Off
Off	On	On	Off	On
Off	On	Off	Off	Off
Off	Off	On	Off	On
Off	Off	Off	Off	Off

We see that the only times current will not flow from S to T is when either both A and C are off (the sixth row), when B and C are off (the fourth row), or when all three switches are off (the eighth row). ■

We often use logical equivalences to reduce complex circuits into simpler ones. The distributive equivalence (see page 53) is often useful. There is another helpful logic equivalence called *absorption,* which can be expressed as $p \vee (p \wedge q) \Leftrightarrow p$. You will be asked to prove the absorption equivalence in the exercises.

Example 3

Reduce the given circuit to an equivalent circuit with fewer switches.

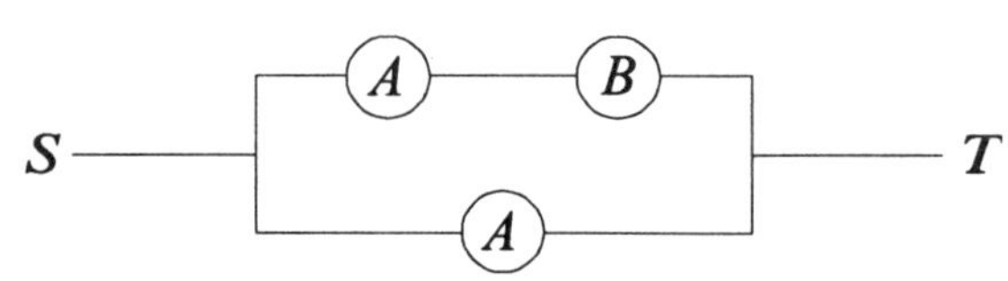

Solution

Our first step is to express the circuit in logic notation. On the upper branch of the circuit, A is in series with B. We represent this branch as $A \wedge B$. The lower branch only contains switch A. It is in parallel with the upper branch. Therefore, we can represent this circuit as $A \vee (A \wedge B)$. But, the absorption equivalence guarantees that $A \vee (A \wedge B)$ is equivalent to A. Hence, the original circuit that contained three switches can be equivalently represented as a circuit that contains only one switch, A.

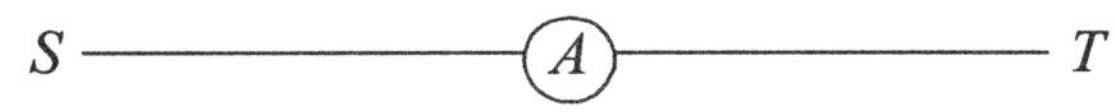

This means that whether or not current flows is only dependent on switch A. If A is on, current flows, otherwise, it does not.

■

Example 4

Reduce the given circuit to an equivalent circuit with fewer switches.

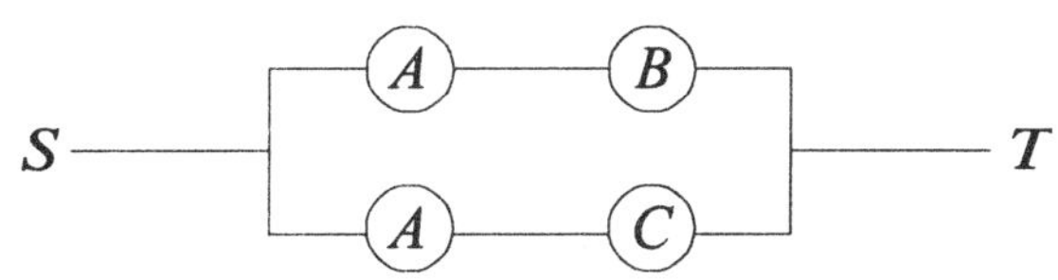

Solution

Both the upper and lower branches have two switches in series. The branches are in parallel with each other. Therefore, we can represent the above circuit as $(A \wedge B) \vee (A \wedge C)$. However, the distributive equivalence guarantees that $(A \wedge B) \vee (A \wedge C)$ and $A \wedge (B \vee C)$ are the same. This means that the circuit can be redrawn with switch A in series with a parallel branch consisting of switches B and C. The resulting reduced circuit is shown below. Notice that it has only three switches.

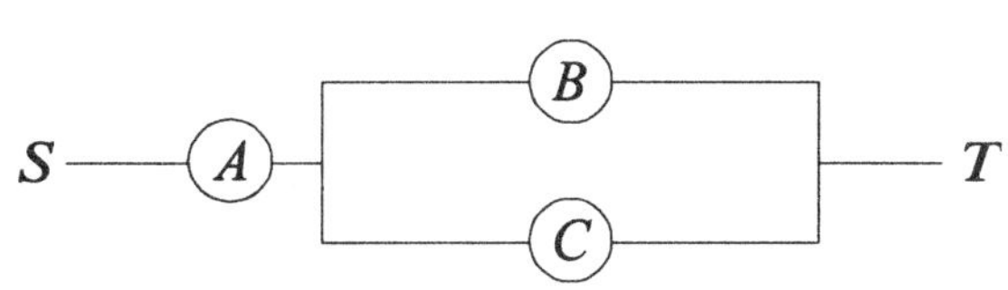

■

Example 5

Reduce the given circuit to an equivalent circuit with fewer switches.

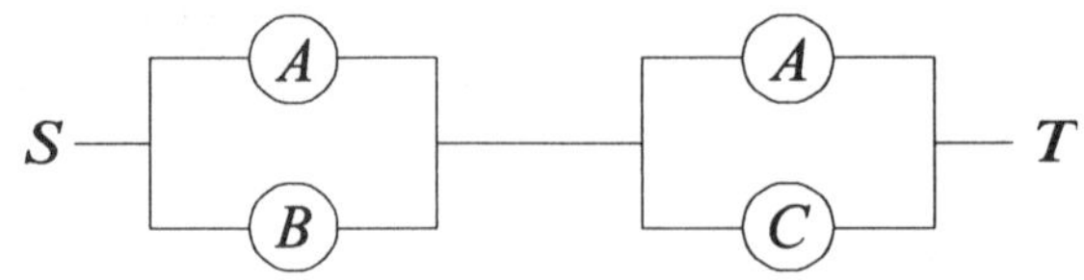

Solution

We have two parallel circuits in series with one another. The first and second parallel circuits may be expressed as $A \vee B$ and $A \vee C$ respectively. Since these two circuits are in series with each other, the entire circuit may be expressed as $(A \vee B) \wedge (A \vee C)$. Now we can use the distributive equivalence to express this as $A \vee (B \wedge C)$. The equivalent reduced circuit is shown below.

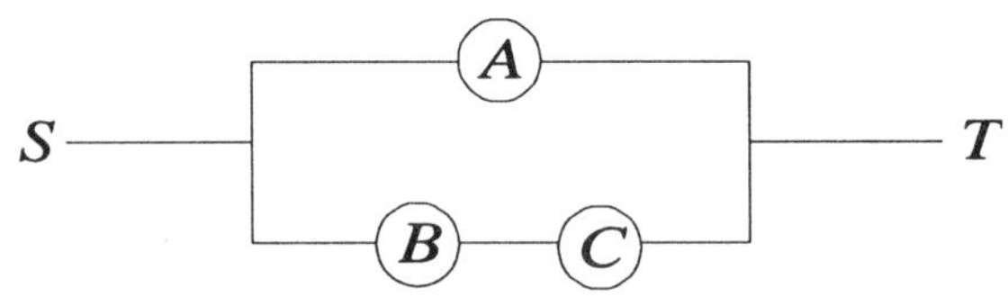

■

In the previous examples, we used either the absorption equivalence alone or the distributive equivalence alone to reduce a circuit. Sometimes, we can apply both laws to the same circuit as we try to reduce it.

Example 6

Reduce the given circuit to an equivalent circuit with fewer switches.

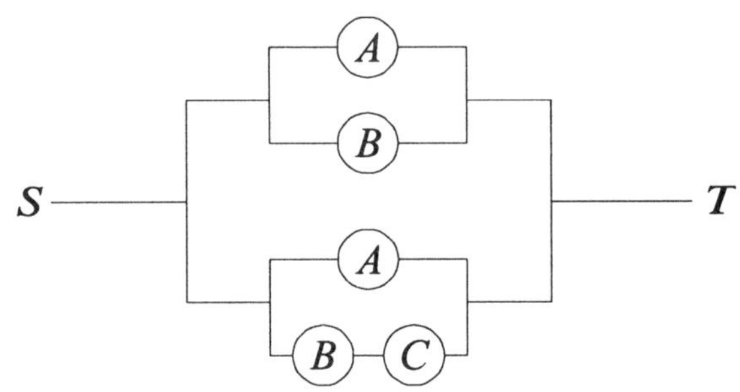

Solution

The upper branch of this parallel circuit can be expressed as the parallel sub-circuit $A \vee B$. The lower branch consists of a parallel sub-circuit, one of whose branches consists of a series circuit. Therefore, the lower branch can be expressed as $A \vee (B \wedge C)$. Hence, the entire circuit can be expressed as $(A \vee B) \vee [A \vee (B \wedge C)]$. If we apply the distributive equivanence to the lower branch, $A \vee (B \wedge C)$, we obtain $(A \vee B) \wedge (A \vee C)$. We can now express the entire circuit as $(A \vee B) \vee [(A \vee B) \wedge (A \vee C)]$. However, the absorption equivalence allows us to express this last statement as $A \vee B$. Our reduced circuit is shown below.

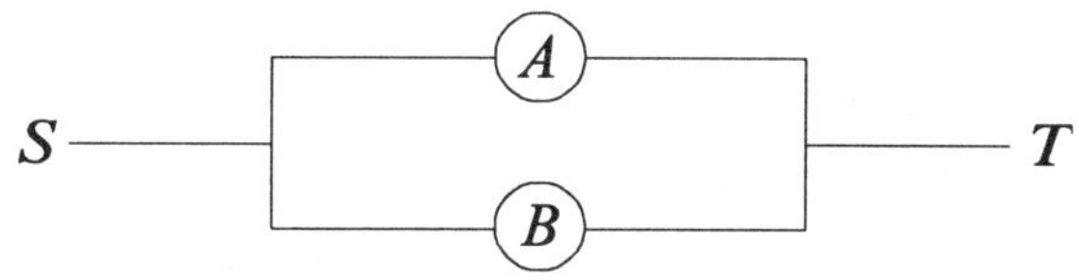

■

In-Class Exercises and Problems for Section 4.6

In-Class Exercises

For Exercises 1–8, use the circuit below to determine whether or not current will flow from *S* to *T*.

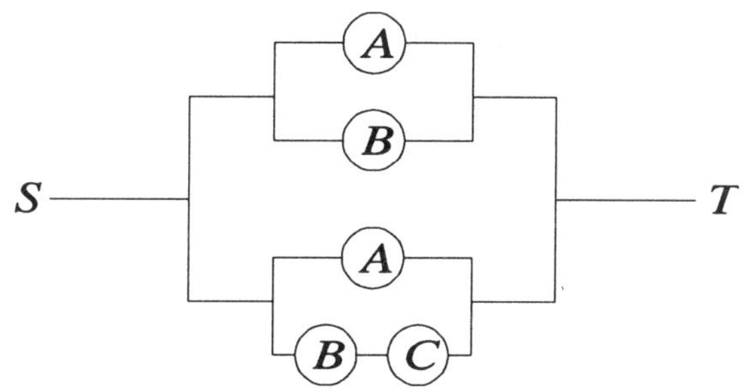

	A	*B*	*C*
1.	On	On	On
2.	On	On	Off
3.	On	Off	On
4.	On	Off	Off
5.	Off	On	On
6.	Off	On	Off
7.	Off	Off	On
8.	Off	Off	Off

For Exercises 9–13, determine if each of the following pairs of circuits are equivalent.

9. 

and

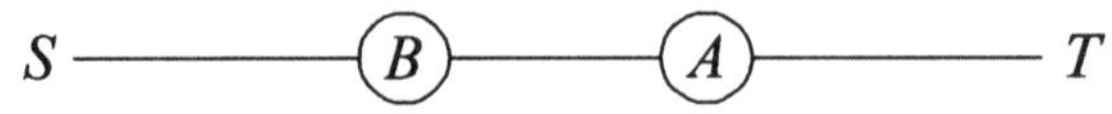

10. 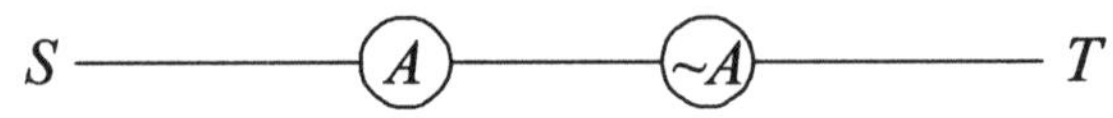

and

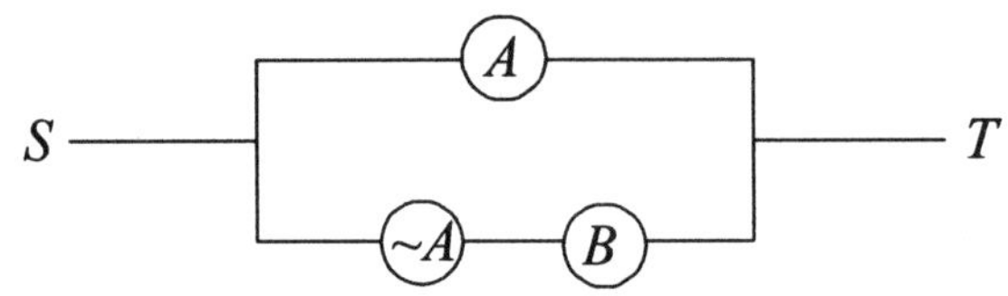

11. 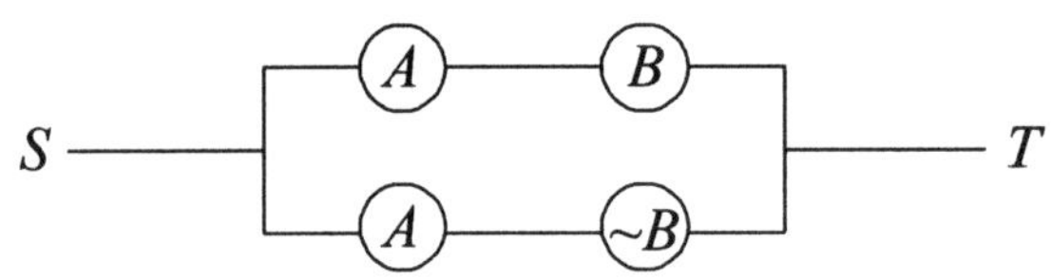

and

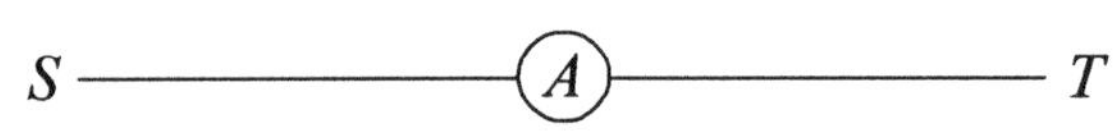

12. 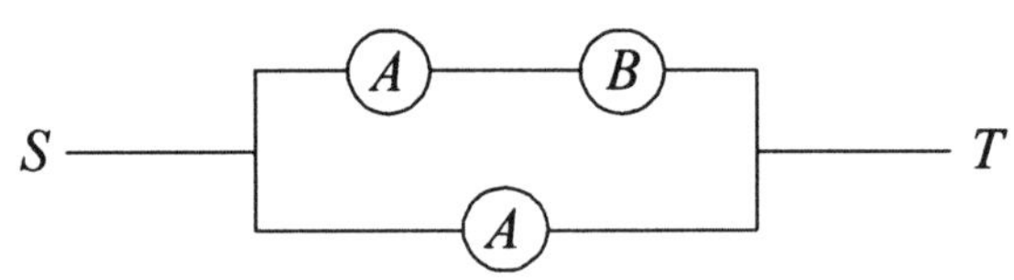

and

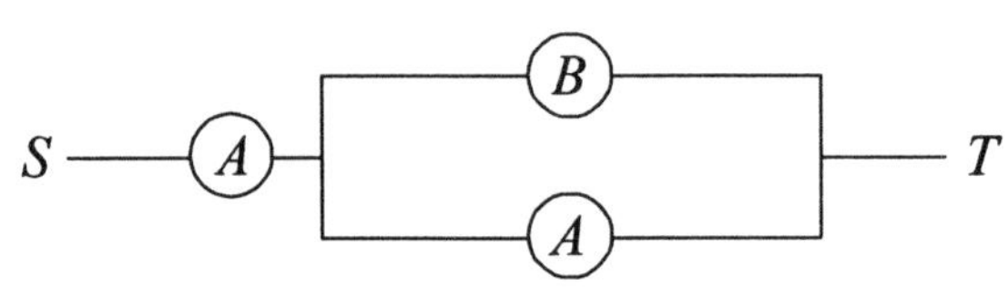

13. 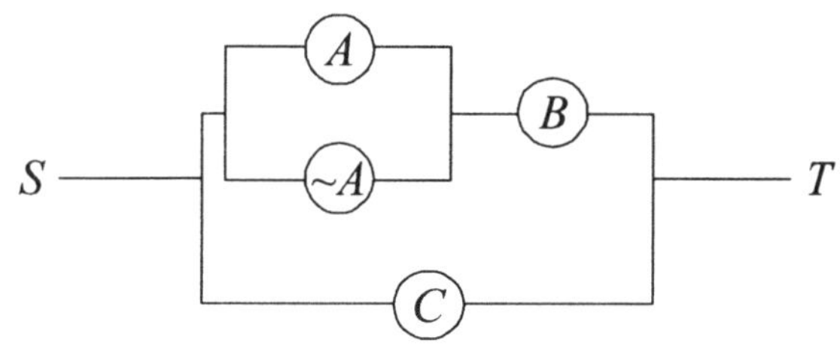

and

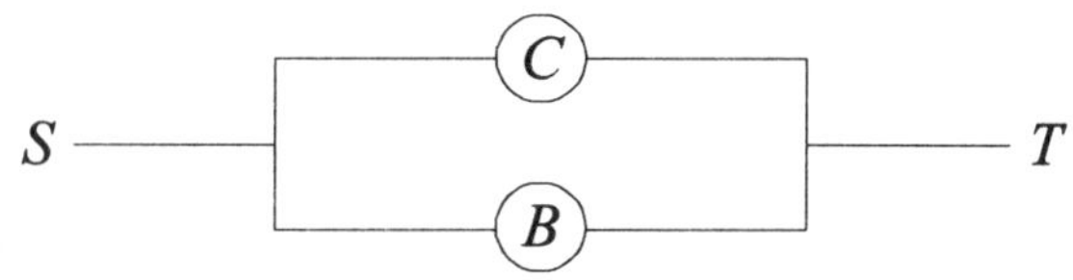

14. Design the simplest single circuit that has current flowing from *S* to *T* when either of the three conditions apply.
   - *A*, *B* and *C* are all on.
   - *A* and *B* are on but *C* is off.
   - *A* and *C* are off but *B* is on.

Problems

For Problems 15–22, use the circuit below to determine whether or not current will flow from *S* to *T*.

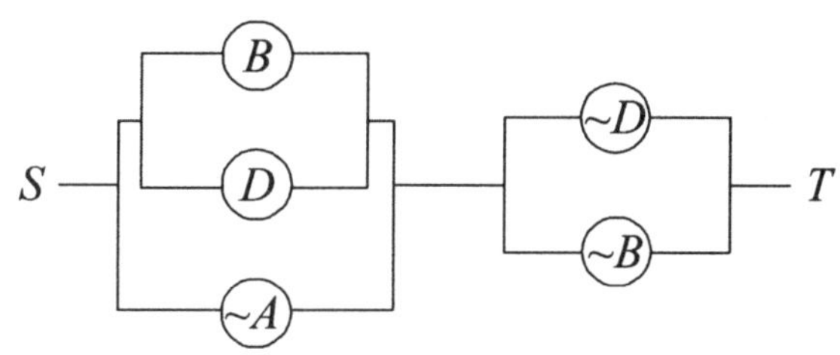

	A	*B*	*D*
15.	On	On	On
16.	On	On	Off
17.	On	Off	On
18.	On	Off	Off
19.	Off	On	On
20.	Off	On	Off
21.	Off	Off	On
22.	Off	Off	Off

For Problems 23–32, represent each circuit using logic notation. Then redesign each circuit into one which has fewer switches. Finally, express your answer in logic notation.

23.
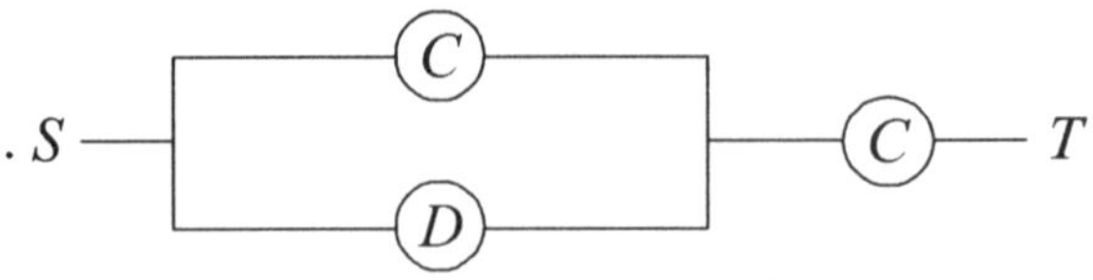

24.
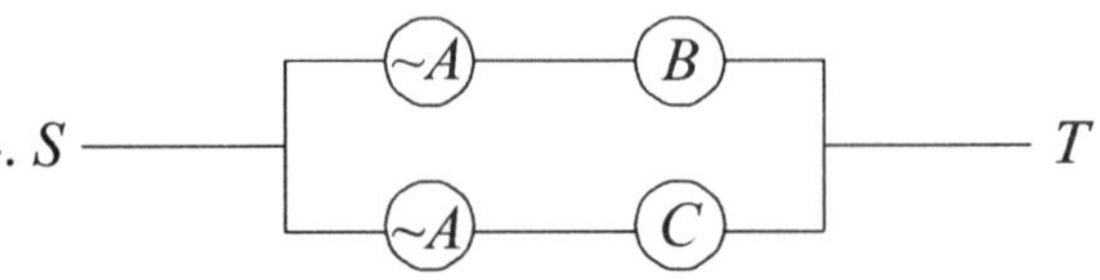

25.
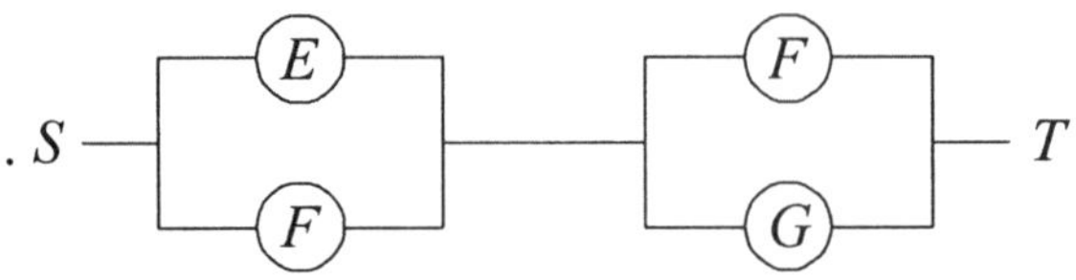

26.
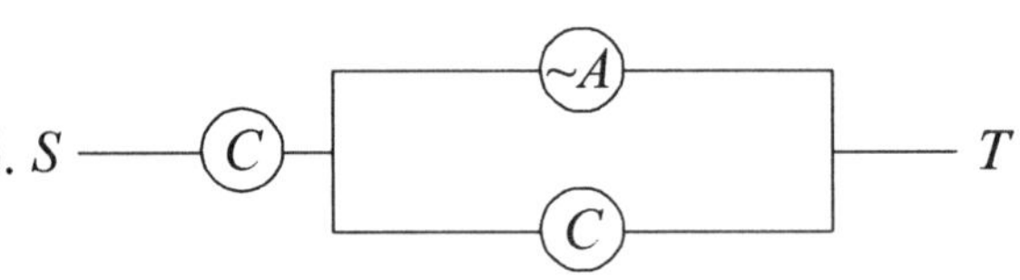

27.
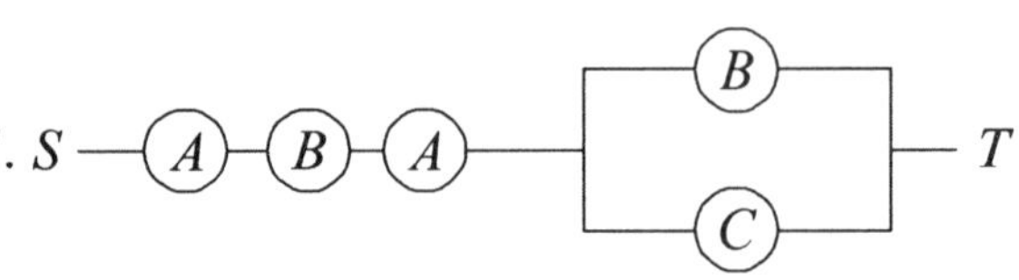

28.
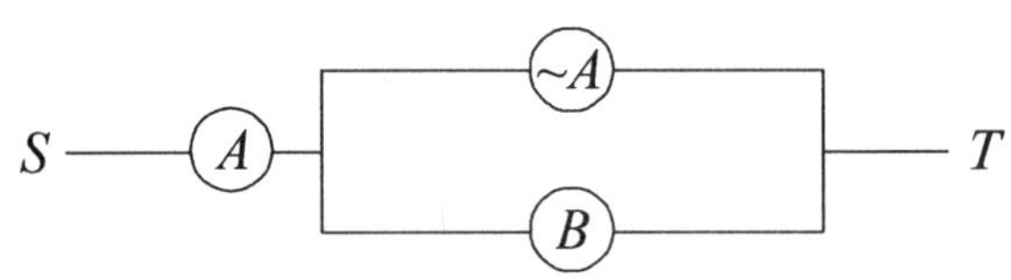

29.
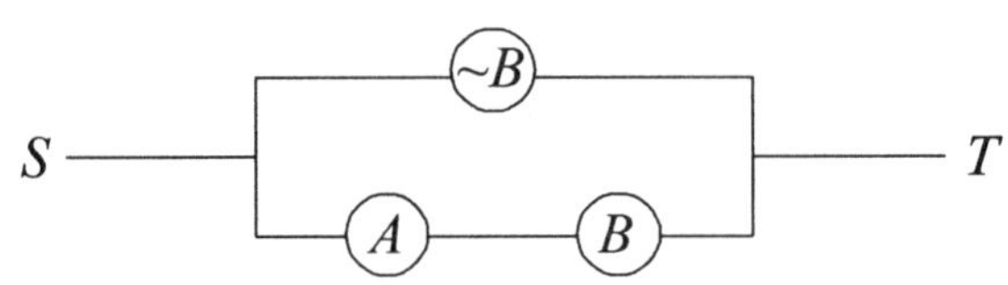

30.
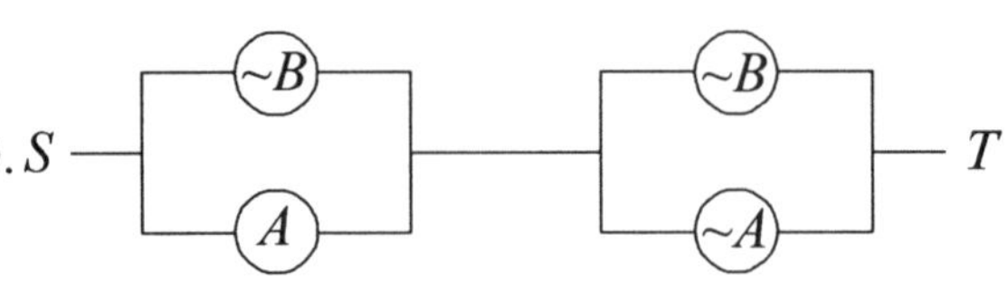

31. 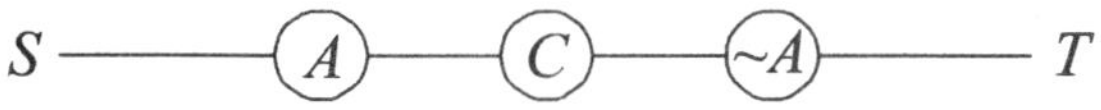

32. 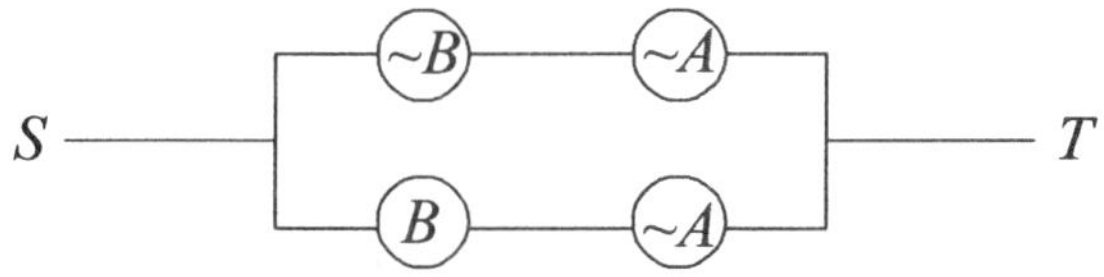

33. Think of a voting machine as having a concealed keypad for each voter. A motion is proposed and each voter votes either yes or no. Each yes vote closes a switch in the circuit and each no vote leaves the switch open. If current can get through the circuit after all the votes are cast, a motion passes and a light goes on at *T*. Suppose that there are three people on a committee and in order for a motion to pass, there must be a unanimous vote. That means persons *A*, *B* and *C* must all vote yes for the motion to pass. The set notation for this is $A \wedge B \wedge C$. The circuit resulting from this notation looks like the one shown below.

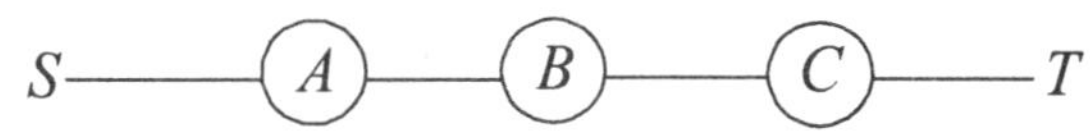

Notice that in this circuit, if anyone votes no, current does not flow through the system. Also notice that this circuit cannot be reduced, since all its components are unique. Suppose three people, *A*, *B*, and *C* vote on a motion, and the motion will pass if a majority of the three voters are in favor of it.

a. Write the three logic expressions for the condition of majority rules.
b. Since any of the three conditions will cause the proposal to pass, which logic connective should be used to join the three expressions?
c. Use the distributive equivalence to simplify the expression obtained in part (b), and then draw the circuit.
d. Suppose the motion will pass if *A* votes yes or both *B* and *C* vote yes. Write the logic expression for this condition.

34. Use a truth table to show that $p \vee (p \wedge q)$ is equivalent to *p*.

Chapter 4 Review

1. Rewrite the following argument, changing all fuzzy statements into *TF* statements. Then, formally prove the argument.

   Either Candice will go on a nice vacation or she will stay home. She will not go on a nice vacation and she will take 5 vacation days next week. If she has a busy week at work this week, then she will not take 5 vacation days next week. Therefore, Candice was not busy at work this week and she will stay home.

2. Let $\tau(p) = 0.65$ and $\tau(q) = 0.24$.
   a. Find $\tau[(p \vee q) \wedge q]$.
   b. What do you notice about the answer to part (a)?
   c. What does this suggest about the statement $(p \vee q) \wedge q$?

3. Gary needs to decide which of five job offers to accept. He will take into consideration how close the job is to his home, the salary, and the job satisfaction. Let *p*: The job offered to Gary is close to home, *q*: The job offered to Gary pays well, and *r*: The job offered to Gary is satisfying.
   Consider the action table shown below. Let $\tau(p) = 0.9$, $\tau(q) = 0.8$, and $\tau(r) = 0.4$. Determine an appropriate final action for the statement "The job offered to Gary pays well and is satisfying" with the statement "The job offered to Gary is close to home."

ACTION TABLE

τ	Action (Take job offer y)
0.1	5
0.2	5
0.4	4
0.6	3
0.8	2
0.9	1

4. Very often, the manager of a baseball team must decide whether or not to remove the starting pitcher. If the pitcher is losing and has thrown over ninety pitches, and a relief pitcher is ready, the

relief pitcher is brought in. If the relief pitcher is not ready, the starting pitcher is left in for one more batter. If the starting pitcher is losing but has not thrown more than ninety pitches and the team is losing by three or more runs, then the relief pitcher is brought in immediately. However, if the team is losing by less than three runs, the starting pitcher is left in for one more inning.

a. Construct a flowchart that depicts the manager's decision-making process.
b. What managerial decision should be made if the starting pitcher threw 60 pitches and is losing by 2 runs?
c. What managerial decision should be made if the starting pitcher threw 95 pitches and the relief pitcher is ready?
d. What managerial decision should be made if the starting pitcher threw 50 pitches and is losing by 5 runs?

5. Supply an appropriate Yes/No question for the flowchart below.

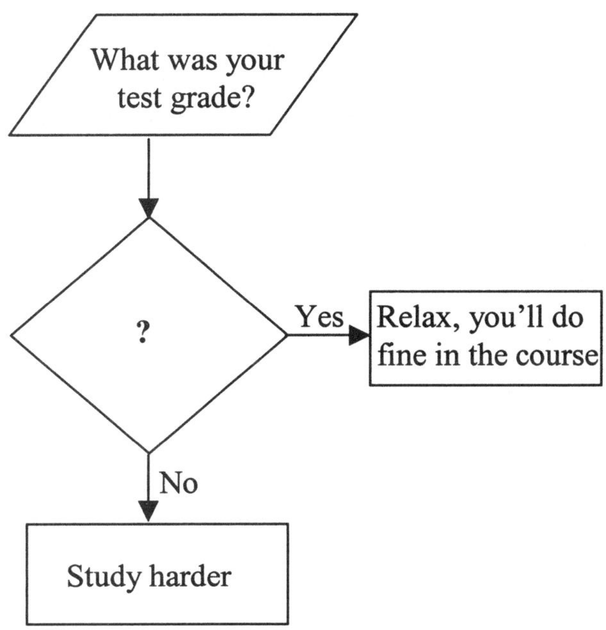

For questions 6–13, use the circuit below to determine whether or not current will flow from S to T.

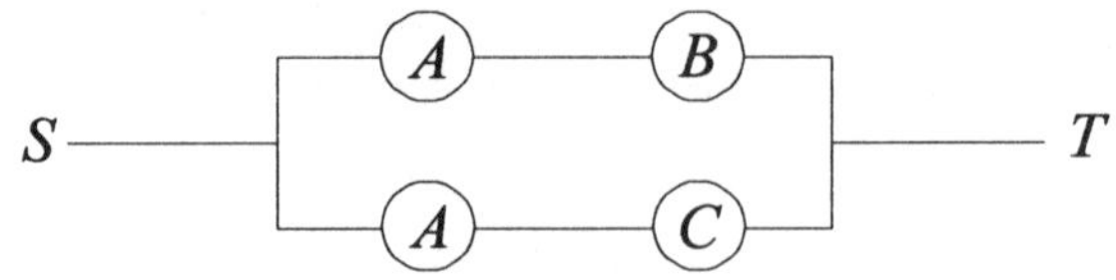

	A	B	C
6.	On	On	On
7.	On	On	Off
8.	On	Off	On
9.	On	Off	Off
10.	Off	On	On
11.	Off	On	Off
12.	Off	Off	On
13.	Off	Off	Off

For questions 14–16, represent each given circuit using logic notation. Then redesign each circuit into one that has fewer switches.

14. 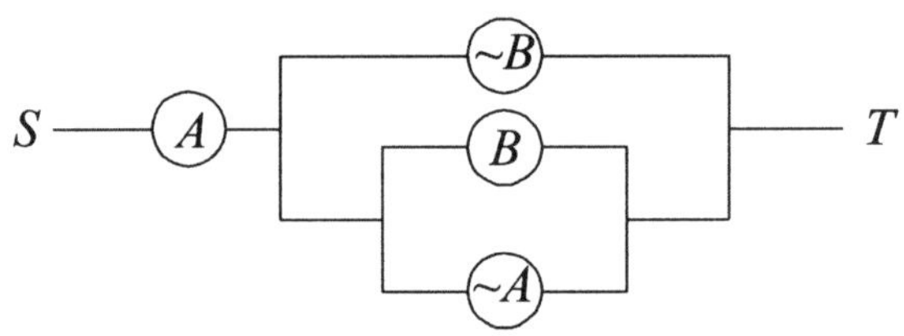

15. 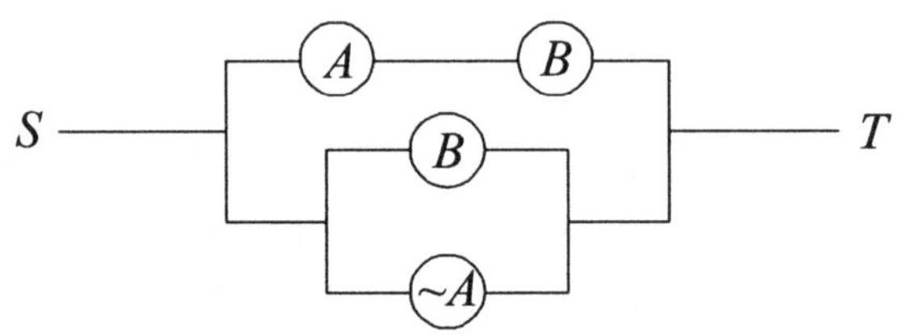

16. 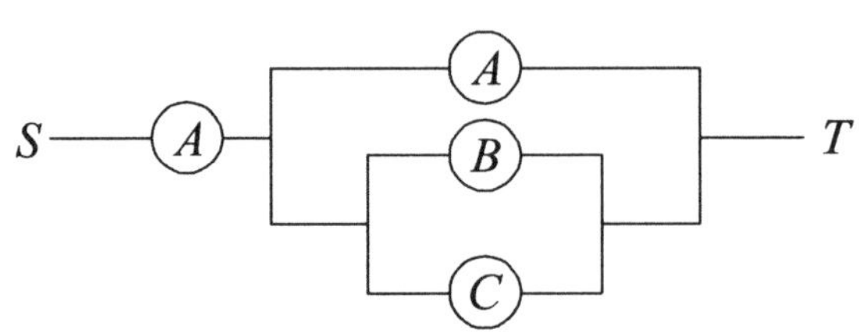

Sample Exam: Chapter 4

1. Which of the following statements would be a reasonable fuzzy interpretation of the statement "John received a 92% on his fuzzy logic test"?
   a. John got an A on his test.
   b. John got the best grade in his class on the test.
   c. John did very well on this test.
   d. John did better on this test than on his last test.

2. Which of the following statements is a *TF* statement?
   a. The trains are not running well today.
   b. The trains are running behind schedule today.
   c. The trains are running smoothly today.
   d. Taking the trains today is your best bet to get to work on time.

3. Let $\tau(p) = 0.35$ and $\tau(q) = 0.22$. What is $\tau(\sim p \vee q)$?
   a. 0.22
   b. 0.35
   c. 0.65
   d. 0.78

4. Let $\tau(p) = 0.17$ and $\tau(q) = 0.68$. Which of the following statements has a degree of truthfulness of 0.32?
   a. $p \rightarrow q$
   b. $\sim p \rightarrow q$
   c. $\sim p \leftrightarrow q$
   d. $p \leftrightarrow q$

5. Rachel went to four shoe stores, and each time had to wait for a salesperson. At the first store she waited 10 minutes, at the second store she waited 15 minutes, at the third store she waited 25 minutes and at the last store she waited only 5 minutes. Which of the following is a reasonable degree of truthfulness for the statement "Rachel had a long wait at the second store"?
   a. 0.25
   b. 0.50
   c. 0.75
   d. 1

6. Michael asked 9 people about his new truck. Six said it was absolutely gorgeous, while 3 said it wasn't 'rugged looking.' Which of the following is a reasonable degree of truthfulness for the statement "Michael's truck is not 'rugged looking' "?
   a. 0.30
   b. 0.33
   c. 0.60
   d. 0.67
7. Consider the following flowchart.

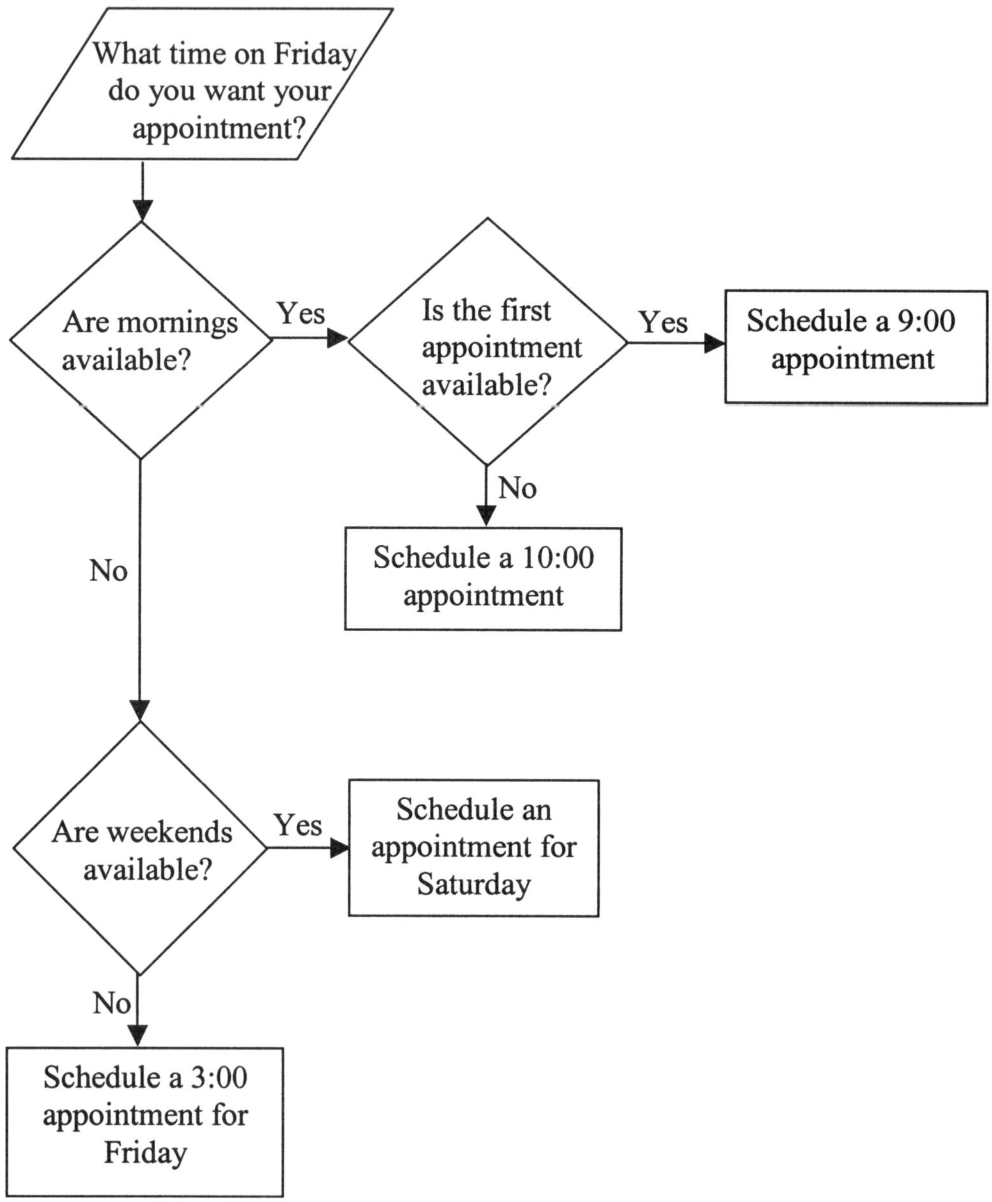

When will the appointment be made if mornings and weekends are NOT available?

a. Friday, 9 am
b. Friday, 10 am
c. Friday, 3 pm
d. Saturday

8. The following is the pricing policy for Dollar Daze. If an item is tagged with a blue ticket, it costs $1.00. If an item is tagged with a red ticket, it costs $2.00. If more than 12 of the same color ticketed items are bought, the price is reduced by 10%. Create a flowchart that one can use to find the subtotal for a customer's purchase.

Consider the following situation for questions 9–10.

Cristina is having a potluck party and must decide what type of disposable plates to buy: white paper plates, coated paper plates, cardboard disposable plates or plastic disposable plates. Everyone that is coming is bringing a dish. Cristina must consider the following:

Dish x is a hot dish.
Dish x will be heavy and juicy.

Consider the Input Table and Action Table below.

INPUT TABLE

Dish x	τ(dish x is hot)	τ(dish x is heavy)	τ(dish x is juicy)
Eggplant	0.7	0.6	0.4
Rice	0.3	0.2	0.1
Lasagna	0.8	0.8	0.6
Apple Pie	0.2	0.4	0.3

ACTION TABLE

τ	Buy y
0.1	1: White Paper Plates
0.2	1: White Paper Plates
0.3	2: Coated Paper Plates
0.4	2: Coated Paper Plates
0.5	3: Plastic Disposable Plates
0.6	3: Plastic Disposable Plates
0.7	3: Plastic Disposable Plates
0.8	4: Cardboard Disposable Plates
0.9	4: Cardboard Disposable Plates

9. Find an appropriate final action for: "Eggplant is a hot dish" with "Eggplant will be heavy and juicy."

10. Find an appropriate final action for: "Rice is a hot dish" with "Rice will be heavy and juicy."

For questions 11–18, use the circuit below to determine whether or not current will flow from *S* to *T*.

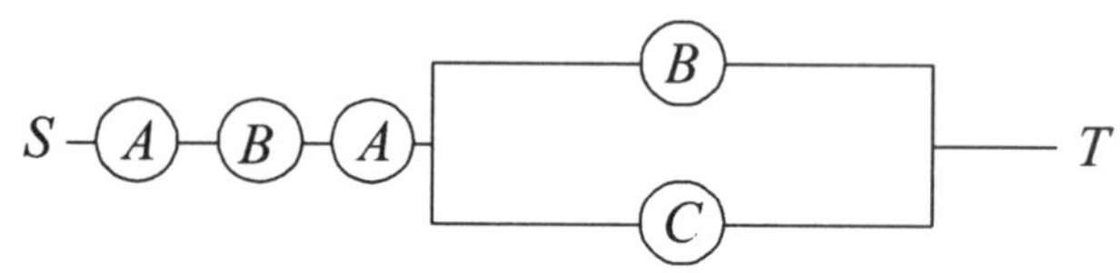

	A	*B*	*C*
11.	On	On	On
12.	On	On	Off
13.	On	Off	On
14.	On	Off	Off
15.	Off	On	On
16.	Off	On	Off
17.	Off	Off	On
18.	Off	Off	Off

19. Represent the circuit below using logic notation. Then redesign it into a circuit that has fewer switches.

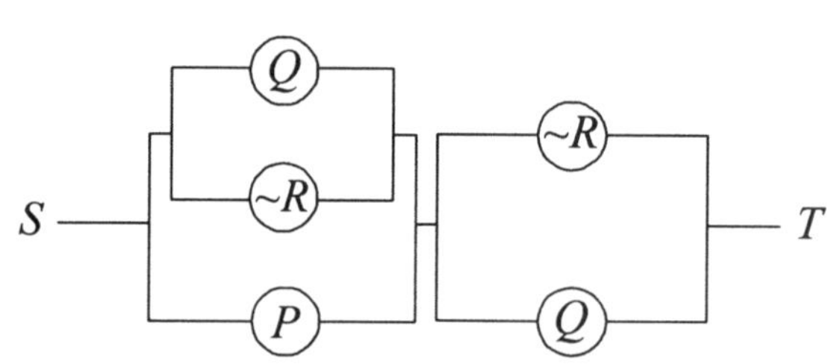

CHAPTER FIVE

SET THEORY

A concept that we are all familiar with is that of a set. We speak of a set of dishes, a set of CDs, or a tool set. The branch of mathematics that is concerned with the general study of sets is set theory. Set theory is important in the study of mathematics, and has many important applications in other areas as well.

5.1 Sets and Notation

Introduction

Without getting too technical, we will think of a *set* as any collection. The objects that are in the set, called *elements,* or *members* of the set, may be numbers, letters, objects, or any other entities. Once we describe a particular set, we must be able to determine whether or not a given entity is or is not an element of that set.

As a matter of notation, braces such as { }, are used to enclose the elements of a set.

Example 1

The set consisting of the first three months of the year may be denoted as {January, February, March}.

■

Just as we allow lowercase letters to represent simple statements in logic, we will use uppercase letters to represent sets. For example, we could say that the set consisting of the first three months of the year is called A. We then would write A = {January, February, March}.

The phrase "is an element of" is expressed by the symbol $\in$, while the phrase "is not an element" is expressed symbolically as $\notin$.

Example 2

If A = {January, February, March}, then March $\in A$, but June $\notin A$.

■

In general, sets can be described in one of three ways. First, we could simply list the elements of a set within set braces, as we did in Example 1. This method is satisfactory as long as the number of elements is sufficiently small. If, however, the set has a large number of elements, listing them all becomes tedious or impossible. Rather, we can often just write a description of the set within set braces.

For instance, if we wanted to list the set of all whole numbers from one to one hundred, we could write {all whole numbers from one to one hundred}, or we could write $\{1,2,3,\ldots,100\}$, where the three dots mean "and continuing in this fashion, up to." The third way a set can be described is using *set builder* notation. This notation is expressed symbolically as $S = \{x|x \text{ has a given property}\}$. The letter x denotes any element in the set, and the vertical line $|$ is read as "such that."

Example 3

If $S = \{1,3,5,7,\ldots, 99\}$, then S is the set of odd numbers from one to ninety-nine. Therefore, $5 \in S$, $97 \in S$, $36 \notin S$, and $103 \notin S$.

■

Example 4

If B = {whole numbers larger than five but less than 1000}, then we know $7.25 \notin B$ and $976 \in B$.

■

Example 5

If $S = \{x | x \text{ is a sports car}\}$, the elements in set S would consist of all sports cars.

■

The Universal Set

We define the *universal set,* or simply, the *universe,* as the set of elements from which we construct all other sets in a particular discussion. We express the universe symbolically by U. If we define $U = \{1, 2, 3, 4, 5\}$, then the numbers one through five are the only numbers that exist for this particular discussion.

Example 6

Let $U = \{1, 2, 3, 4, 5\}$. Could the set $A = \{1, 3, 5, 7\}$ be constructed from this universe?

Solution

Since the element seven is not in the universe, this set cannot be constructed from the given universe.

■

The Null Set

The *null set,* is a set that contains no elements. This set is also referred to as the *empty set*. The null set is expressed symbolically as either $\varnothing$ or $\{ \}$. At first, you may not be aware of the need to be able to talk about an empty set, but if you think for a moment, you will realize that such sets are ubiquitous. The set of all female presidents of the United States as of 2006 is an empty set, as is the set of all positive numbers less than zero.

Example 7

Find the set of all numbers that are multiples of 100 and odd.

Solution

Since all multiples of 100 are even, there are no multiples of 100 that are odd. Therefore, the set of all such numbers is empty, i.e. $\varnothing$.

■

Equal Sets

Two sets, A and B, are *equal sets* if every element of A is in B, and conversely, if every element in B is in A. Using some of our logic connectives, we can define the equality of two sets, A and B as $(A = B) \Leftrightarrow [[(x \in A) \rightarrow (x \in B)] \wedge [(x \in B) \rightarrow (x \in A)]]$.

The biconditional equivalence allows us to define the equality of two sets in a more compact way: $(A = B) \Leftrightarrow [(x \in A) \leftrightarrow (x \in B)]$.

Example 8

If $A = \{1,2,3\}$ and $B = \{3,2,1\}$, is $A = B$?

Solution

Since every element in set A is in set B and every element in set B is in set A, $A = B$.

■

Example 9

Let $A = \{\text{all whole numbers greater than 1 and smaller than 20}\}$. Let $B = \{2,3,4,5,6,7,8,9,10,11,12,13,14,15,16,17,18,19,20\}$. Is $A = B$?

Solution

Every element in set A is in set B. However, there is an element in set B that is not in set A, namely the number 20. Therefore, set A is not equal to set B. This is expressed symbolically as $A \neq B$.

■

Subsets

Set A is a *subset* of set B, expressed symbolically as $A \subseteq B$, if and only if every element in set A is an element of set B. This definition can be stated symbolically as $(A \subseteq B) \Leftrightarrow [(x \in A) \rightarrow (x \in B)]$. Notice that as a consequence of this definition, every set is a subset of itself. Furthermore, if $A = B$ then $A \subseteq B$ and $B \subseteq A$.

Example 10

If $A = \{1,3,5,6,7,8,9\}$ and $B = \{1,5,6\}$, is A a subset of B? Is B a subset of A?

Solution

Since there is at least one element in A that is not an element of B, A is not a subset of B. However, every element of B is an element of A. Therefore, $B \subseteq A$.

■

Example 11

Let $A = \{1,2,4,5\}$ and $B = \{2,5,7,9\}$. Is A a subset of B? Is B a subset of A?

Solution

There are elements in A that are not in B. There are also elements in B that are not in A. Hence, neither set is a subset of the other.

■

One of the most counterintuitive notions in set theory is that the null set is a subset of any set, A. To see why this is true, we need to recall that if this were not the case, there would be a least one element in $\varnothing$ that could not be found in A. This is not possible, since the empty set contains no elements.

Example 12

List all the subsets of $A = \{5, 6\}$.

Solution

The subsets of A are $\{5,6\}, \{5\}, \{6\}$, and $\varnothing$.

■

Example 13

Find all the subsets of $B = \{7, 8, 9\}$.

Solution

The subsets of B are $\{7,8,9\}, \{7\}, \{8\}, \{9\}, \{7,8\}, \{7,9\}, \{8,9\}$, and $\varnothing$.

■

Proper Subsets

Set A is called a *proper subset* of set B, expressed symbolically as $A \subset B$, if and only if set A is a subset of set B, but set A does not equal set B. This definition can be stated symbolically as $(A \subset B) \Leftrightarrow [[(x \in A) \rightarrow (x \in B)] \wedge (A \neq B)]$.

As a consequence of this definition, it should be noted that if A is not a subset of B, then there is at least one element in A that cannot be found in B.

Example 14

List all the proper subsets of $A = \{5, 6\}$.

Solution

The proper subsets of A are $\{5\}, \{6\}$, and $\varnothing$.

■

Since every set is a subset of itself, the empty set (which has no elements) has one subset, which is the set itself. If a set has one element, it must have two subsets, itself and the empty set. In Examples 12 and 13, we saw that if a set has two elements, it has four subsets and if a set has three elements it has eight subsets.

The following table summarizes these results.

Number of elements in A	Number of subsets of A
0	1
1	2
2	4
3	8
n	2^n

It appears that as we add an element to a set, the number of subsets doubles from the previous set. This result can be stated with an equation that expresses the number of subsets as a function of the number of elements. If a set A has n elements, then A has 2^n subsets.

As an equation, we have:

$$\text{number of subsets of } A = 2^n,$$
and
$$\text{number of proper subsets of } A = 2^n - 1,$$

where n is the number of elements in set A.

Example 15

Can a set have exactly 67 subsets?

Solution

The number of subsets of a given set must be a power of two. Since 67 is not a power of two, there is no such set.

■

The Power Set

The *power set* is the set of all subsets of a given set. If $A = \{1, 2, 3\}$, the power set of *A*, denoted as $\mathscr{P}(A)$, contains $2^3 = 8$ elements, and $\mathscr{P}(A) = \{\{1\}, \{2\}, \{3\}\, \{1,2\}, \{1,3\}, \{2,3\}, \{1,2,3\}, \varnothing\}$. Notice that the elements of a power set are sets themselves.

Example 16

If $B = \{1, 2, 3, 4, 5\}$, how many elements are there in $\mathscr{P}(B)$, the power set of *B*?

Solution

The power set consists of all subsets of *B*. Since *B* has 5 elements, its power set has $2^5 = 32$ elements.

■

In-Class Exercises and Problems for Section 5.1

In-Class Exercises

In Exercises 1–5, represent each set by listing its elements.

1. The set of whole numbers less than 20 that are divisible by 4.
2. $\{x \mid x \text{ is a multiple of 5 between 18 and 42}\}$.
3. The set of all two or three letter words that can be formed by using the letters "e", "n" or "o".
4. The set of all even numbers that can be written using the digits 1, 3, or 5.
5. The set of days of the week that do not contain the letter "s".

In Exercises 6–8, describe each set with an English sentence.

6. {0, 2, 4, 6, 8, 10}
7. {31, 33, 35, 37, 39}
8. {a, b, c, d}

For Exercises 9–19, suppose that $U = \{0, 1, 2, 3, 4, 5, 6, 7, 8, 9, 10\}$, $A = \{1, 3, 5, 7, 9\}$, $B = \{0, 2, 4, 6, 8, 10\}$, $C = \{1, 2, 5, 7, 9\}$, $D = \{1, 3, 7, 9\}$, $E = \{8, 2, 10\}$, and $F = \{9, 7, 1, 3\}$.

9. Determine if each statement is true or false.

a. $A = C$	e. $D = F$	i. $\varnothing \subset C$
b. $B \subseteq U$	f. $E \subseteq B$	j. $F \subset A$
c. $A \subset D$	g. $\varnothing \in D$	k. $F \subset D$
d. $7 \in B$	h. $F \subseteq D$	l. $D \subset A$

10. Find the number of elements of C.
11. Find the number of subsets of C.
12. Find the number of proper subsets of C.
13. Find the number of elements in $\mathscr{P}(C)$.
14. Find the number of subsets of B
15. Find the number of proper subsets of B.
16. Find the number of elements in $\mathscr{P}(D)$.
17. Find the number of proper subsets of D.
18. Find the number of elements in $\mathscr{P}(A)$.
19. List $\mathscr{P}(E)$.

Problems

For Problems 20–25, represent each given set by listing its elements.

20. The set of all Presidents of the United States before Thomas Jefferson.
21. The set of whole numbers greater than 3 and less than 10.
22. The set of months that have an "r" in their spelling.
23. The set of all three-letter English words that can be formed using the letters "d", "g", "o", and "c", without repetition.
24. The set of states in the Union that begin with the word "New".
25. The set of negative numbers larger than 3.

For Problems 26–31, describe each set in English.

26. $\{1,3,5,7\}$

27. $\{1,3,5,7,...,21\}$

28. $\{3,6,9,12,...,36\}$

29. $\{1,4,9,16,25,36\}$

30. $\{\{1\},\{2\},\{1,2\},\{\ \}\}$

31. $S = \{x \mid x \text{ is a day of the week beginning with the letter } z\}$

For Problems 32–41, suppose $A = \{1,3,5,6,7\}$, $B = \{1,5,6\}$, $C = \{1,3,5,7,9,11\}$, and $D = \{1,3,7,9\}$. Determine if the given statement is true or false.

32. $A \subseteq B$

33. $B \subseteq A$

34. $D \subset A$

35. $A \subseteq C$

36. $1 \in B$

37. $\{1,3\} \subset D$

38. $\{1,3\} \in D$

39. $\varnothing \subseteq B$

40. $\varnothing \in B$

41. $B \subseteq D$

For Problems 42–46, suppose $A = \{1,3,5,6\}$, $B = \{1,5,6\}$, and $C = \{1,3,5,7,9,11\}$. Find:

42. The number of elements in B.

43. The number of subsets of B.

44. The number of elements in $\mathscr{P}(A)$.

45. The number of proper subsets of C.

46. The number of subsets of C less the number of proper subsets of C.

47. Is the following a correct definition for $A = B$? Explain your reasoning.

$$(A = B) \Leftrightarrow [[(x \in A) \rightarrow (x \in B)] \wedge [(x \notin A) \rightarrow (x \notin B)]]$$

The number of elements in a set is called its *cardinality*. Thus, the cardinality of set $A = \{1,5,6\}$ is 3, and the cardinality of set $B = \{a,b,c,\ldots,z\}$ is 26. Sets that have the same cardinality are said to be *equivalent*. For Problems 48-52, answer true or false for each question.

48. If two sets are equal they are equivalent.
49. If two sets are equivalent they are equal.
50. If A is a subset of B, the cardinality of A must be less than the cardinality of B.
51. The set $\{\varnothing\}$ has cardinality zero.
52. The cardinality of $A = \{4,5,\{1,2\},6\}$ is four.

5.2 Connectives

Introduction

In logic, we quickly discovered the need to unite *TF* statements. In this section, we will learn to combine sets. The similarities between negations, conjunctions, disjunctions, implications, and biconditionals and their set theory counterparts should become apparent.

Complementation

The operation of *complementation* in set theory is analogous to the operation of negation in logic. If set A is drawn from some universe, then the complement of A, denoted A', is the set of all elements of the universe that are not in A. We can express this definition symbolically as $(x \in A') \Leftrightarrow [(x \notin A) \wedge (x \in U)]$.

Example 1

If $U = \{1,3,5,7,8,9\}$ and $A = \{1,5,7,9\}$ then $A' = \{3,8\}$. Notice that the numbers 2, 4, and 6 are not in the universe, so they cannot appear in A'.

■

Example 2

If $U = \{1,3,5,7,8,9\}$ and $A = \{1,3,5,7,8,9\}$ then $A' = \varnothing$.

■

Example 2 points out that, in general, the complement of the universal set is the empty set. Conversely, the complement of the empty set is the universal set.

Membership Tables

A *membership table* in set theory is analogous to a truth table in logic. It provides a way to show that an element is or is not a member of a set. For example, the membership table for complementation is shown below.

A	A'
$\in$	$\notin$
$\notin$	$\in$

The table shows us that if an element is in set A, then it is not in A', and if an element is not in A then it must be in A'. The $\in$ symbol behaves just like the truth value of "true" and $\notin$ behaves just like the truth value of "false."

Notice that this membership table looks very much like the negation truth table in logic. We will construct membership tables for each of the connectives we encounter.

Intersection

If two sets, A and B, are drawn from some universe, U, then the *intersection* of A and B, denoted $A \cap B$, is the set of all elements that are in both A and B. We state this definition symbolically as $[x \in (A \cap B)] \Leftrightarrow [(x \in A) \wedge (x \in B)]$.

If two sets do not share any elements, they are said to be *disjoint.* Since disjoint sets do not share any elements, their intersection is the null set.

Example 3

Suppose $A = \{34, 56, 57, 66, 94\}$ and $B = \{35, 56, 66, 95\}$. Find $A \cap B$.

Solution

The elements that are in both sets are 56 and 66. Therefore, $A \cap B = \{56, 66\}$.

■

Let's construct a membership table for the intersection of two sets, A and B. If an element is in both sets, it belongs in the intersection, otherwise, it does not. This is illustrated in the membership table below.

A	B	$A \cap B$
$\in$	$\in$	$\in$
$\in$	$\notin$	$\notin$
$\notin$	$\in$	$\notin$
$\notin$	$\notin$	$\notin$

Notice that this membership table is analogous to the truth table for the conjunction connective. Even the symbol we use for intersection looks quite similar to the symbol for conjunction. This should help you remember how the intersection connective behaves.

Example 4

Let $U = \{1,...,10\}$, $A = \{1,...,6\}$, $B = \{7,8\}$, and $C = \{6,7,8\}$. Find $A \cap B$ and $A' \cap C$.

Solution

Set A and set B are disjoint. Therefore, $A \cap B = \varnothing$. Since A' is the set of all elements in the universe that are not in A, $A' = \{7,8,9,10\}$. Therefore, $A' \cap C = \{7,8\}$.

■

Union

If two sets, A and B, are drawn from some universe, U, then the *union* of A and B, denoted $A \cup B$, is the set of all elements that are either in A or B. We can state this definition symbolically as $[x \in (A \cup B)] \Leftrightarrow [(x \in A) \vee (x \in B)]$.

Example 5

Let $A = \{1,2,3\}$ and $B = \{2,3,4\}$. Find $A \cup B$.

Solution

The set of all elements that are in either A or B is $A \cup B = \{1,2,3,4\}$.

■

Let's construct a membership table for the union of two sets, A and B. If an element is in either set, it belongs in the union, otherwise, it does not. This is illustrated in the membership table below.

A	B	$A \cup B$
$\in$	$\in$	$\in$
$\in$	$\notin$	$\in$
$\notin$	$\in$	$\in$
$\notin$	$\notin$	$\notin$

Again, notice that this membership table is analogous to the truth table for disjunction. Even the symbol we use for union looks quite similar to the symbol for disjunction. This should help you remember how the union connective behaves.

Example 6

If $U = \{1,2,3,4,5,6,7,8,9,10\}$, $A = \{1,2,4,5,7,8\}$, $B = \{1,7,8,9,10\}$, and $C = \{4,8,9,10\}$, find $(A' \cup B) \cap C$.

Solution

We proceed by first resolving $A' \cup B$. We know that $A' = \{3,6,9,10\}$, so $A' \cup B = \{1,3,6,7,8,9,10\}$. Hence, $(A' \cup B) \cap C = \{8,9,10\}$. ■

Set Difference

If two sets, A and B are drawn from some universe, then the *set difference*, $A - B$ is defined to be those elements in A but not in B. Stating the definition of set difference symbolically, we have $[x \in (A - B)] \Leftrightarrow [(x \in A) \wedge (x \notin B)]$. Notice that the definition states that x is not in B. This means that x must be in B'. Therefore, we can also write the definition as $[x \in (A - B)] \Leftrightarrow [(x \in A) \wedge (x \in B')]$.

The membership table for set difference is constructed by recognizing that if an element is in the set difference $A - B$, it first must be in set A, but *cannot* be in set B.

A	B	$A - B$
$\in$	$\in$	$\notin$
$\in$	$\notin$	$\in$
$\notin$	$\in$	$\notin$
$\notin$	$\notin$	$\notin$

Example 7

Let $U = \{1,2,3,4,5\}$, $A = \{1,4,5\}$, and $B = \{1,3,4\}$. Find $A - B$.

Solution

We remove from A all those elements that are found in set B. The elements that remain in A comprise $A - B$. Therefore, $A - B = \{5\}$. ■

Example 8

Let $U = \{1,2,3,4,5\}$, $A = \{1,4,5\}$, and $B = \{1,3,4\}$. Find $B - A$.

Solution

We remove from B all those elements that are found in set A. The elements that remain in B comprise the set $B - A$. Therefore, $B - A = \{3\}$. ■

As Examples 7 and 8 point out, unlike the operations of union and intersection, which are commutative, set difference is not. In general, $A - B \neq B - A$.

Example 9

If $U = \{1, 2, ..., 7\}$, $A = \{1, 2, 3, 7\}$, $B = \{1, 2, 3, 5, 6\}$, and $C = \{1, 2, 3, 7\}$, find $(A \cap B) - C$.

Solution

First, we resolve the intersection inside the parentheses. The set $A \cap B = \{1, 2, 3\}$. Now, we remove from this intersection any elements that are found in C. Therefore, $(A \cap B) - C = \varnothing$.

■

Example 10

If $U = \{1, 2, ..., 7\}$, $A = \{1, 2, 3, 7\}$, $B = \{1, 2, 3, 5, 6\}$, and $C = \{2, 3, 7\}$, find $C - (A \cap B)$.

Solution

Again, $A \cap B = \{1, 2, 3\}$. This time, we remove all elements in this intersection from C. Therefore, $C - (A \cap B) = \{7\}$.

■

Example 11

If $U = \{\text{whole numbers from 1 to 10}\}$, $A = \{2,4,6,8,10\}$, $B = \{1,9\}$, and $C = \{1,4,5,8,9\}$, find $[(A \cap C)' - B]'$.

Solution

The solution to this more complicated problem is resolved in small steps, beginning with the contents of the inner parentheses. First, we need to find all elements common to sets A and C. We find that $A \cap C = \{4, 8\}$. Second, we find the complement of $A \cap C$. This is the set $(A \cap C)' = \{1, 2, 3, 5, 6, 7, 9, 10\}$. Next, remove from $(A \cap C)'$ any elements found in B, so $(A \cap C)' - B = \{2, 3, 5, 6, 7, 10\}$. Finally, we obtain the complement of $(A \cap C)' - B$, that is, all elements not in $(A \cap C)' - B$, to obtain $[(A \cap C)' - B]' = \{1, 4, 8, 9\}$.

■

In-Class Exercises and Problems for Section 5.2

In-Class Exercises

In Exercises 1–20, assume $U = \{1, 2, 3, 4, 5, 6, 7\}$, $A = \{1, 3, 5, 7\}$, $B = \{2, 4, 6\}$, $C = \{1, 2, 5, 6\}$, and $D = \{2, 3, 4\}$. Find the elements in each of the given sets.

1. $C \cup D$
2. $C \cap D$
3. $C - D$
4. $D - C$
5. $B \cup D'$
6. $D - A'$
7. $(A \cup B)'$
8. $(A \cap B) \cup B'$
9. $(D \cap A)' - B$
10. $(C' \cup A)' - D$
11. $(A \cap D') \cup C'$
12. $(A \cap C) - (A \cap D)$
13. $(B \cap D) \cup (B - D)$
14. $[(A \cup C)' \cap (C - D)]'$
15. $(A - B) \cup (C - D')$
16. $[(A' \cap D) - C']'$
17. $(C \cup B') - (A' \cap D)$
18. $C' - [(B \cup D) \cap (D - A')]$
19. $[(A' - C)' \cap (B \cup C')] - (A \cup D)$
20. $[(A - C) \cap (B - D)'] \cup [(C - A) \cap (D - B)']$

Problems

For Problems 21–30, assume $U = \{1, 2, 3, ..., 10\}$, $A = \{1, 3, 4, 7, 10\}$, $B = \{7, 8, 10\}$, $C = \{3, 5, 7\}$, and $D = \{2, 3, 8, 9\}$. Find the elements of each set.

21. $A \cap C$
22. $A \cup B'$
23. $D' - B$
24. $A \cap D'$
25. $A' \cap B'$
26. $A \cap B \cap C \cap D$
27. $(A \cap B') \cup (C' \cap D)$
28. $(A - C) - D$
29. $[(A' \cap C)' - (B' \cup D)]'$
30. $(D - C') \cap (B - A')$

For Problems 31–40, suppose that $U = \{\text{the months of the year}\}$. If set $A = \{x | x \text{ has 30 days}\}$, set $B = \{x | x \text{ has an } r \text{ in its spelling}\}$, set $C = \{\text{April, May, June}\}$, and set $D = \{\text{May,...,October}\}$, find

31. $A \cap B$
32. $A \cup B'$
33. $D - B$
34. $A \cap D$
35. $A' \cap B'$
36. $A \cap B \cap C \cap D$
37. $C \cap D$
38. $(A - C) - D$
39. $[(A' \cap C)' - (B' \cup D)]'$
40. $U - A - B - C - D$

5.3 Venn Diagrams

A *Venn diagram* is a pictorial way of representing sets and the relationships that exist among them. A Venn diagram consists of a rectangle, the interior of which represents the universe. Within the rectangle, circular regions represent the sets under discussion.

Example 1

Suppose in a given universe, set B is a proper subset of set A. Draw a possible Venn diagram for this relationship.

Solution

Inside a rectangle, we draw two circles, one entirely within the other.

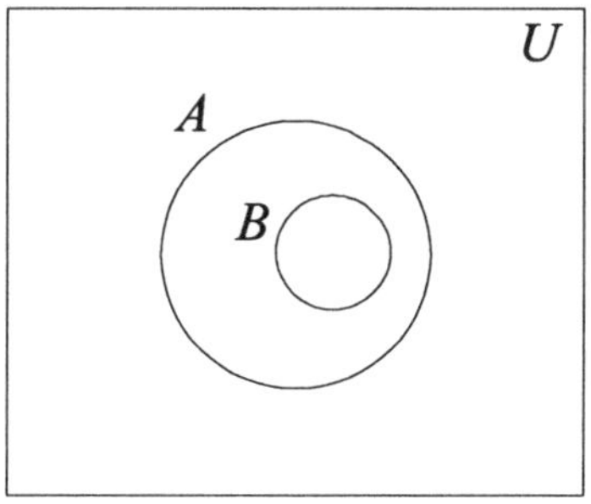

■

Example 2

Assume that in some universe, sets A and B share some, but not all of their elements. Draw a Venn diagram that depicts this condition.

Solution

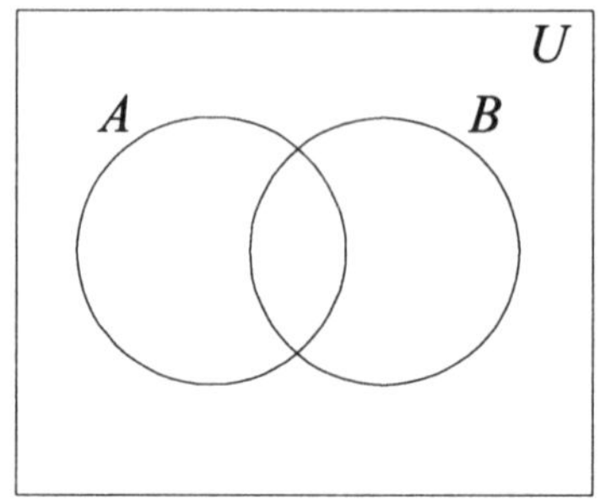

■

Example 3

Assume that in some universe, sets A and B share some, but not all of their elements, and that C is a proper subset of $A \cap B$. Draw a Venn diagram that depicts this condition.

Solution

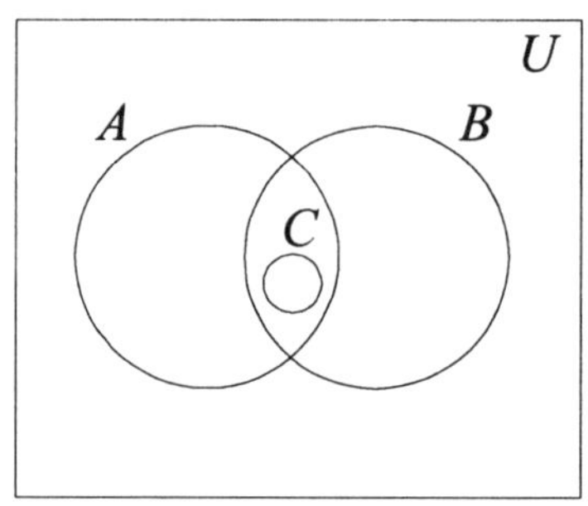

■

Example 4

Suppose sets A and B are disjoint, and C is a proper subset of A. Draw a Venn diagram that depicts this condition.

Solution

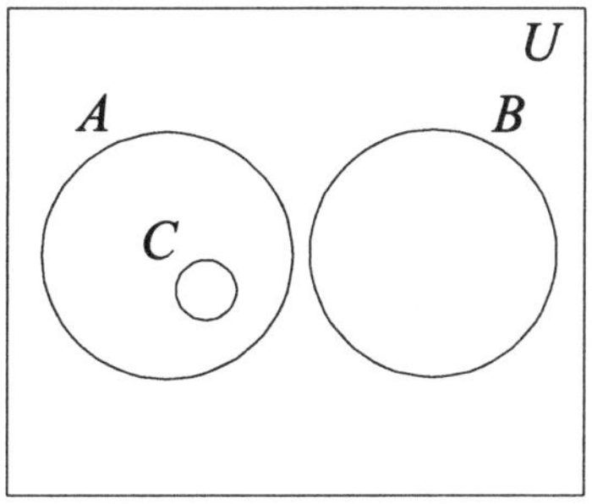

■

Example 5

Let $U = \{1, 2, 3, 4, 5, 6\}$, $A = \{1, 2, 3\}$, $B = \{2, 3, 4\}$. Place the elements into their proper regions in a Venn diagram.

Solution

Since sets A and B share elements, we will draw two intersecting circles and place the elements 2 and 3 within the intersection. The element 1 is not in B but is in A. Similarly, the element 4 is not in A but is in B. The elements 5 and 6 are in neither set.

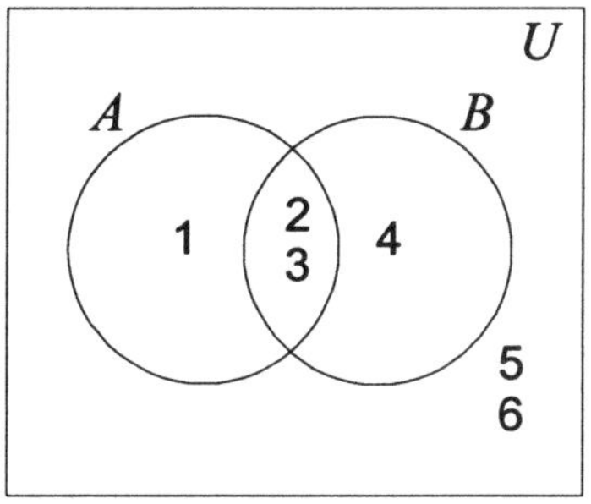

■

Example 6

Suppose that $U = \{1, 2, 3, \ldots, 10\}$, $A = \{1, 3, 4, 7, 8\}$, $B = \{3, 4, 9, 10\}$, and $C = \{1, 2, 4, 10\}$. Place the elements into their proper regions in a Venn diagram.

Solution

Since there are three sets, we must have three circles. Each circle shares some of its elements with each of the other two circles. Therefore, each circle must intersect the other two circles. The element 4 is in all three sets, so it must be placed in their common intersection. The element 3 is in sets A and B, but it is not in set C. The element 1 is in sets A and C but not in set B. The element 10 is in sets B and C, but not in set A. The elements 5 and 6 are not in any of the three sets. Finally, the elements 7 and 8 are only in A, the element 9 is only in B and the element 2 is only in C. These results are shown in the following Venn diagram.

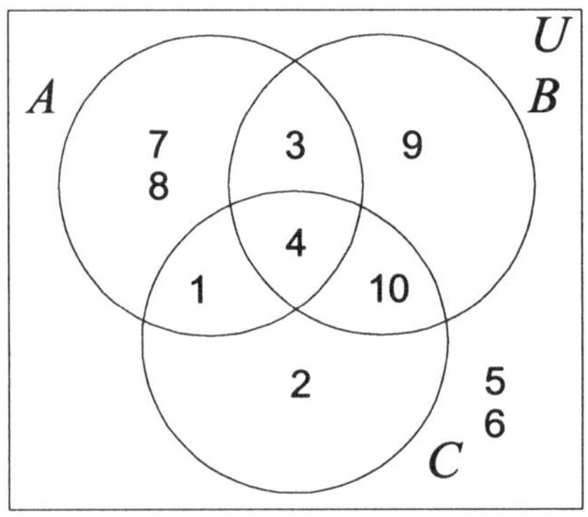

■

The previous example illustrates the need to be able to identify and name the various regions in any Venn diagram. We will begin by returning to the situation in which we have only two sets in some universe.

In the Venn diagram below, we have four distinct regions. We have labeled these regions with Roman numerals so that we may refer to the regions in a concise manner. The order in which we label the regions is immaterial, however if the regions are labeled as shown below, the rows of a corresponding membership table for the Venn diagram will correspond to the numbered regions.

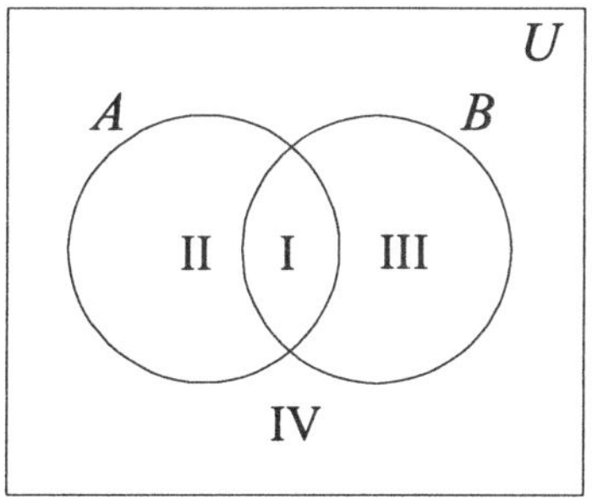

Notice that the region $A \cap B$ is denoted by the Roman numeral I, while $A \cup B$ consists of the *three* regions I, II, and III. The region B' consists of regions II and IV.

Example 7

Name the region or regions in the above Venn diagram that represent $A \cap B'$.

Solution

In order to be in $A \cap B'$, an element must be in set A and outside set B. This is region II.

■

Example 8

Name the region or regions in the above Venn diagram that represent $(B - A)'$.

Solution

First, we need to find $B - A$. This set consists of all elements in B that are not in A. This corresponds to region III. Then we take the complement of this region, so our answer consists of regions I, II, and IV.

■

With three intersecting circles, there are eight distinct regions. Again, the order in which we assign the numbering is immaterial, but if the regions are labeled as shown below, each of the rows of a corresponding membership table for the Venn diagram will correspond to the numbered regions.

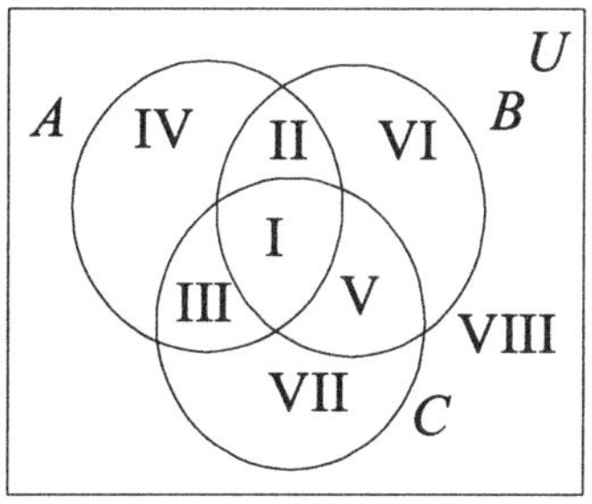

Note that the intersection of all three sets, i.e., $A \cap B \cap C$, is represented by region I. This is the region that is common to all three sets. Note also that the intersection of *any* two of the three sets consists of two regions. For example, $A \cap B$ is represented by regions I and II. It is also important to realize that each of the three sets A, B, and C consists of four regions. For example, set A is represented by regions I, II, III and IV.

If we wish to identify an element that is *only* in set A, we must look in region IV, while an element that is only in set B is found in region VI. Region VII consists of elements only in set C.

Example 9

Name the numbered region in terms of sets A, B, and C.

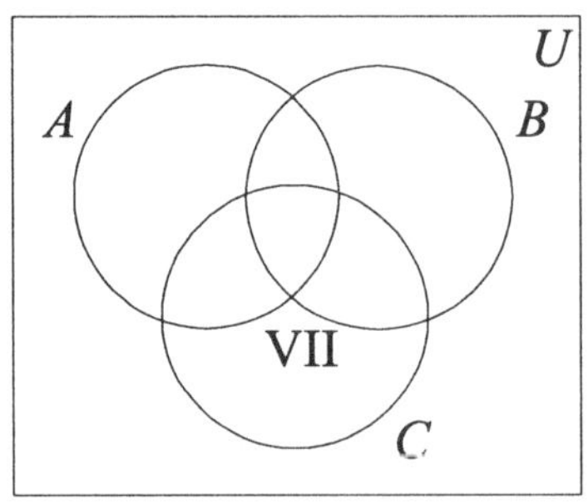

Solution

We begin by noting that region VII is in set C, but C consists of more than just region VII. The regions that C shares with both A and B have been removed from C. Therefore one way to express region VII is $C-(A \cup B)$. Another way to name region VII is $(C-A)-B$. ■

Example 10

Use a membership table to decide which region or regions in the Venn diagram below belong to $(A \cup B)' - C$.

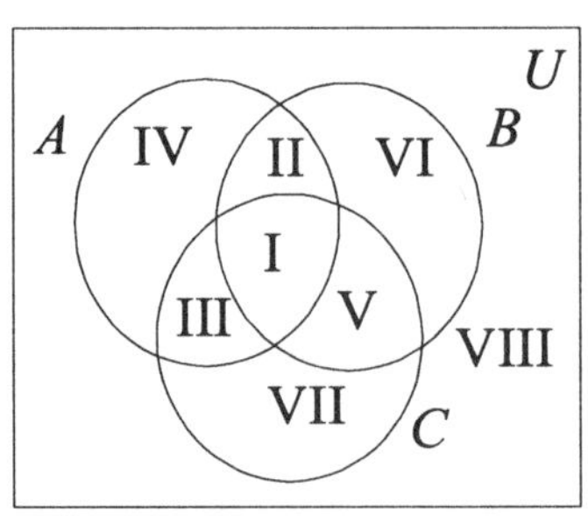

Solution

Since there are three sets under discussion, our membership table will have eight rows. We will have columns for each set, for $A \cup B$, $(A \cup B)'$, and finally for $(A \cup B)' - C$. If an element is in either A or B, it belongs to $A \cup B$, and therefore does not belong to $(A \cup B)'$. Due to the nature of the set difference operation, we are interested only in those elements that are in $(A \cup B)'$, but are not in C. This occurs in row eight of the membership table.

Region	A	B	C	$A \cup B$	$(A \cup B)'$	$(A \cup B)' - C$
I	$\in$	$\in$	$\in$	$\in$	$\notin$	$\notin$
II	$\in$	$\in$	$\notin$	$\in$	$\notin$	$\notin$
III	$\in$	$\notin$	$\in$	$\in$	$\notin$	$\notin$
IV	$\in$	$\notin$	$\notin$	$\in$	$\notin$	$\notin$
V	$\notin$	$\in$	$\in$	$\in$	$\notin$	$\notin$
VI	$\notin$	$\in$	$\notin$	$\in$	$\notin$	$\notin$
VII	$\notin$	$\notin$	$\in$	$\notin$	$\in$	$\notin$
VIII	$\notin$	$\notin$	$\notin$	$\notin$	$\in$	$\in$

Therefore, the only elements that are in $(A \cup B)' - C$ are those elements that are not in A nor B nor C. Hence, the only region that describes $(A \cup B)' - C$ in the previous Venn diagram is region VIII. Notice that the membership table's eighth row shows the elements that are in neither A, nor B, nor C.

■

In-Class Exercises and Problems for Section 5.3

In-Class Exercises

For Exercises 1–5, construct a Venn diagram for each situation. Be sure to place the elements in their correct regions.

1. $U = \{1,3,5,7,9,11\}$, $A = \{3,5,7\}$, $B = \{3\}$
2. $U = \{1,4,9,16,25,36\}$, $C = \{1,4,16,36\}$, $D = \{4,16\}$, $E = \{9,36\}$
3. $U = \{2,4,6,8,10,12,14\}$, $F = \{2,4,6\}$, $G = \{8,12,14\}$, $H = \{12\}$
4. $U = \{1,2,3,4,5,6,7,8,9,10\}$, $A = \{1,3,4,7,8\}$, $B = \{3,4,6,9\}$
   $C = \{1,3,9,10\}$

5. $U = \{2,3,5,7,11,13,17,19\}$, $W = \{2,5,7,13\}$, $X = \{3,5,11,17\}$
$Y = \{2,7\}$, $Z = \{11,17\}$

For Exercises 6–9, construct a Venn diagram that satisfies the conditions stated.

6. Sets D, E, and F are proper subsets of U. $D \cap E \neq \varnothing$, and neither is a subset of the other. Also, $D \cap F = \varnothing$, $E \cap F = \varnothing$.
7. Sets A, B, C and D are proper subsets of U. $C \subset A$, $D \subset B$, and $A \cap B = \varnothing$.
8. Sets A, B, and C are proper subsets of U. $B \subset A$, $C \subset A$, and $B \cap C = \varnothing$.
9. Sets A, B, C and D are proper subsets of U. $A \cap B \neq \varnothing$, $C \subset A$, $D \subset (A \cap B)$, and $B \cap C = \varnothing$. Neither A nor B is a subset of the other.

Use the Venn diagram below for Exercises 10–19. Name the region or regions described.

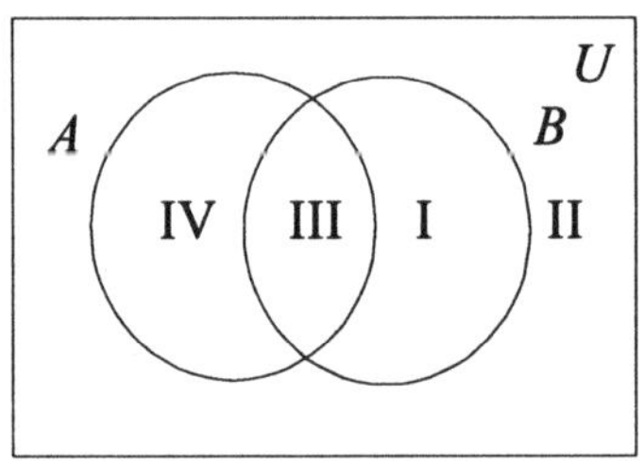

10. A
11. B
12. A'
13. B'
14. $A \cap B$
15. $A - B$
16. $(A \cap B)'$
17. $A' \cup B$
18. $(A \cap B') \cup (B - A)$
19. $(A - B')' - (A' \cap B)$

Use the Venn diagram below for Exercises 20–34. Name the region or regions described.

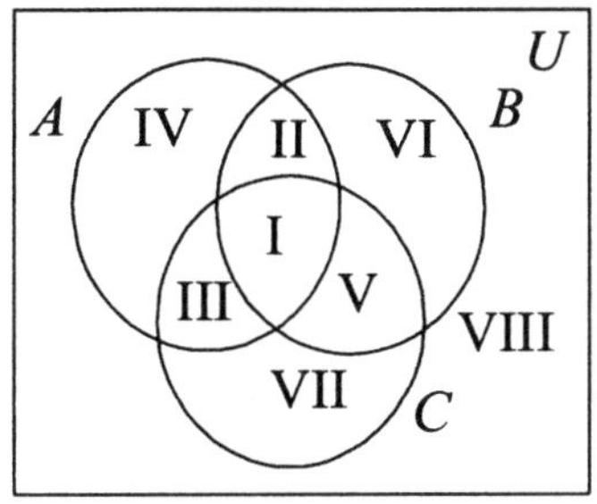

20. B
21. C'
22. $B \cap C$
23. $A' \cup C$
24. $B - C$
25. $(A - B) \cap C$
26. $(A \cap B \cap C)' - B'$
27. $(A \cup C') - (A \cap B)$
28. $(A' \cup B') \cap (B \cup C)$
29. $[B' - (A \cap C')]'$
30. $[C - (A - B)]' \cap (A' \cup B')$
31. Only set *A*
32. Set *B* but not set *A*
33. Set *B* and set *C* but not set *A*
34. Set *A* or set *C* but not in both.

For each given Venn diagram in Exercises 35–38, name the designated region or regions in terms of sets *A*, *B* and *C*. There are many correct answers for each Venn diagram.

35.

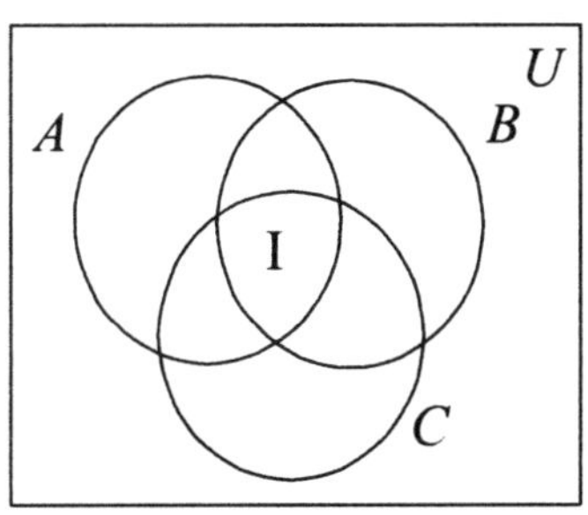

37.

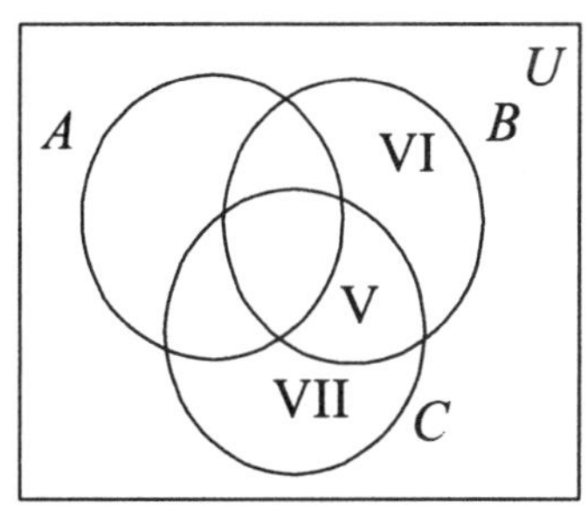

36.

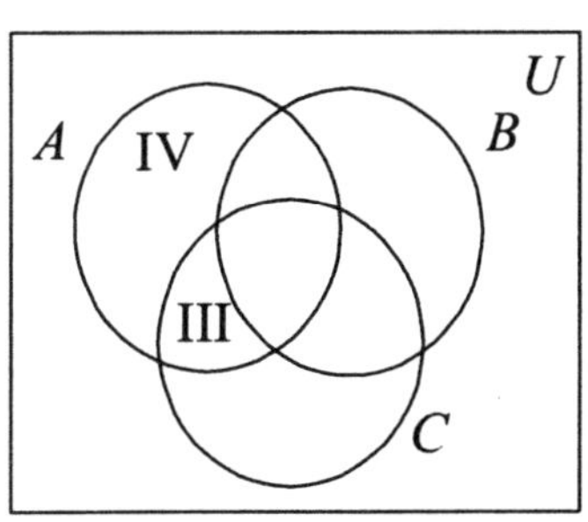

38.

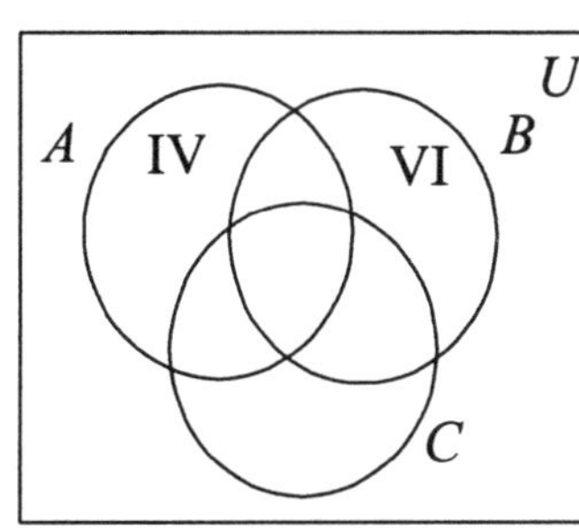

Problems

For Problems 39–44, construct a Venn diagram for each situation. Be sure to place the elements in their correct regions.

39. $U = \{10,11,12,13,14\}$, $A = \{11,12\}$, $B = \{10,11,14\}$.

40. $U = \{1,2,...,6\}$, $A = \{1,2,3\}$, $B = \{4,5,6\}$.

41. $U = \{2,4,6,8,10\}$, $A = \{2,4,6,10\}$, $B = \{4,10\}$, $C = \{8\}$.

42. $U = \{1,3,5,...,11\}$, $A = \{1,3,5\}$, $B = \{3,5,7\}$, $C = \{1,3,5,7,9\}$.

43. $U = \{1,2,3,5,8,13,21\}$, $A = \{1,2,3,5\}$, $B = \{2,3,8\}$, $C = \{3\}$.

44. $U = \{1,2,3,...,10\}$, $A = \{2,3,7,10\}$, $B = \{2,3\}$, $C = \{3,5,7\}$.

For Problems 45–49, construct a Venn diagram that satisfies the conditions stated.

45. Sets A, B, and C are proper subsets of U. $A \subset B$, $B \subset C$.

46. Sets A, B, and C are proper subsets of U. $A \subset B$, $B \cap C \neq \varnothing$, neither B nor C is a subset of the other, $A \cap C = \varnothing$.

47. Sets A, B, and C are proper subsets of U. $A \cap B = \varnothing$, $C \subset B$.

48. Let A, B, and C be three proper subsets of U with the following conditions: $A \cap B \neq \varnothing$, neither A nor B is a subset of the other, $C \subset (A \cap B)$.

49. Sets A, B, and C are proper subsets of U. $A \cap B \neq \varnothing$, $B \cap C \neq \varnothing$, $A \cap C = \varnothing$. No set is a subset of any other set.

For Problems 50–59, use the Venn diagram below. Name the region or regions described.

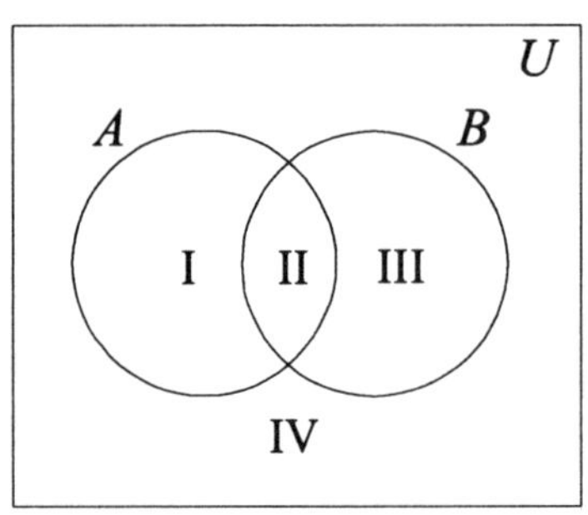

50. A
51. $A \cap B$
52. $A \cup B$
53. $(A \cup B)'$
54. $A' \cap B'$
55. $A - B'$
56. $(A' \cap B')'$
57. $(A \cup B) - (A \cap B)$
58. $(A' \cap B')' \cap (A' \cap B)$
59. $[(A \cap B') \cap (B \cap A')]'$

For Problems 60–79, use the Venn diagram below. Name the region or regions described.

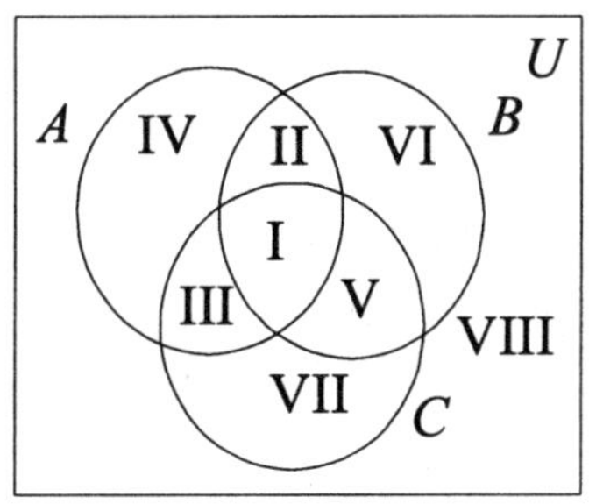

60. A
61. $A \cap B$
62. $A \cup B$
63. $(A \cup B)'$
64. $A' \cap B'$
65. $A - B'$
66. $(A' - B')'$
67. $(A \cup B) \cap C$
68. $(A - C) \cap B'$
69. $(A \cup C) - B$
70. $A' \cap B' \cap C'$
71. $(A - B) - C$
72. $A - (B - C)$
73. $[(A \cap B) \cap C]'$
74. $(A - B) \cup (A - C) \cup (B - C)$
75. $(A - B) \cap (A - C) \cap (B - C)$
76. Sets A and B, but not set C
77. Only set C
78. Sets B or C but not set A
79. Set B but not set A nor set C

For Problems 80–83, use a membership table to decide which region or regions in the Venn diagram below belong to the given set.

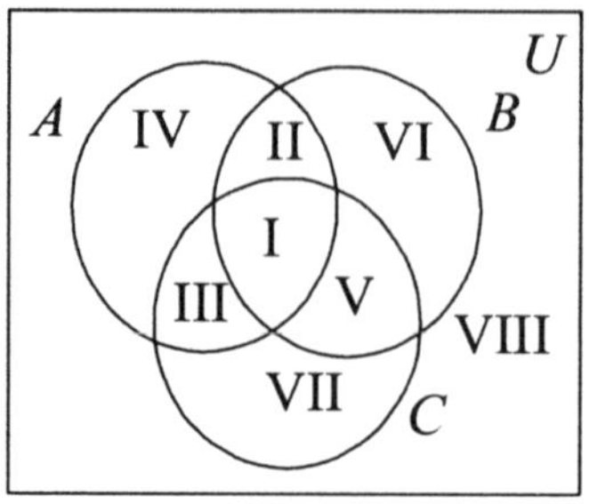

80. $(A \cap B')' \cap C$

81. $(A' - B) \cap C'$

82. $(A \cap C') - B$

83. $(B - C)' \cap (A \cup B)$

For Problems 84–91, name the designated region or regions in terms of sets *A*, *B* and *C*. There may be more than one correct way to name the region.

84.

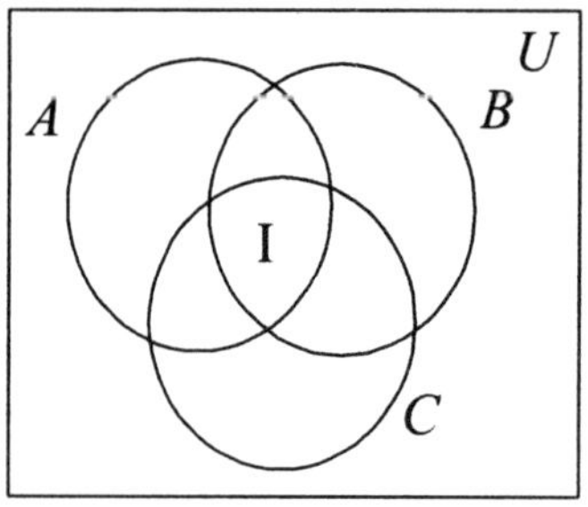

85.

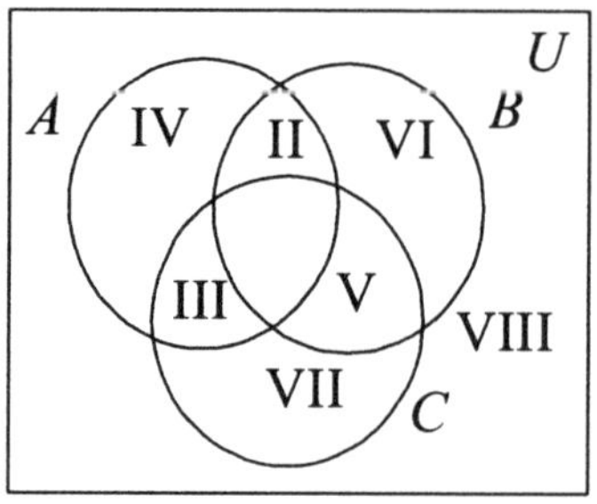

86.

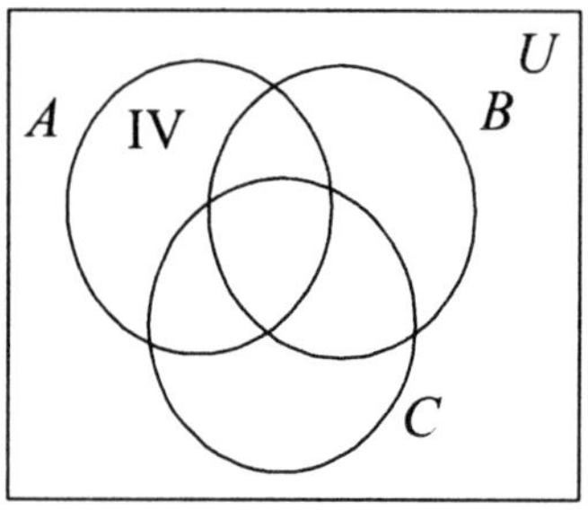

87.

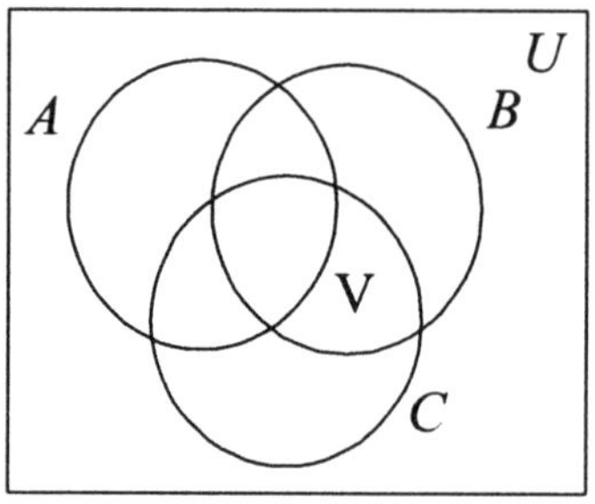

88.

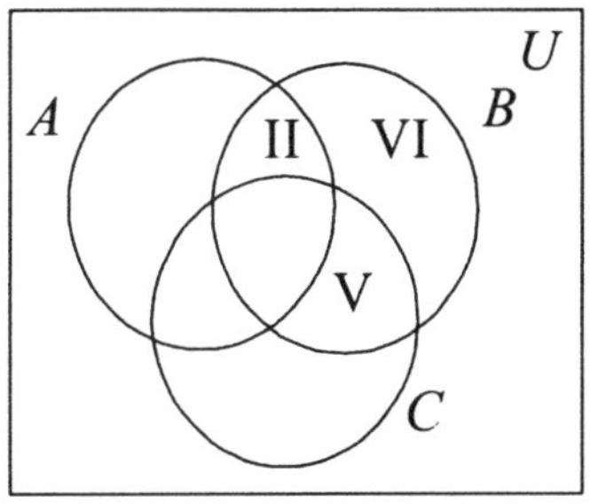

89.

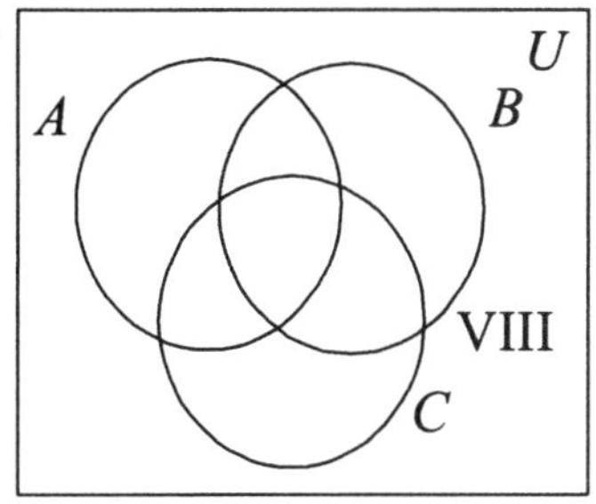

90.

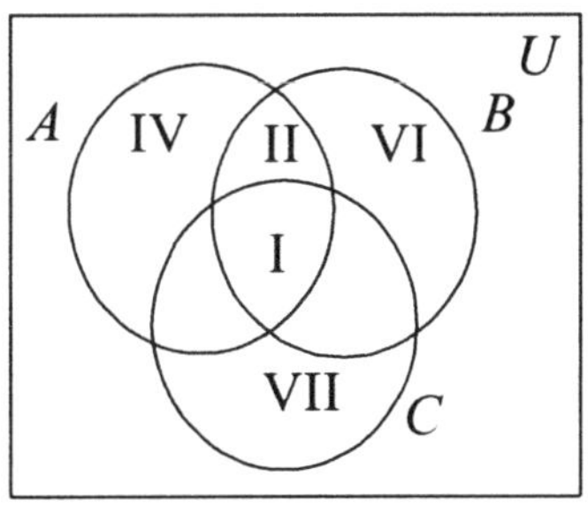

91.

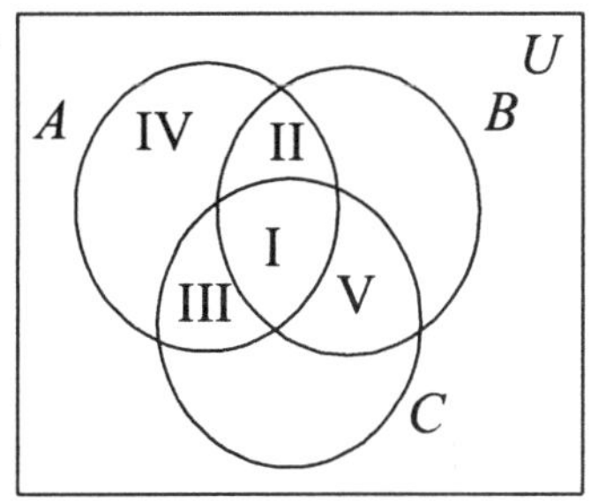

5.4 Laws of Equal Sets

Introduction

Many of the equivalences that we studied in Chapter One have identical counterparts in set theory. In this section, we will use membership tables and Venn diagrams to verify these set equivalences. A summary of the laws of equal sets can be found on page 288.

Double Complement Law

The double negation equivalence in logic was expressed symbolically as $\sim(\sim p) \Leftrightarrow p$. In set theory, the corresponding equivalence is expressed symbolically as $(A')' = A$, and is called the *double complement law*. Let's use a membership table to show that the sets on left and right sides of the equal sign are the same.

A	A'	$(A')'$	$A=(A')'$
$\in$	$\notin$	$\in$	T
$\notin$	$\in$	$\notin$	T

We see that the first and third columns always have the same elements. Therefore, $(A')' = A$.

DeMorgan's Law

DeMorgan's equivalences state that $\sim(p \wedge q) \Leftrightarrow (\sim p \vee \sim q)$, and $\sim(p \vee q) \Leftrightarrow (\sim p \wedge \sim q)$. In set theory, these equivalences correspond to $(P \cap Q)' = (P' \cup Q')$ and $(P \cup Q)' = (P' \cap Q')$ respectively, and are called *DeMorgan's laws.*

Example 1

Use a membership table to show that $\sim(a \wedge b) \Leftrightarrow (\sim a \vee \sim b)$ is true for sets.

Solution

The corresponding form of DeMorgan's equivalence in set theory is expressed as $(A \cap B)' = A' \cup B'$. The membership table is constructed below.

A	B	A'	B'	$A \cap B$	$(A \cap B)'$	$A' \cup B'$	$(A \cap B)' = A' \cup B'$
$\in$	$\in$	$\notin$	$\notin$	$\in$	$\notin$	$\notin$	T
$\in$	$\notin$	$\notin$	$\in$	$\notin$	$\in$	$\in$	T
$\notin$	$\in$	$\in$	$\notin$	$\notin$	$\in$	$\in$	T
$\notin$	$\notin$	$\in$	$\in$	$\notin$	$\in$	$\in$	T

Since the sixth and seventh columns are identical, we have shown that $(A \cap B)' = A' \cup B'$.

■

Distributive Law

Example 2

Use a Venn diagram to verify a distributive law for sets. In particular, verify that $A \cup (B \cap C) = (A \cup B) \cap (A \cup C)$.

Solution

The regions that correspond to $B \cap C$ are I and V. Set A consists of regions I, II, III, and IV. Therefore, $A \cup (B \cap C)$ consists of regions I, II, III, IV, and V.

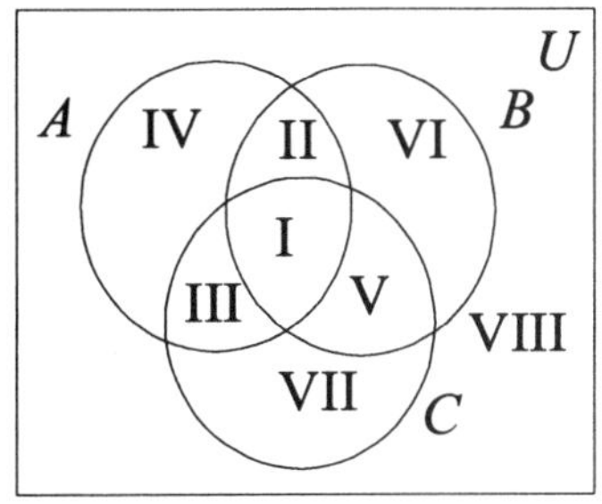

The set $A\cup B$ consists of regions I, II, III, IV, V, and VI, while set $A\cup C$ consists of regions I, II, III, IV, V, and VII. Therefore, $(A\cup B)\cap(A\cup C)$ consists of regions I, II, III, IV, and V. These are exactly the same regions we obtained for $A\cup(B\cap C)$. Hence, $A\cup(B\cap C)=(A\cup B)\cap(A\cup C)$.

■

Absorption Law

In set theory, as well as in electronic circuit theory, we often refer to the *absorption law*. One form of the absorption law is expressed symbolically as $A\cap(A\cup B)=A$.

Example 3

Use a membership table to verify the absorption law shown above.

Solution

A	B	$A\cup B$	$A\cap(A\cup B)$	$A\cap(A\cup B)=A$
$\in$	$\in$	$\in$	$\in$	T
$\in$	$\notin$	$\in$	$\in$	T
$\notin$	$\in$	$\in$	$\notin$	T
$\notin$	$\notin$	$\notin$	$\notin$	T

Since the first and fourth columns are identical, $A\cap(A\cup B)=A$.

■

There are other Laws of Equal Sets that are also worth mentioning. The *Idempotent Law* simply states that the union (or intersection) of any set with itself is exactly the set itself. The *Inverse Law* states that the intersection of a set with its complement is empty, while the union of a set with its complement is the universe. The *Complement Law* states that the complement of the empty set is the universe, while the complement of the universe is the empty set. The *Identity Law* tells us how to combine a set with the universal set and the null set using the operations of union and intersection. *Set difference* allows us to write a difference of two sets as an intersection. All eleven of the Laws of Equal Sets are summarized in the table below.

Laws of Equal Sets

1. Double Complement Law	$(A')' = A$
2. Commutative Law	$A \cap B = B \cap A$ $A \cup B = B \cup A$
3. Associative Law	$A \cap (B \cap C) = (A \cap B) \cap C$ $A \cup (B \cup C) = (A \cup B) \cup C$
4. Distributive Law	$A \cap (B \cup C) = (A \cap B) \cup (A \cap C)$ $A \cup (B \cap C) = (A \cup B) \cap (A \cup C)$
5. DeMorgan's Law	$(A \cap B)' = A' \cup B'$ $(A \cup B)' = A' \cap B'$
6. Absorption Law	$A \cap (A \cup B) = A$ $A \cup (A \cap B) = A$
7. Idempotent Law	$A \cup A = A$ $B \cap B = B$
8. Inverse Law	$A \cap A' = \varnothing$ $A \cup A' = U$
9. Complement Law	$\varnothing' = U$ $U' = \varnothing$
10. Identity Law	$A \cap U = A$ $A \cup U = U$ $A \cap \varnothing = \varnothing$ $A \cup \varnothing = A$
11. Set Difference Law	$A - B = A \cap B'$

Simplifying Sets

The Laws of Equal Sets can be used to simplify compound statements about sets.

Example 4

Simplify the set $A \cap (A' \cup B)$.

Solution

First, use the distributive law to obtain

$$A \cap (A' \cup B) = (A \cap A') \cup (A \cap B).$$

The inverse law allows us to express $(A \cap A')$ as $\varnothing$. Therefore,

$$A \cap (A' \cup B) = (A \cap A') \cup (A \cap B) = \varnothing \cup (A \cap B).$$

Then, using the identity law, $\varnothing \cup (A \cap B) = (A \cap B)$. ■

Example 5

Use laws of equal sets to verify that $(A \cup B) \cap (A' \cap B)' = A$.

Solution

First use DeMorgan's Law to obtain $(A' \cap B)' = A \cup B'$. Now we can write

$$(A \cup B) \cap (A' \cap B)' = (A \cup B) \cap (A \cup B').$$

Using the distributive law we obtain

$$(A \cup B) \cap (A \cup B') = A \cup (B \cap B').$$

Since $(B \cap B') = \varnothing$, we have $A \cup (B \cap B') = A \cup \varnothing = A$. Therefore, we have $(A \cup B) \cap (A' \cap B)' = A$. ■

Often, we choose to simplify a set using statements and reasons, much like we did in writing formal proofs.

Example 6

Write a formal proof for the to show the relationship in Example 5 is true.

Solution

Statement	Reason
1. $(A \cup B) \cap (A' \cap B)'$	Given
2. $(A \cup B) \cap (A \cup B')$	DeMorgan's Law
3. $A \cup (B \cap B')$	Distributive Law
4. $A \cup \varnothing$	Inverse Law
5. A	Identity Law

■

In-Class Exercises and Problems for Section 5.4

In-Class Exercises

For Exercises 1–2, supply the correct Law of Equal Sets that justifies the reasoning from one step to the next.

1. $A-(B'\cap A)$ given
   $A\cap(B'\cap A)'$ *a.*
   $A\cap(B\cup A')$ *b.*
   $(A\cap B)\cup(A\cap A')$ *c.*
   $(A\cap B)\cup\varnothing$ *d.*
   $A\cap B$ *e.*

2. $[A\cup(A\cap B)]\cup(B'-A')$ given
   $A\cup(B'-A')$ *a.*
   $A\cup(B'\cap A)$ *b.*
   $(A\cup B')\cap(A\cup A)$ *c.*
   $(A\cup B')\cap A$ *d.*
   $A\cap(A\cup B')$ *e.*
   A *f.*

For Exercises 3–5, use a Venn diagram to verify that each of the following statements are true.

3. $(A\cap B)'-C=(A'\cup B')\cap C'$
4. $A\cap(C-D)'=(A-C)\cup(A-D')$
5. $(R\cup S)-(C-S)=S\cup(R-C)$

For Exercises 6–10, simplify each set.

6. $B\cap(B'\cup C)$
7. $A'\cup(B\cap B')$
8. $(H'\cup H)\cap B'$
9. $(A\cup B)\cap(A\cup B')$
10. $(A\cup B)\cup A'$

Problems

For Problems 11–14, use a membership table to verify each of the following Laws of Equal Sets.

11. $A \cap B = A - B'$
12. $(A \cup B)' = A' \cap B'$
13. $A \cap (B \cup C) = (A \cap B) \cup (A \cap C)$
14. $A \cup (A \cap B) = A$

For Problems 15–19, use the Venn diagram below with the technique shown in Example 2 to verify each of the following Laws of Equal Sets.

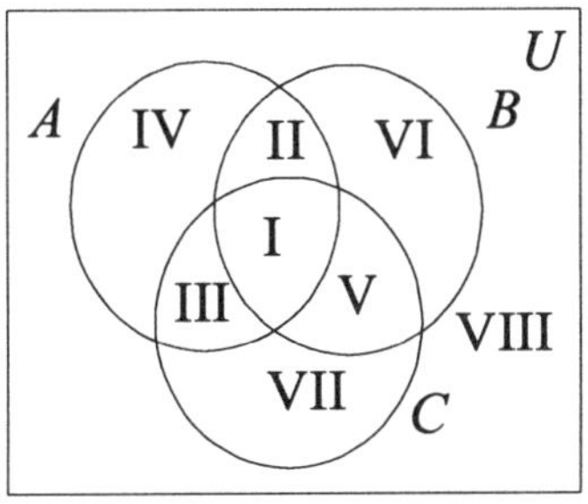

15. $A - B = A \cap B'$
16. $(A \cup B)' = A' \cap B'$
17. $(A \cup B) \cup C = A \cup (B \cup C)$
18. $A \cap (B \cup C) = (A \cap B) \cup (A \cap C)$
19. $A \cup (A \cap B) = A$

For Problems 20–22, supply the Law of Equal Sets that justifies the reasoning from one step to the next.

20.	$(P - R)' \cup R'$	given
	$(P \cap R')' \cup R'$	*a.*
	$(P' \cup R) \cup R'$	*b.*
	$P' \cup (R \cup R')$	*c.*
	$P' \cup U$	*d.*
	U	*e.*

21.	$S' \cap (P \cap S')'$	given
	$S' \cap (P' \cup S)$	*a.*
	$(S' \cap P') \cup (S' \cap S)$	*b.*
	$(S' \cap P') \cup \varnothing$	*c.*
	$S' \cap P'$	*d.*
	$S' - P$	*e.*

22.	$[(C \cap U) - (A' \cup \varnothing)]' \cap C$	given
	$[C - (A' \cup \varnothing)]' \cap C$	*a.*
	$(C - A')' \cap C$	*b.*
	$(C \cap A)' \cap C$	*c.*
	$(C' \cup A') \cap C$	*d.*
	$C \cap (C' \cup A')$	*e.*
	$(C \cap C') \cup (C \cap A')$	*f.*
	$\varnothing \cup (C \cap A')$	*g.*
	$C \cap A'$	*h.*
	$C - A$	*i.*

For Problems 23-32, simplify each set.

23. $A' \cup (B \cap B')$
24. $(C' \cup C) \cap D'$
25. $A \cap (A' \cup C)$
26. $(E \cup A') \cap (E \cup A)$
27. $B' \cap (B \cap A')'$
28. $G \cap (G' \cap H')$
29. $(C' \cap D) \cup (C \cup D)'$
30. $[(R \cap R')' \cap (R - S)] \cap S$
31. $P \cap (P - R)'$
32. $\{[C' \cap (C' \cup D)] \cap C\}'$

Chapter 5 Review

For questions 1–5, use set notation to represent each given set by listing its elements.

1. The set of vowels found in the word "rhythm."
2. The set of all continents.
3. The set of all whole numbers between 9 and 10.
4. The set of all months with 30 days.
5. The set of all planets in our solar system.

For questions 6–10, assume that $U = \{1,2,3,4,5,6,7,8,9\}$, $R = \{2,5,8,9\}$, $S = \{1,2,7,8\}$, $T = \{3,7,9\}$ and $W = \{4,8\}$. Find the elements in each of the given sets.

6. $(R \cap S') - W$
7. $(T' \cup S) \cap (R - S)$
8. $(S \cup R')' \cap (T \cap W)$
9. $(R \cap S \cap W) - (R' \cup S')'$
10. $(R - W')' - (R \cup S \cup T \cup W)'$

For questions 11–18, assume that the universe is $U = \{1, 2, 3, ..., 10\}$. Let set $A = \{2, 4, 6, 8\}$, set $B = \{2, 4, 6\}$, set $C = \{1, 3, 5, 7\}$, and set $D = \{4, 6, 8\}$. Indicate whether each of the following statements is true or false.

11. $A - B = D - B$
12. $B - A = C \cap D$
13. $A \cup C = U$
14. $(A - D) \subset B$
15. $A' \cap C' \neq \varnothing$
16. $B \cup D = A \cap U$
17. $A - (B \cup C) = A - B$
18. $B \cap C' = A \cap D$

19. Let the universe be $U = \{1,2,3,4,5,6,7,8,9,10\}$, $D = \{1,5\}$, $E = \{2,3,7,8,10\}$, $F = \{1,3,5,8,9\}$, $G = \{7,10\}$, and $H = \{8\}$. Construct a Venn diagram for this situation. Be sure to place each element in its proper region.

20. Construct a Venn diagram that satisfies the following conditions: Sets *A*, *B*, *C*, *D* and *E* are all proper subsets of *U*, $A \cap C = \varnothing$, $A \cap B \neq \varnothing$, $B \cap C \neq \varnothing$, $D \subset (A - B)$, and assume $E \subset [B - (A \cup C)]$.

For questions 21–29, let $M = \{\text{m,a,t,h}\}$ and $E = \{\text{e,n,g,l,i,s,h}\}$.

21. Find the number of elements in *M*.

22. Find the number of elements in $\mathscr{P}(M)$.

23. Find the number of elements in $\mathscr{P}(E)$.

24. Find the number of proper subsets of *E*.

25. List all the elements in $\mathscr{P}(M \cap E)$.

26. Is $a \in M$?

27. Is $\{g\} \subset E$?

28. Is $\{\text{mat}\} \subseteq M$?

29. Is $\varnothing \in (M \cup E)$?

30. Given the following Venn diagram, place an "X" in the region or regions that represents $(A \cap B') \cup B$.

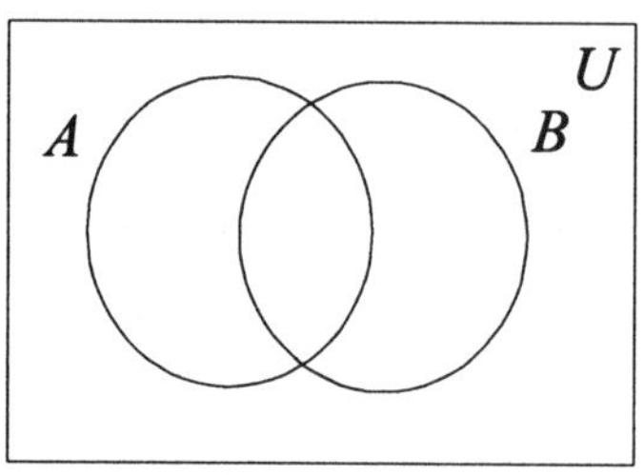

31. Given the following Venn diagram, place an "X" in the region or regions that represents $(A - B)' \cap (C \cup A)$.

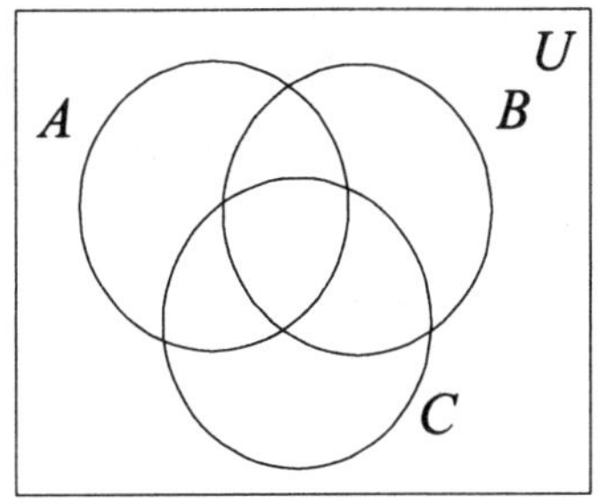

For questions 32–33, consider the Venn diagram below. There are many correct answers for each part.

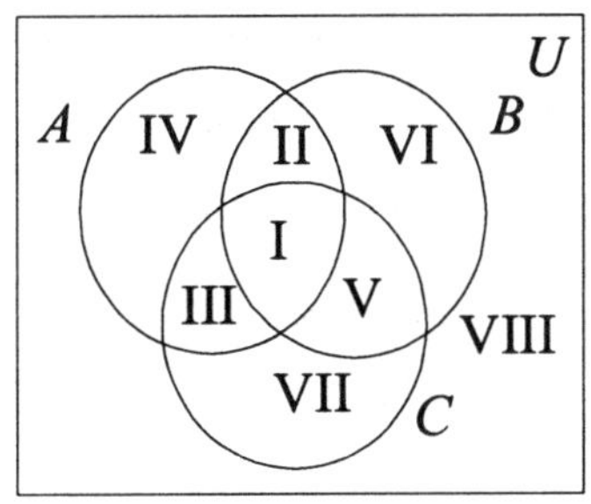

32. Name the numbered region or regions described by the following conditions.

    a. $(A \cap B) - C$

    b. $(A \cup B) \cap (A - C)$

    c. $[(A \cap B') \cup B]'$

    d. $[(B - C) \cup (C - A)']'$

33. Name the designated region or regions in terms of sets *A, B* and *C*. There may be more than one correct way to name the region.

    a. Regions II and V.

    b. Region VI.

    c. Regions I, IV and VII.

    d. Region III.

    e. Regions VII and VIII.

34. Use a membership table to determine if the compound statement $[A\cup(A\cup B')']-C=(A\cap C')\cup(B\cap C')$ is true.

35. Match each numbered statement with the appropriate Law of Equal Sets.

i. $R'\cup(R'\cap S)=R'$	a. DeMorgan's Law
ii. $R'\cup(P\cup S')=R'\cup(S'\cup P)$	b. Distributive Law
iii. $(R\cup S')'\cap P=(R'\cap S)\cap P$	c. Commutative Law
iv. $R'\cup(P\cup S')=(R'\cup P)\cup S'$	d. Absorption Law
v. $R'\cup(P\cap S')=(R'\cup P)\cap(R'\cup S')$	e. Associative Law

36. Describe all the relationships that are exhibited in the Venn diagram below.

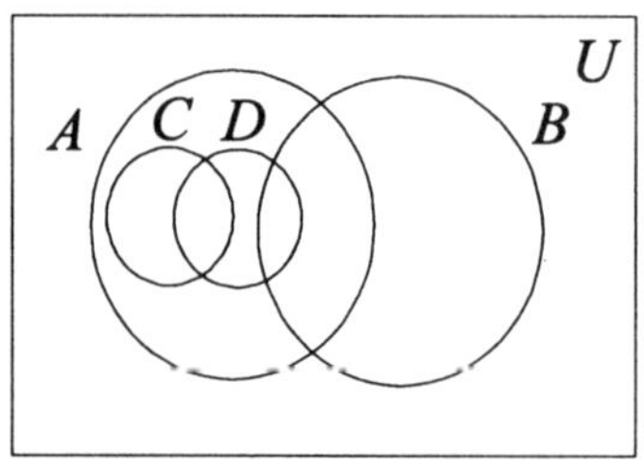

37. For each of the following compound statements, supply the Law of Equal Sets that justifies each step.

a. 1. $(R-S')'-R'$	1. given
2. $(R\cap S)'-R'$	2.
3. $(R'\cup S')-R'$	3.
4. $(R'\cup S')\cap R$	4.
5. $R\cap(R'\cup S')$	5.
6. $(R\cap R')\cup(R\cap S')$	6.
7. $\varnothing\cup(R\cap S')$	7.
8. $R\cap S'$	8.
9. $R-S$	9.

b. 1. $[A\cup(A-B)]\cup[A-(B\cup A')]$ 1. given
2. $[A\cup(A\cap B')]\cup[A-(B\cup A')]$ 2.
3. $A\cup[A-(B\cup A')]$ 3.
4. $A\cup[A\cap(B\cup A')']$ 4.
5. $A\cup[A\cap(B'\cap A)]$ 5.
6. $A\cup[A\cap(A\cap B')]$ 6.
7. $A\cup[(A\cap A)\cap B')]$ 7.
8. $A\cup(A\cap B')$ 8.
9. A 9.

38. Use the Laws of Equal Sets to verify each of the following statements.
 a. $(A-B)\cup(A-C')=A-(B\cap C')$
 b. $[P'\cap(P'\cup S)]-(P'\cap Q)=(P\cup Q)'$
 c. $(A\cup C)'-[A'\cup(A'\cap B)]=\varnothing$

Sample Exam: Chapter 5

For questions 1-10, assume that $U=\{1,2,3,4,5,6,7,8,9,10\}$, $S=\{3,6,9\}$, $P=\{1,2,4,7,10\}$, $Q=\{2,3,4,6,9\}$, and $R=\{1,6,9,10\}$.

1. Find the elements in $P'\cap Q$.
2. Find the elements in $(S-Q)'$.
3. Find the elements in $(R\cup S)-(R\cap S)$.
4. Find the elements in $Q'-(R\cap P)$.
5. Find the elements in $(Q\cup S)'\cap P$.
6. How many elements are in $\mathscr{P}(Q\cap S)$?
7. Does $(P\cap S')=P\cup(P\cap R')$?
8. Is $\{369\}\in\mathscr{P}(S)$?
9. Which of the following is a true statement?
 a. $(Q-S)\subset(P\cap Q)$
 b. $(Q-S)\subseteq(P\cap Q)$
 c. $(Q-S)\cap(P\cap Q)=\varnothing$
 d. $(Q-S)'\cup(P\cap Q)=\varnothing$
 e. none of these

10. Which of the following is a true statement?
    a. $\varnothing \subset S$
    b. $\varnothing \in \mathscr{P}(R)$
    c. $P \cap S = \varnothing$
    d. all of the above
    e. none of the above

For questions 11–20, indicate whether the statement is true or false.

11. If $x \in (A \cap B)'$ then $x \in A'$ or $x \in B'$.
12. $(B \subset D) \rightarrow (B \subseteq D)$.
13. $[x \in (C - D)] \Leftrightarrow [(x \in C) \wedge (x \notin D)]$
14. $(C \subseteq A) \Leftrightarrow (A \subseteq C)$.
15. If $P \subseteq Q$ then $P = Q$.
16. If $M = \{1,2,5,8\}$ and $N = \{8,2,5,1\}$ then $M \subset N$.
17. $(R = S) \Leftrightarrow [\,[(x \in R) \rightarrow (x \in S)] \vee [(x \in S) \rightarrow (x \in R)]]$
18. If $x \in (S - W)$ then $x \in W'$.
19. $G = U - G'$.
20. If $x \in [P \cap (P \cup Q)]$ then $x \in Q$.
21. For the given Venn diagram, sets *A*, *B*, and *C* are proper subsets of the universal set.

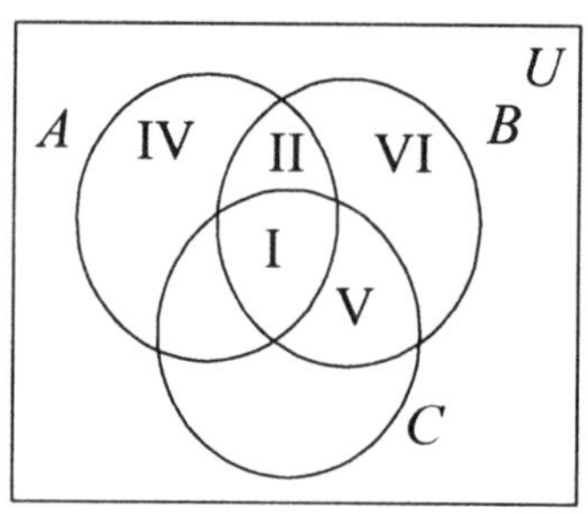

Which of the following represents the numbered regions?
    a. $(A \cup B) - C$
    b. $B \cup C'$
    c. $B \cup [A - (B \cup C)]$
    d. $(A \cup B) - (A \cap C)$
    e. none of these

22. Let $U = \{1,2,3,5,7,11,13,17,19,23,29\}$, $A = \{1,2,3,5,13,17\}$, $B = \{1,2\}$, $C = \{3,5,7,11,19,23\}$, and $D = \{7,11,23\}$. Construct a Venn diagram that satisfies the given conditions. Be sure to place all of the elements in their correct regions.

23. Use a membership table to determine if the compound statement $P-(R\cap S)'=(P-R')\cup(P-S')$ is true.

24. For the given Venn diagram, sets A, B, P and W are proper subsets of the universal set.

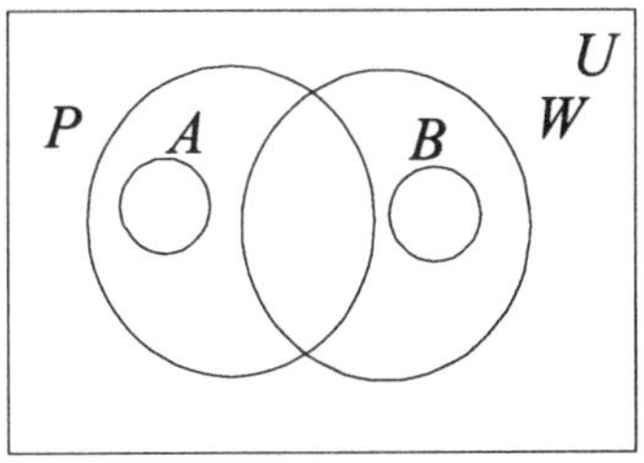

Which of the following is a true statement?

a. $A\cap W\neq\varnothing$
b. $P\cap W=\varnothing$
c. $(x\in P)\rightarrow(x\in A)$
d. $B\subset(P\cup W)$
e. none of these

For questions 25–33, supply the Law of Equal Sets that justifies each step.

$C\cap[(C-A)'\cup A]$	given
$C\cap[(C\cap A')'\cup A]$	25.
$C\cap[A\cup(C\cap A')']$	26.
$C\cap[A\cup(C'\cup A)]$	27.
$C\cap[A\cup(A\cup C')]$	28.
$C\cap[(A\cup A)\cup C']$	29.
$C\cap(A\cup C')$	30.
$(C\cap A)\cup(C\cap C')$	31.
$(C\cap A)\cup\varnothing$	32.
$C\cap A$	33.

34. Use the Laws of Equal Sets to verify that the compound statement $W\cap(S\cap W)'=W-S$ is true.

CHAPTER SIX

APPLICATIONS OF LOGIC AND SET THEORY

Logic is at the center of many applications such as simple networks, while set theory can be used to solve survey problems, basic probability, and even logical proofs. In this chapter we will examine these applications.

6.1 Testing the Validity of Arguments Using Venn Diagrams

Since the operations of negation, disjunction and conjunction in logic are analogous respectively to complementation, union and intersection in set theory, it is not surprising that there should be a way to test the validity of arguments using set theory. Venn diagrams provide a means for doing this.

A four step algorithm is used to test for validity.

1. Construct a Venn diagram with as many intersecting circles as there are variables in the argument, and label all the regions.
2. Write the set theory counterparts of the premises and the conclusion of the argument. In the case of conditional statements, first use the conditional equivalence to express the conditional statement as a disjunctive statement.
3. Identify the region(s) in the Venn diagram containing the intersection of all the premises.
4. Identify the region(s) in the Venn diagram containing the conclusion.
5. If the intersection of the premises is a subset of the conclusion, the argument is valid. If not, the argument is invalid.

Example 1

Use a Venn diagram to show that the following argument is valid.

$$\begin{array}{l} p \vee q \\ \sim p \\ \hline q \end{array}$$

Solution

First, construct a Venn diagram with two intersecting circles, and label the regions.

Next, write the set theory counterpart of the given argument. We now have the following Venn diagram and argument.

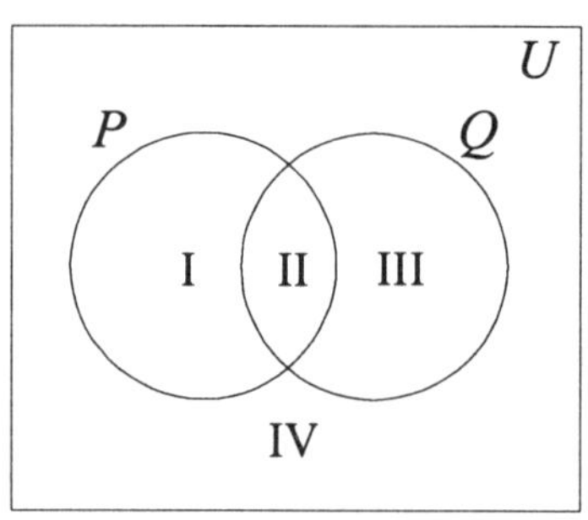

$$\begin{array}{l} P \cup Q \\ P' \\ \hline Q \end{array}$$

The first premise corresponds to regions I, II, and III. The second premise corresponds to regions III and IV. The conclusion corresponds to regions II and III. The intersection of the premises is region III which is a subset of the conclusion region. Hence, the argument is valid.

■

Example 2

Use a Venn diagram to show that the following argument is valid.

$$\begin{array}{l} p \to (q \vee r) \\ \sim q \\ \underline{\sim r} \\ \quad \sim p \end{array}$$

Solution

Our Venn diagram will have three intersecting circles. Using the conditional equivalence, $p \to (q \vee r)$ can be written as $\sim p \vee (q \vee r)$. We now have the following Venn diagram and argument.

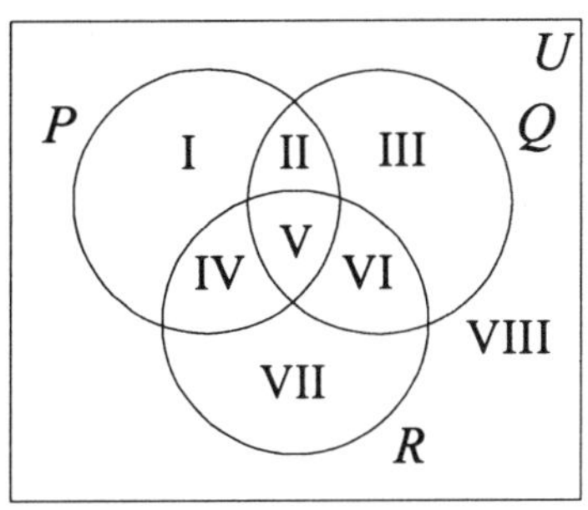

$$\begin{array}{l} P' \cup (Q \cup R) \\ Q' \\ \underline{R'} \\ \quad P' \end{array}$$

The first premise is represented by regions II through VIII, the second premise by regions I, IV, VII, and VIII, and the third premise by regions I, II, III, and VIII. The conclusion is represented by regions III, VI, VII, and VIII. The intersection of these premises occurs only in region VIII. Since this is a subset of the conclusion region, the argument is valid.

■

Example 3

Use a Venn diagram to show that the following argument is invalid.

$$\begin{array}{l} p \\ \underline{p \to (q \vee r)} \\ \quad r \end{array}$$

Solution

Our Venn diagram will have three intersecting circles. Using the conditional equivalence, $p \to (q \vee r)$ can be written as $\sim p \vee (q \vee r)$. Next, write the set theory counterpart of the given argument. We now have the following Venn diagram and argument.

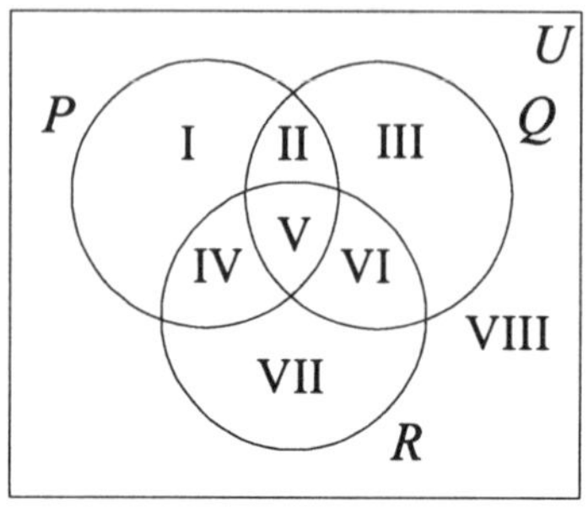

$$\begin{array}{c} P \\ P' \cup (Q \cup R) \\ \hline R \end{array}$$

The first premise is represented by regions I, II, IV, and V, the second premise by regions II, III, IV, V, VI, VII, and VIII. The intersection of the premises is regions II, IV, V. The conclusion is represented by regions IV, V, VI, and VII. Since the intersection of the premises is not a subset of the conclusion, the argument is invalid.

■

In-Class Exercises and Problems for Section 6.1

In-Class Exercises

For Exercises 1–8, use a Venn diagram to test the validity of each argument.

1. $$\begin{array}{c} a \vee \sim b \\ \sim b \to c \\ \sim a \\ \hline \sim c \end{array}$$

2. $$\begin{array}{c} p \to q \\ p \\ \sim q \vee r \\ \hline r \end{array}$$

3. $$\begin{array}{c} r \to s \\ s \to w \\ \hline r \to w \end{array}$$

4. $$\begin{array}{c} \sim(\sim a \vee b) \\ a \to c \\ \hline c \wedge a \end{array}$$

5. $\sim b$

$a \rightarrow b$

$c \rightarrow a$

$\overline{c \wedge b}$

6. $\sim (r \rightarrow s)$

$s \rightarrow p$

$\sim p$

$\overline{r \rightarrow p}$

7. $p \vee \sim q$

$\sim p$

$\sim q \rightarrow \sim r$

$\overline{\sim (r \vee p)}$

8. $\sim (n \wedge \sim p)$

$m \rightarrow n$

$\sim p$

$\overline{\sim m \vee p}$

Problems

For Problems 9–16, use a Venn diagram to test the validity of each argument.

9. $(p \rightarrow q) \vee r$

$\sim r$

$\overline{p \wedge \sim q}$

10. $(p \vee q) \rightarrow r$

$\sim p \rightarrow q$

$\overline{r}$

11. $\sim p \vee r$

$p \vee r$

$\overline{r}$

12. $(p \vee q) \rightarrow r$

$\sim p \rightarrow \sim q$

$\overline{r}$

13. $(p \rightarrow q) \rightarrow r$

$\sim p \rightarrow q$

$\overline{r \vee p}$

14 $(p \rightarrow q) \rightarrow r$

$\sim p \vee q$

$\overline{r \vee q}$

15. $\sim a$

$a \vee \sim b$

$\sim b \rightarrow c$

$\overline{c \rightarrow a}$

16. w

$x \rightarrow z$

$\sim w \vee x$

$\overline{z \rightarrow w}$

6.2 Venn Diagrams and Survey Problems

One of the most useful applications of set theory is solving what have come to be known as *survey problems*. Typically, such problems involve splitting the results of a survey into disjoint sets in order to quickly arrive at solutions to questions that are not easily answered by considering the information in the survey as it is initially presented.

Example 1

Suppose there is a meeting of the softball and basketball teams. All in attendance play either softball or basketball. There are 28 people who play softball and 33 who play basketball. How many people were in attendance at this meeting?

Solution

At first, one would guess that there were $28 + 33 = 61$ people present. This situation is represented by the Venn diagram below.

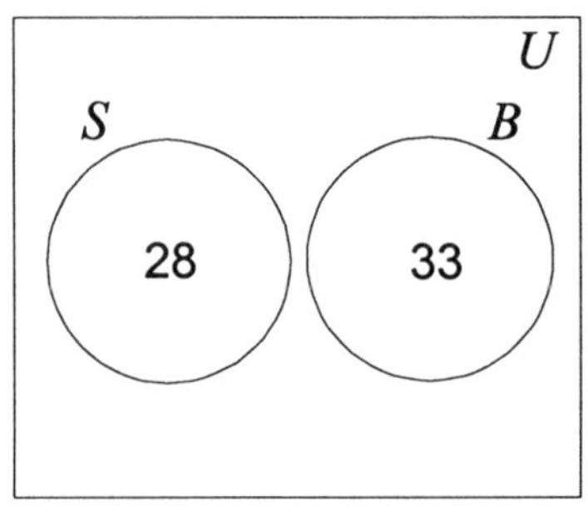

However, this solution may not be correct. Consider the possibility that all 28 softball players also play basketball. If this were the case, a total of only 33 people would be needed to satisfy the conditions given.

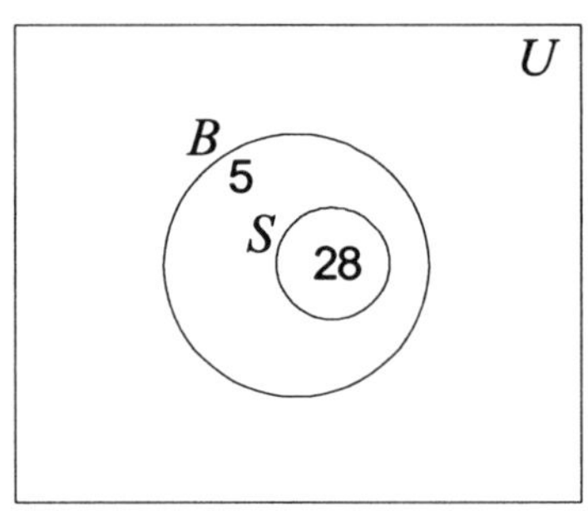

Armed only with the given information, the best we can do is say that there were between 33 and 61 people present.

■

Example 2

Suppose we are now told that there are exactly 8 people who play both softball and basketball. How can we refine our previous answer given this new information?

Solution

The new information guarantees that the two sets of athletes are not disjoint, and neither is a subset of the other. In fact, we are told that there are eight athletes in the intersection of the two sets.

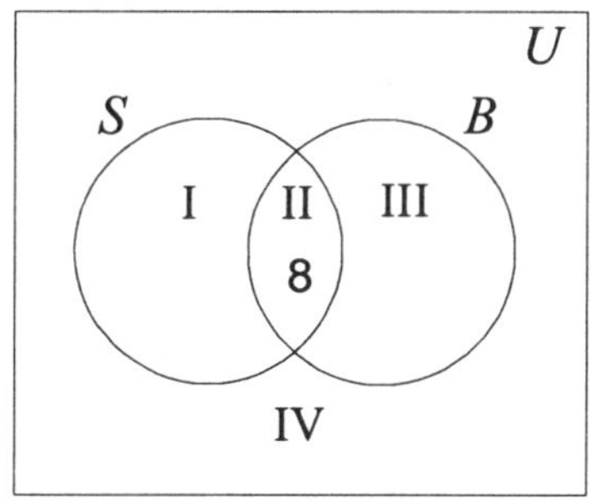

Softball players are distributed in regions I and II. Since there are 28 softball players, and eight of them have been already accounted for in region II, there must be exactly 20 in region I. Similarly, the 33 basketball players are distributed in regions II and III. Since eight of them have been already accounted for in region II, there must be exactly 25 in region III.

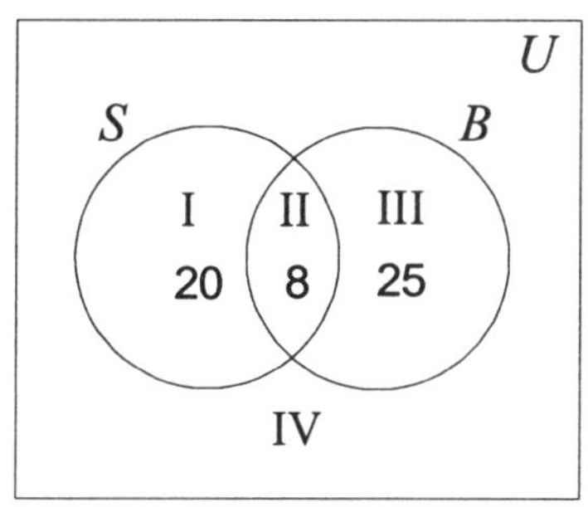

We now see that there were exactly $20 + 8 + 25 = 53$ people in attendance at the meeting.

■

Example 3

In a restaurant parking lot there were 15 black cars, 20 sedans and 5 black sedans. In all, there were 41 cars in the lot.

a. How many cars were black, but not sedans?

b. How many sedans were not black?

c. How many cars were neither black nor sedans?

Solution

We construct a Venn diagram with two intersecting circles. We know there are 5 black sedans, so we place 5 cars in region II. Sedans are distributed in regions I and II. Knowing there are 20 sedans, and five of them have been already accounted for in region II, there must be exactly 15 in region I. Similarly, the 15 black cars are distributed in regions II and III. Since five of them have been already accounted for in region II, there must be exactly 10 in region III.

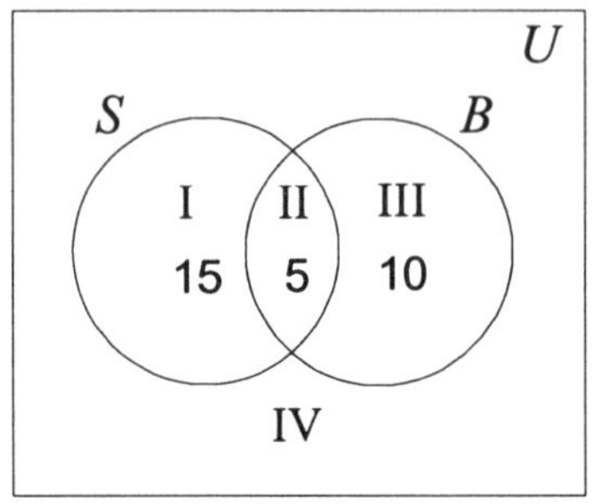

So far, we have accounted for 30 cars. The information given states that there were 41 cars in the lot, so the remaining 11 cars must be in rcgion IV.

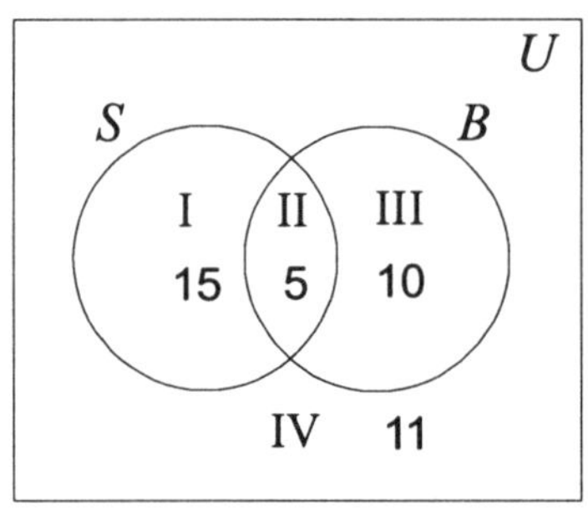

In order to answer part (a), we need to identify the elements that are in *B* but not in *S*. These are in region III. Hence, there are 10 cars that are black, but not sedans. For part (b), we need to identify the elements that are in *S* but not in *B*. These are in region I. This means that 15 cars are sedans, but not black. Finally, the cars that are neither black nor sedans are in region IV. Therefore, there were 11 cars that were neither black nor sedans.

■

Example 4

At a sorority party attended by 46 people, it was discovered that 26 people took a mathematics course, 28 took an English course, and 30 took a psychology course. There were 15 people who took both mathematics and English, 21 who took both English and psychology, and 18 who took both mathematics and psychology. Ten people took all three courses.

a How many students took at least one of the courses?

b. How many students took none of the three courses?

c. How many students took exactly one of the courses?

d. How many students took mathematics and English, but not psychology?

Solution

We begin by drawing three mutually intersecting circles and labeling them in the usual way.

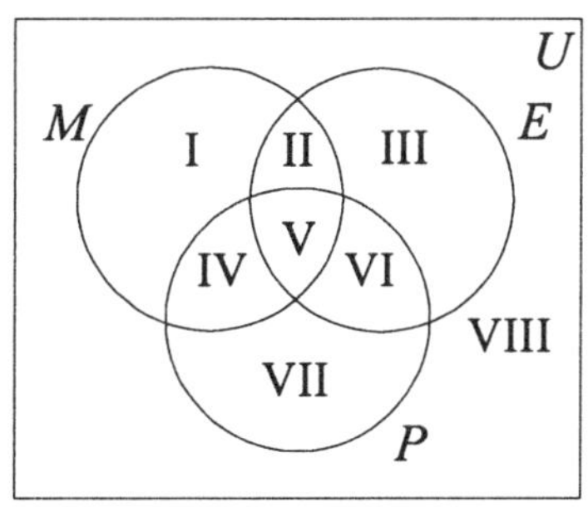

The 46 people must be divided into the eight disjoint regions. We might try to begin with the fact that 26 people took mathematics, but the mathematics set consists of four disjoint regions and at this point, it is unclear how the 26 people distribute themselves into these four regions.

However, the people who took all three courses can be assigned to exactly one region, namely region V. Next, consider those people who can be assigned to exactly two sets, that is, the people who took either mathematics and English, English and psychology, or mathematics and psychology. First, there were 15 people who took mathematics and English. The regions that make up mathematics and English are II and V, so the 15 people must be distributed between these two regions. Since there are already 10 people in region V, we put the remaining 5 people in region II. This is shown next.

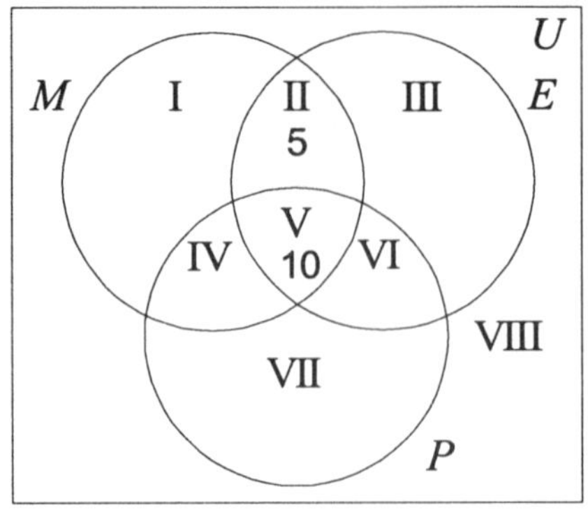

We now continue with the 21 students who took English and psychology. The regions that make up English and psychology are V and VI. Since there are already 10 people in region V, we put the remaining 11 people in region VI.

Similarly, regions IV and V describe the students who took mathematics and psychology. Again, there are already 10 people in region V, so the remaining 8 people are in region IV.

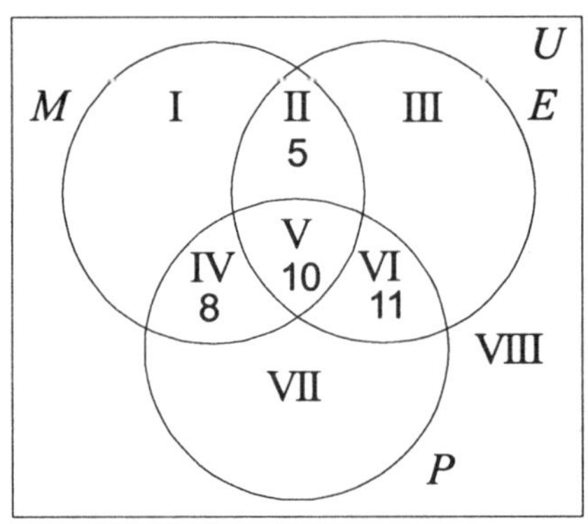

Now we are able to turn our attention to those people who took mathematics. The 26 people who took mathematics must be assigned to regions I, II, IV, and V. Since regions II, IV, and V already account for 23 people, there are exactly 3 people left for region I. In the same way regions III and VII have 2 people and 1 person, respectively. Summing all the people in the disjoint regions considered thus far (all except region VIII), accounts for 40 people. Since there were 46 people at the party, there are 6 people left for region VIII. Now we can begin to answer the questions posed.

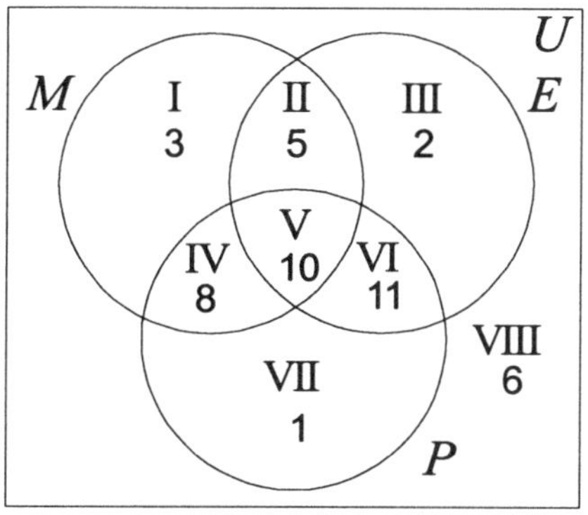

Part (a) asked how many people took at least one course. These are the people in regions I through VII. There are 40 people distributed in these regions.

The answer to part (b) is obtained by recognizing that the people who took none of the three courses are found in region VIII. There are six people who took none of the three courses.

For part (c), we are interested in only regions I, III, and VII, since all other regions within the given sets are in two or three sets. Therefore, there are 6 people who took only one of the three courses.

In part (d), we are interested in the set $M \cap E \cap P'$. This is region II. Hence, five people took mathematics and English, but not psychology. ■

Example 5

At a doctor's office, the following data was gathered over the course of one week by the office staff. There were 535 patients who saw a doctor. Ninety of these patients were treated for heart disease, 65 were treated for diabetes, and 105 were treated for emphysema. Thirty-five patients were treated for both heart disease and diabetes, 70 were treated for both heart disease and emphysema, and 40 were treated for both emphysema and diabetes. Exactly 30 people were treated for all three conditions.

a. How many patients were treated for conditions other than heart disease, diabetes, or emphysema?
b. How many patients were treated for only one of the three conditions mentioned?
c. How many patients were treated for at least two of the conditions mentioned?

Solution

Begin by placing 30 in region V, which contains the elements of $H \cap D \cap E$. Now assign people to regions shared by two sets. Regions II and V make up the set $H \cap D$. We know that there are 35 people distributed in these two regions, and 30 are already in region V. Therefore, 5 people belong in region II. Reasoning in this fashion for the sets $H \cap E$ and $D \cap E$ we find that region IV has 40 people and region VI has 10 people. Since there are 90 people in set H who must be distributed over regions I, II, IV, and V, we assign 15 people to region I. Similarly, we find there are 20 people in region III, and 25 people in region VII. We have accounted for 145 people in regions I through VII. Therefore, there must be 390 people in region VIII.

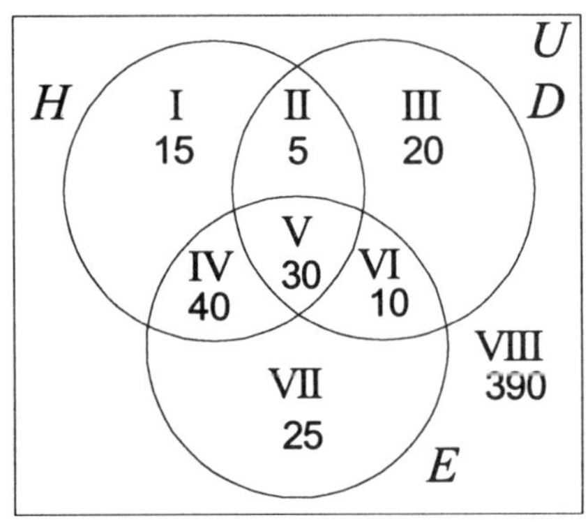

Now we can begin to answer the questions asked. For part (a) we are being asked for $(H \cup D \cup E)'$, which is region VIII. Therefore, there were 390 people who were treated for something other than heart disease, diabetes, or emphysema.

In part (b), the people who were treated for only one condition are found in regions I, III, and VII. Hence, a total of 60 people were treated for only one of the three conditions mentioned.

To answer part (c), we look to regions II, IV, V, and VI. Thus, there were 85 people who were treated for at least two of the conditions mentioned.

■

In-Class Exercises and Problems for Section 6.2

In-Class Exercises

1. Twenty-five people went to dinner. There were two choices of vegetables: carrots and string beans. Three people ordered both, ten had only carrots and eight had neither string beans nor carrots. How many had only string beans?

2. There were 60 people at a party. Thirty were Irish, 30 were Polish and 35 were German. Seven were all three. Thirteen were Polish and German and not Irish. Eight were Polish and Irish and not German. Ten were Irish and German and not Polish.
   a. How many were only Irish?
   b. How many were neither Irish, nor Polish nor German?

3. In a certain high school, there are 50 students in the high school band. Twenty students can play the clarinet, 10 students can play the saxophone and six students can play the oboe. Five students indicated that they can play both the clarinet and saxophone, 2 indicated that they can play both the clarinet and oboe, while 1 student can play both the saxophone and oboe. No student can play all three instruments.
   a. How many students cannot play any of these instruments?
   b. How many students can play the clarinet or oboe?
   c. How many students cannot play the clarinet and saxophone?
   d. How many students cannot play the saxophone or oboe?

4. In a certain college, every freshman was required to take at least one of the following courses: social science, physical science or humanities. A total of 380 students took a humanities course, 420 students took a social science course and 500 students took a physical science course. One hundred forty students took both a social science course and a physical science course, while 180 students took both a social science and a humanities course. Also, 120 students took both a physical science course and a humanities course. One hundred students took all three types.
   a. How many freshman are there at this college?
   b. How many freshman are taking only humanities courses?

c. How many freshman are taking only one of the required courses?
d. How many freshman are taking a social and a physical science course, but not a humanities course?
e. How many freshman are taking a social science or physical science course, but not a humanities course?

5. A survey was conducted in a school cafeteria to determine what foods students like to eat. Altogether, 400 students indicated that they liked hamburgers, 280 liked frankfurters and 340 liked pizza. Some of these students also indicated that they preferred more than one of the choices. Fifty said they would eat hamburgers and frankfurters but not pizza and 130 students would eat only pizza. Two hundred students would eat only hamburgers while 120 would eat both hamburgers and frankfurters. There were 700 students in the cafeteria at the time of the survey.
   a. How many liked at least one of the three foods?
   b. How many liked none of the three foods?
   c. How many liked only frankfurters?
   d. How many liked hamburgers or frankfurters, but not pizza?
   e. How many liked hamburgers and frankfurters, but not pizza?
   f. How many liked pizza or frankfurters but not both?
   g. How many liked only one of these foods?
   h. How many liked only two of these foods?

6. A survey was taken to see what kind of chocolate candy some people preferred. Twelve people liked Snickers, fourteen people liked Kit Kat and 13 people preferred Milky Way. Eight people liked both Snickers and Milky Way, 10 liked Snickers and Kit Kat but 2 liked only Kit Kat. Four people didn't like chocolate and 7 liked them all.
   a. How many people participated in the survey?
   b. How many like only Milky Way?
   c. How many liked Kit Kat and Milky Way but not Snickers?
   d. How many liked exactly two kinds of chocolate?
   e. How many like Snickers or Milky Ways but not Kit Kat?

7. A student designed a survey for her statistics course. One question on the survey was to determine the number of people who saw Friday the 13th, Scream, and Insomnia. After surveying

55 students, she determined the following: 17 had seen Scream and 23 had seen Insomnia. Six had seen Friday the 13th and Scream, 10 had seen Scream and Insomnia, and 7 had seen only Insomnia. Two students indicated they had seen all three movies and twenty hadn't seen any of these movies.

a. How many students had seen Friday the 13th but neither of the other two movies?
b. How many students had seen exactly one of these movies?
c. How many students had seen at least two of these movies?

Problems

For Problems 8–14, use a Venn diagram to help solve each of the survey problems.

8. This past semester, a local community college offered vaccinations for measles and mumps to their students, faculty and staff, totaling 20,000 individuals. The Health Center reported that 2,400 of these individuals received the vaccination for both measles and mumps, 14,000 received the vaccination for measles, and 8,200 received the vaccination for mumps.
   a. How many members of the college community did not receive either vaccination this past semester?
   b. How many members of the college community received only the measles vaccination?
9. Anne and Sterling each made a list of their favorite television shows. Anne's list had nine shows and Sterling's list had seven shows. When they compared their lists, they found that a total of 12 different shows had been mentioned.
   a. How many shows were on both of their lists?
   b. How many shows were on Anne's list but not on Sterling's list?
   c. How many shows were on Sterling's list but not on Anne's list?
10. One hundred middle school children belong to a YMCA. Kaysha volunteers her time to run an after school homework assistance center for students needing help with math, social studies, and English. Half of the students need help in their math homework, one-quarter need help in their social studies homework, and 40% need help with their English homework.

Eleven students need help with both math and social studies and 7 students need help with both math and English homework. While 2 students require help with all three subject areas, 25 need help only in English.

a. How many students need only social studies help?
b. How many students don't require Kaysha's help?

11. In the 2002 winter Olympic games, 19 countries won gold medals, 19 countries won silver medals, and 20 countries won bronze medals. Fifteen countries won both gold and silver medals, 14 countries won both gold and bronze medals, and 16 countries won both silver and bronze. Twelve countries won gold, silver, and bronze medals.
a. How many countries won only silver medals?
b. How many countries won gold and silver but no bronze medals?
c. How many countries won gold or bronze but no silver medals?
d. How many countries won at least one medal?

12. Seventy-five faculty members are in the Mathematics, Statistics, and Computer Processing Department. Each faculty member teaches at least one course. Twenty-nine faculty members teach statistics courses, 44 faculty members teach computer courses, 30 faculty members teach both mathematics and computer courses, 28 faculty members teach both mathematics and statistics courses, 3 faculty members teach computers and statistics courses, and 2 faculty members teach all three.
a. How many faculty members teach only mathematics?
b. How many faculty members do not teach statistics?
c. How many faculty members teach only one subject area?

13. A computer printout for the employees of a company indicated that 300 employees grossed over $40,000 while 600 employees grossed under $50,000. Since there are only 700 employees in this company, some of them must have been counted twice.
a. How many employees fell into neither category?
b. How many grossed over $40,000 but under $50,000?
c. How many grossed under $40,000?
d. How many grossed over $50,000?

14. A marketing firm surveyed homes in a community and obtained the following information: 1,850 homes had color television sets, 150 homes had VCRs, and 260 had computers. While every home with a VCR had a color television, there were 10 homes with computers, but no televisions. Fifty homes had all three items while 350 homes did not have any of the items.
    a. How many homes had color televisions, but had neither VCRs nor computers?
    b. How many had color televisions and VCRs but no computers?
    c. How many had color televisions or computers but no VCRs?

6.3 Sets and Probability

Introduction

A useful application of set theory is in an area of mathematics known as probability. We shall illustrate the nature of probability by considering several examples.

Example 1

To determine which football team will kick off to begin the game, a coin is tossed in the air. How likely is it that the coin will land "heads" up?

Solution

To answer the question, we try to determine what fraction of time we would expect to see a head facing up if we tossed the coin many times. This fraction is called the probability of heads facing up. We say that the probability of getting heads is ½. This is because, if we tossed the coin lots of times, we'd expect to get heads one-half the time.

■

Probability Definition

In order to deal with more complicated problems, we need a more precise way to define probability. The *probability* that an event E will occur is defined as

$$P(E) = \frac{\text{the number of ways event } E \text{ may occur}}{\text{the total number of possible outcomes}}.$$

Our set theory notation gives us an easy way to evaluate the probability fraction for a particular event. We define a *sample space* to be the set of all possible outcomes that may occur when we perform an experiment. The sample space corresponds to the universe in set theory. In Example 1, the experiment is tossing a coin, and the sample space would be the set $S = \{\text{heads, tails}\}$. The event whose probability we wish to determine would be the set $E = \{\text{heads}\}$. The outcome "heads" is called a *sample point* in the sample space. Notice that in general, E must be a subset of S. Since the number of ways event E can occur is the same as the number of elements in set E, and the total number of possible outcomes is the same as the number of elements in S,

$$P(E) = \frac{\text{the number of elements in set } E}{\text{the number of elements in the entire sample space } S}.$$

In Example 1,

$$P(E) = P(\text{head}) = \frac{\text{the number of elements in set } E}{\text{the number of elements in set } S} = \frac{1}{2}.$$

Example 2

John Zakie has complained that when course registration is done in alphabetical order, the people whose last names start with "A" always get first choice. In order to make the registration process fairer, it was decided to hold an alphabet lottery to pick the letter of the last name with which to begin the registration process. Since there are 26 letters in the alphabet, 26 slips of paper, each containing a different letter of the alphabet, are placed in a jar and then one is selected. We assume that each slip of paper has an equal opportunity of being chosen. How likely is it that the letter "z" will be the first one chosen? How likely is it that the first letter chosen will be a vowel?

Solution

The sample space is the set consisting of the letters of the alphabet, i.e., $S = \{a, b, c, \ldots, z\}$, and $E = \{\text{the letter } z \text{ is chosen}\}$. We know that

$$P(E) = \frac{\text{the number of elements in set } E}{\text{the number of elements in the entire sample space}}.$$

Since there are 26 elements in the sample space, and only one way to satisfy the event " z will be chosen" we have

$$P(z \text{ will be chosen}) = \frac{1}{26}.$$

To determine how likely is it that the first letter chosen will be a vowel we count the number of elements in this event. There are five vowels in the alphabet. They form the set $E = \{a, e, i, o, u\}$. Then,

$$P(E) = \frac{\text{the number of elements in set } E}{\text{the number of elements in the entire sample space}},$$

so,

$$P(\text{a vowel is selected}) = \frac{5}{26}.$$

■

Example 3

A die is a small cube with dots on each of its six faces, numbered 1 through 6. Suppose a fair die is rolled. What is the probability that the side facing up is the number 5? What is the probability that it is a number less than 3? What is the probability that it is an even number? Note that a *fair* die is one such that each of its faces is equally likely to show if the die is rolled. Unless otherwise noted, all dice we discuss will be assumed fair.

Solution

For the die, the sample space is $S = \{1, 2, 3, 4, 5, 6\}$, so the total number of elements in the sample space is 6. The set representing a roll of 5 has just one point in it, the number 5 itself, so $E_1 = \{5\}$. Therefore,

$$P(5) = \frac{1}{6}.$$

The set that contains numbers less than 3 is $E_2 = \{1, 2\}$. Therefore,

$$P(\text{less than } 3) = \frac{2}{6} = \frac{1}{3}.$$

The set that contains only even numbers is $E_3 = \{2, 4, 6\}$. Using the same reasoning as before,

$$P(\text{even}) = \frac{3}{6} = \frac{1}{2}.$$

■

Example 4

An urn contains seven colored chips: four green chips, two red chips and one blue chip. One chip is selected at random. Determine $P(\text{green})$, $P(\text{red})$ and $P(\text{blue})$.

Solution

The number of elements in the sample space is the total number of chips, 7. The subset of green chips has 4 elements in it; the subset of red chips has 2 elements; and the subset of blue chips has only 1 element. This gives us

$$P(\text{green}) = \frac{4}{7},\ P(\text{red}) = \frac{2}{7},\ \text{and } P(\text{blue}) = \frac{1}{7}.$$

■

Mutually Exclusive Events

In Example 4, there are only three possible events that can occur if we select a chip at random. The chip must be green, red, or blue. These events are all *mutually exclusive,* that is, they cannot happen at the same time. For example, a chip cannot be both red and green at the same time. Moreover, these three events represent all possible outcomes. If we add the probabilities of all three mutually exclusive events, we obtain

$$P(\text{green}) + P(\text{red}) + P(\text{blue}) = \frac{4}{7} + \frac{2}{7} + \frac{1}{7} = 1.$$

This illustratcs an important result in probability: *The sum of the probabilities of all the mutually exclusive events in a sample space is always one.*

Complementary Probability

If $P(E)$ represents the probability that event E occurs, we let $P(E')$ represent the probability that event E will not occur. We call $P(E')$ the *complementary probability* of event E.

Example 5

Consider a deck of 52 ordinary playing cards. One card is selected at random. Determine the probability of not getting an ace.

Solution

The total number of elements in the sample space is 52. Notice that it is easier to find the number of cards that are aces than it is to count the cards that are not aces. Since there are 4 aces,

$$P(\text{ace}) = \frac{4}{52}.$$

The number of cards that are not aces is $52 - 4 = 48$. Therefore,

$$P(\text{not an ace}) = P(\text{ace}') = \frac{48}{52}.$$

■

We observe that the event "ace" and the event "not an ace" are mutually exclusive events. If a card is selected, either it will or will not be an ace. Thus, $P(\text{ace}) + P(\text{not an ace}) = 1$. This was illustrated in Example 5 where we saw that

$$P(\text{ace}) + P(\text{ace}') = \frac{4}{52} + \frac{48}{52} = 1.$$

The solution to Example 5 illustrates a formula that can be used to compute the complementary probability of any event E. Since $P(E) + P(E') = 1$, we can write

$$P(E') = 1 - P(E).$$

Example 6

A die is rolled. Find the probability that the number rolled is not greater than four.

Solution

The total number of elements in the sample space is 6. Two of these are numbers that are greater than four, so

$$P(\text{not greater than 4}) = 1 - P(\text{greater than 4}) = 1 - \frac{2}{6} = \frac{4}{6} = \frac{2}{3}.$$

■

Probability and Percents

Sometimes we may not know the total number of elements in a sample space, but we may be given the relationship among the various events in terms of percentages.

Example 7

At a certain movie theater, 70% of the women are wearing shoes, 20% are wearing sneakers, and 10% are wearing boots. If a woman is chosen at random, find the probability that she is wearing shoes, the probability that she is wearing sneakers, and the probability that she is not wearing boots.

Solution

Since percent means "for every hundred", we can treat the sample space as if it contains 100 elements. The fact that seventy percent of the women are wearing shoes indicates that 70 of every 100 women are wearing shoes, so $P(\text{shoes}) = \dfrac{70}{100} = 0.7$. Since 20% of the women are wearing sneakers, $P(\text{sneakers}) = \frac{20}{100} = 0.2$. Finally, since 10% of the women are wearing boots, the probability that a woman is wearing boots is $P(\text{boots}) = 0.1$. Therefore, the probability a woman is not wearing boots is $P((\text{boots})') = 1 - P(\text{boots}) = 1 - 0.1 = 0.9$. ■

In-Class Exercises and Problems for Section 6.3

In-Class Exercises

1. Twenty-six blocks, each containing one letter of the alphabet, are placed in a carton. One of the blocks is picked at random. Find the probability that the block is
   a. the letter "x".
   b. the letter "p".
   c. not the letter "q".
   d. not a vowel.
   e. a letter contained in the word "dog".
   f. not a letter contained in the word "frog".
   g. neither an "a" nor a "b".

2. A die is rolled. Find:
   a. $P(2)$.
   b. $P(1) + P(2) + P(3) + P(4) + P(5) + P(6)$.
   c. $P(2 \text{ or } 3)$.
   d. $P(\text{odd})$.
   e. $P(\text{not } 4)$.
   f. $P((5)')$.
   g. $P(\text{neither 5 nor 6})$.
   h. $P(7)$.
   i. $P(\text{less than 4})$.
   j. $P(\text{not less than 3})$.
   k. $P((\text{greater than 5})')$.
   l. $P((\text{less than 9})')$.

3. The results for a television survey at 9 pm on a certain night were as follows: 20% of the homes surveyed were tuned to

NBC, 10 % were tuned to ABC, 5% were tuned to CNN and 15% were tuned to FOX. The survey company believes that this reflects the viewing habits of the entire nation. If a home is called at random and one television is on in that home, find the probability that the television is tuned to

a. CNN.
b. either NBC or FOX.
c. one of the four networks cited in the survey.
d. none of the four networks.

4. A card is selected from an ordinary deck of fifty-two cards. Find

   a. $P(\text{club})$.

   b. $P(\text{not a club})$.

   c. $P(\text{not a black card})$.

   d. $P(\text{not a king})$.

   e. $P(\text{the six of spades})$.

   f. $P(\text{a queen or a king})$.

   g. $P(\text{neither a queen nor a king})$.

   h. $P(\text{the card is less than five})$, assuming that the ace counts as a one.

Problems

5. An urn contains seven colored chips: two red, four green and one blue. Find the probability that a chip selected at random is

   a. red.
   b. green.
   c. blue.
   d. not red.
   e. not blue.
   f. neither green nor blue.
   g. white.
   h. red or green.

6. A certain die is "loaded" to favor the numbers 1, 2, 3 and 4. That is, $P(1) = P(2) = P(3) = P(4) = \frac{1}{5}$, but $P(5) = P(6) = \frac{1}{10}$. Find

   a. $P(1) + P(2) + P(3) + P(4) + P(5) + P(6)$.

   b. $P(\text{not a } 2)$.

c. P(not a 4).

d. P(not a 5).

7. In a particular mathematics class, 30% of the students are freshmen, 20% are sophomores, 40% are juniors and 10% are seniors. Find the probability that a student chosen at random from this class is
   a. a junior.
   b. a freshman.
   c. not a senior.
   d. a sophomore or above.
   e. neither a freshman nor a junior.

8. A deck of ordinary playing cards is shuffled thoroughly. If a single card is drawn at random, find the probability that the card is
   a. an ace of diamonds.
   b. a black ace.
   c. a king.
   d. a picture card.
   e. a number greater than 5 but less than 8.
   f. not the queen of hearts.
   g. not a jack.
   h. either an ace or a king.
   i. a red card.
   j. not a diamond.
   k. an even-numbered card.
   l. not an odd-numbered card, if the ace is considered a one.

9. A computer is programmed to generate a random whole number from 1 to 12. Find
   a. $P(7)$.
   b. $P((3)')$.
   c. P(an even number).
   d. $P(19)$.
   e. $P((13)')$.
   f. $P(2 \text{ or } 12)$.
   g. $P((1 \text{ or } 10)')$.
   h. P(a number less than 10).
   i. P((a number greater than 5)$'$).

6.4 Venn Diagrams and Probability

Introduction

We have already seen that many of the ideas of probability can be explained using set theory notation. Using the notions of union, intersection, and Venn diagrams, we now expand our previous results to find probabilities of compound events.

We begin our discussion by considering situations that require finding the probability that either of two mutually exclusive events will occur. Events that are mutually exclusive can be represented by disjoint sets in a Venn diagram.

Example 1

Suppose a die is rolled. What is the probability that a number less than 4 or greater than 5 is rolled?

Solution

Our sample space consists of the set of all the possible numbers we can roll on the die. Thus, $S = \{1, 2, 3, 4, 5, 6\}$. Letting L and G be subsets of S, where L represents the set of numbers less than 4, and G represents the set of numbers greater than 5, we have $L = \{1, 2, 3\}$ and $G = \{6\}$. We observe that L and G have no elements in common, i.e., $L \cap G = \varnothing$. Sets L and G represent mutually exclusive events. A Venn diagram representing this situation is shown below.

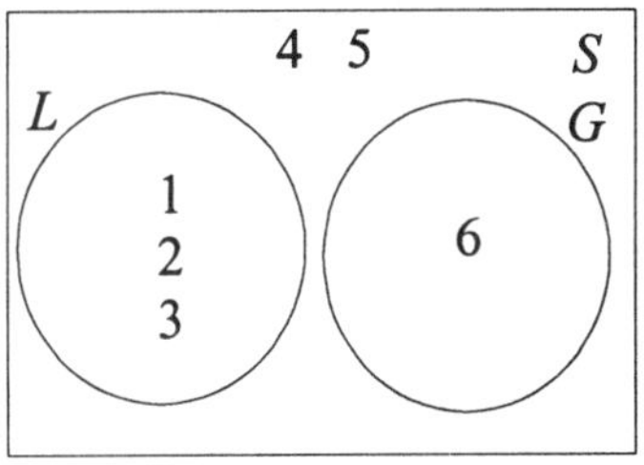

Since L has 3 elements and S has 6 elements,

$$P(\text{a number less than 4}) = P(L) = \frac{3}{6} = \frac{1}{2}.$$

Also, since G only contains one element,

$$P(\text{a number greater than 5}) = P(G) = \frac{1}{6}.$$

The union of the two sets consists of the elements that are in either L or G, that is, $L \cup G$. In this case, $L \cup G = \{1, 2, 3, 6\}$. Therefore,

$$P(\text{a number less than 4 or greater than 5}) = P(L \cup G) = \frac{4}{6} = \frac{2}{3}.$$

■

We observe that since set L and set G are disjoint, we have $P(L \cup G) = P(L) + P(G)$. This result can be generalized as follows:

Given any two mutually exclusive events A and B, that is, $A \cap B = \varnothing$, *it follows that* $P(A \cup B) = P(A) + P(B)$.

Example 2

Suppose we select a card at random from an ordinary deck of 52 playing cards. What is the probability that the card chosen is either an 8 or a picture card?

Solution

Letting the sample space S be the set of all possible outcomes, we find that S contains 52 elements. If we let E represent the set of eights, we find 4 elements belonging to E, namely the eight of hearts, the eight of diamonds, the eight of clubs, and the eight of spades. If T is the set of picture cards, then T contains the kings, the queens and the jacks in all four suits, so T contains 12 elements. No card is both an eight and a picture card, so events E and T are mutually exclusive.

Since $P(E) = \frac{4}{52}$ and $P(T) = \frac{12}{52}$, and $E \cap T = \varnothing$, we have

$$P(\text{an eight or a picture card}) = P(E) + P(T) = \frac{4}{52} + \frac{12}{52} = \frac{16}{52} = \frac{4}{13}.$$ ■

Example 3

In a certain mathematics class, the grades on a particular examination were 3 A's, 5 B's, 12 C's, 5 D's and 5 F's. Find the probability that a student selected at random from this class received either an A or a D.

Solution

The total number of possible elements in the sample space is $3+5+12+5+5=30$. Since events A and D are mutually exclusive,

$$P(A \text{ or } D) = P(A \cup D) = P(A) + P(D) = \frac{3}{30} + \frac{5}{30} = \frac{8}{30} = \frac{4}{15}.$$ ■

The Addition Theorem

If event A and event B are not mutually exclusive, then the probability of obtaining event A or event B is a little more complicated.

Example 4

Suppose a die is rolled. What is the probability that the number rolled is greater than 2 and an even number? What is the probability that the number rolled is greater than 2 or an even number?

Solution

There are six elements in the sample space, $S = \{1,2,3,4,5,6\}$. If we let the set G represent the event "a number greater than 2 is rolled", then $G = \{3,4,5,6\}$. If we let set E represent the event "an even number is rolled", then $E = \{2,4,6\}$. The events are not mutually exclusive, i.e., $G \cap E \neq \varnothing$. In fact, their intersection is $G \cap E = \{4,6\}$, which are the numbers that are both greater than 2 and even. The corresponding Venn diagram is shown below.

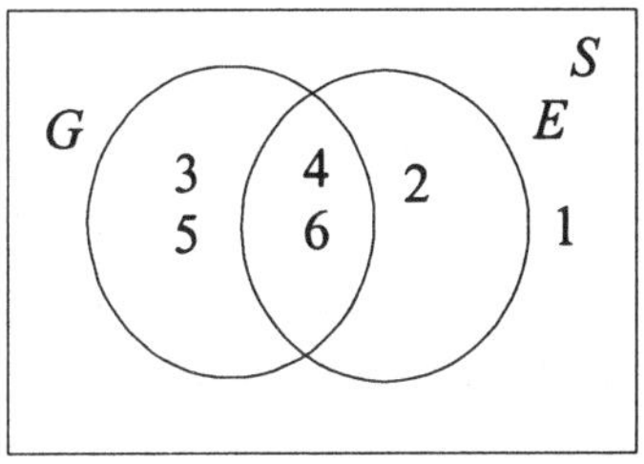

Observe that

$$P(\text{greater than 2 and an even number}) = P(G \cap E) = \frac{2}{6} = \frac{1}{3}.$$

We must now determine the probability that the roll of a die produces a number that is greater than 2 *or* even. Since the events G and E are not mutually exclusive, we cannot simply add $P(G)$ to $P(E)$ in order to find $P(G \cup E)$. If we did, the elements of $G \cap E$ would be counted twice, once as elements of G and then again as elements of E. To correct this problem, we subtract $P(G \cap E)$ from the sum of $P(G)$ and $P(E)$. Symbolically, we have

$$P(G \cup E) = P(G) + P(E) - P(G \cap E).$$

Thus,

$$P(G \cup E) = \frac{4}{6} + \frac{3}{6} - \frac{2}{6} = \frac{5}{6}.$$

We can check this result in the Venn diagram. The number of elements in the universe is 6. The elements that are either greater than 2 or even are members of $G \cup E$, and we count 5 of them. Since there are 6 elements in the universe,

$$P(G \cup E) = \frac{5}{6}.$$

■

As a result of the reasoning used in Example 4, we get the general *Addition Theorem* for the probability of event A or event B. *Given two events, A and B,*

$$P(A \text{ or } B) = P(A \cup B) = P(A) + P(B) - P(A \cap B).$$

Example 5

Suppose we select a card at random from an ordinary deck of 52 playing cards. Determine:

a. P(the card is an ace or a heart),
b. P(the card is a five or a queen),
c. P(the card is a picture card and black).

Solution

a. There are 4 aces in a deck of cards. There are 13 cards in the heart suit. These events are not mutually exclusive, because there is one card, namely the ace of hearts, that satisfies both events. If we let A represent the event "ace" and H represent the event "hearts", the addition theorem gives

$$P(A \cup H) = P(A) + P(H) - P(A \cap H)$$

$$= \frac{4}{52} + \frac{13}{52} - \frac{1}{52} = \frac{16}{52} = \frac{4}{13}.$$

b. There are 4 fives in the deck and there are 4 queens. There is no card that is both a five and a queen, so these events are mutually exclusive. Using F for "fives" and Q for "queens", we can still apply the addition theorem, giving us

$$P(F \cup Q) = P(F) + P(Q) - P(F \cap Q) = \frac{4}{52} + \frac{4}{52} - 0 = \frac{8}{52} = \frac{2}{13}.$$

c. If we let T be the set of picture cards and B be the set of black cards, we wish to find $P(T \cap B)$. We simply count the number of elements in the intersection $T \cap B$, which contains the jack, queen, and king of spades and the jack, queen, and king of clubs.

Since $T \cap B$ contains 6 elements, $P(T \cap B) = \frac{6}{52} = \frac{3}{26}$.

■

Using DeMorgan's Law

We have seen how some of the rules of probability are related to the set theory concepts of union, intersection and complementation.

Remember that $P(A \cap B)$ represents the probability that an event satisfies outcome A and outcome B simultaneously, while $P(A \cup B)$ represents the probability that A or B will occur. Recall that the complementary probability, the probability that event A does *not* happen, is given by

$$P(A') = 1 - P(A).$$

When complementary probability is combined with union or intersection, we can simplify the computations needed to solve the problem with the help of DeMorgan's Law. The two forms of DeMorgan's Law of Equal Sets are given by

$$(A \cap B)' = A' \cup B' \text{ and } (A \cup B)' = A' \cap B'.$$

Example 6

A card is selected from a deck of playing cards. Find the probability that the card is not a king or not a spade.

Solution

If we let K represent the set of kings, and S the set of spades, then the set of cards that are not kings or not spades is $K' \cup S'$. Thus, we are looking for $P(K' \cup S')$. DeMorgan's Law allows us to rewrite $P(K' \cup S')$ as $P((K \cap S)')$. Using the idea of complementary probability, $P((K \cap S)') = 1 - P(K \cap S)$. We know that there is only one card that is both a king and a spade, namely the king of spades.

Therefore, $P(K \cap S) = \frac{1}{52}$. Putting this all together, we have

$$\begin{aligned} P(\text{not a king or not a spade}) &= P(K' \cup S') = P((K \cap S)') \\ &= 1 - P(K \cap S) = 1 - \frac{1}{52} = \frac{51}{52}. \end{aligned}$$

■

Example 7

What is the probability that a randomly selected card is not a club and not a picture card?

Solution

Using C for the set of clubs and T for the set of picture cards, we obtain

$$\begin{aligned} P(\text{not a club and not a picture card}) &= P(C' \cap T') = P((C \cup T)') \\ &= 1 - P(C \cup T). \end{aligned}$$

From the addition theorem,

$$P(C \cup T) = P(C) + P(T) - P(C \cap T)$$
$$= \frac{13}{52} + \frac{12}{52} - \frac{3}{52} = \frac{22}{52}.$$

Therefore,

$$P(C' \cap T') = P((C \cup T)') = 1 - P(C \cup T)$$
$$= 1 - \frac{22}{52} = \frac{52}{52} - \frac{22}{52} = \frac{30}{52} = \frac{15}{26}.$$

■

In-Class Exercises and Problems for Section 6.4

In-Class Exercises

For Exercises 1–5, assume a card is selected at random from an ordinary deck of 52 playing cards. Determine if the two given events are mutually exclusive.

1. The card has on odd number on it; the card is a picture card.
2. The card is a queen; the card is a picture card.
3. The card is a club; the card is a spade.
4. The card is a diamond; the card is red.
5. The card is a 5; the card is a king.

Use the following information to answer Exercises 6-7. There are freshman students, honors students and teachers in a library study room. In all, there are ten people. Let F represent the set of freshman students, H represent the set of honors students, and T represent the set of teachers.

6. Draw a Venn diagram to illustrate the relationship that exists among these three sets if Chris, Jan and Mike are freshmen honors students; Lou, Val and Nan are freshmen non-honors students; Rachel and Will are non-freshmen honors students; and Kate and Pat are teachers.
7. Suppose one person is selected at random from the study room. Use your Venn diagram to determine the probability that the person selected is:

a. a freshman.
b. an honors student.
c. a freshman and an honors student.
d. a freshman or an honors student.
e. a teacher.
f. a teacher and a freshman.
g. a teacher or a freshman.
h. not an honors student.
i. neither an honors student nor a freshman.
j. not an honors student and not a teacher.

For Exercises 8–17, assume E, F, and G are three events from a sample space of all possible outcomes. Let $P(E)=\frac{3}{17}$, $P(F)=\frac{10}{17}$, $P(G)=\frac{7}{17}$, $P(E\cap F)=\frac{1}{17}$, $P(F\cap G)=\frac{4}{17}$, and $P(E\cap G)=0$. Apply the rules of probability to find the following probabilities. Venn diagrams are not necessary.

8. $P(E\cup F)$
9. $P(F\cup G)$
10. $P(E\cup G)$
11. $P(E')$
12. $P(F')$
13. $P(G')$
14. $P((E\cap F)')$
15. $P(E'\cup F')$
16. $P(E'\cap F')$
17. $P(E'\cap G')$

For Exercises 18-29, consider a bingo machine containing seventy-five balls numbered 1 through 75. Determine the probability that the number on the first ball picked is:

18. 30.
19. less than 30.
20. greater than 30.
21. less than 30 or greater than 30.
22. even.
23. odd.
24. less than 10.
25. odd and less than 10.

26. odd or less than 10.
27. not odd and not less than 10.
28. not even and does not end in 3.
29. not even and not greater than 60.

For Exercises 30-39, assume one card is chosen from a half-deck of playing cards that only contains the red cards: the diamonds and the hearts. Assuming that the ace card acts like the number one, find the probability that the card is:

30. a king or a queen.
31. a king or a picture card.
32. a king and a picture card.
33. a diamond or a heart.
34. a diamond and a heart.
35. an even number less than 9.
36. an even number or a number less than 9.
37. a diamond and a picture card.
38. not a diamond or not a picture card.
39. not a heart and not a picture card.

Problems

Use your understanding of mutually exclusive events to answer Problems 40–41.

40. A die is rolled. Which events are mutually exclusive?
    a. The number is even; the number is odd.
    b. The number is a 3; the number is odd.
    c. The number is greater than 4; the number is less than 5.
    d. The number is less than 3; the number is greater than 4.
41. A man and a woman are selected from a room full of people. Which events are mutually exclusive?
    a. The woman has brown hair; the man has brown hair.
    b. The woman has brown hair; the woman has blue eyes.
    c. The woman has blue eyes; the woman has brown eyes.
    d. The man has a blonde hair; the man has black hair.

Draw Venn diagrams to help you complete Problems 42–43.

42. A card is selected at random from the 13 hearts in a deck of cards: A, 2, 3, 4,…,10, J, Q, and K. In the Venn diagram, let T represent the elements that are picture cards and E represent the elements that are even-numbered cards. Find:
    a. P(a picture card).
    b. P(an even-numbered card).
    c. P(a picture card or an even-numbered card).
    d. P((a picture card or an even-numbered card)′).
    e. P(not a picture card and not an even-numbered card).

43. Ten chips are placed into a container. Four are colored red and six are blue. The four red chips are numbered 1 through 4, while the six blue chips are numbered 5 through 10. In the Venn diagram, let B be the subset of elements that are blue and let O be the subset of elements that are odd. Find:
    a. $P(B)$.
    b. $P(O)$.
    c. $P(B \cap O)$.
    d. $P(B \cup O)$.
    e. $P((B \cup O)')$.
    f. $P((B \cap O)')$.
    g. $P(B' \cup O')$.
    h. $P(B' \cap O')$.

Do Problems 44–47 by applying the rules of probability. It is not necessary to draw Venn diagrams.

44. A and B represent two events that are mutually exclusive. Event A can be expected to occur 30% of the time and event B is expected 40% of the time. Find each of the following probabilities.
    a. $P(A')$
    b. $P(B')$
    c. $P(A \cup B)$
    d. $P(A \cap B)$

e. $P((A \cap B)')$

f. $P((A \cup B)')$

g. $P(A' \cup B')$

h. $P(A' \cap B')$

45. Suppose $P(Q) = \frac{5}{31}$, $P(R) = \frac{7}{31}$, and $P(Q \cap R) = \frac{3}{31}$. Find the value of each of the following:

a. $P(Q')$.

b. $P(R')$.

c. $P(Q \cup R)$.

d. $P((Q \cup R)')$.

e. $P((Q \cap R)')$.

f. $P(Q' \cup R')$.

g. $P(Q' \cap R')$.

46. Suppose that one hundred slips of paper numbered from 1 to 100 are placed into a bag. Find the probability that the number on a slip of paper picked at random is:

a. greater than 50.
b. greater than 50 or less than 21.
c. not greater than 50 and not less than 21.
d. greater than 50 or a multiple of 10.
e. not greater than 50 and not a multiple of 10.
f. an odd number or a number ending in 4.
g. not an odd number and not a number ending in 4.
h. an even number or a multiple of 5.
i. not an even number and not a multiple of 5.

47. From an ordinary deck of fifty-two playing cards one card is chosen at random. Find the probability that the card is:

a. black or a heart.
b. black or a picture card.
c. a queen or a spade.
d. black and not a 2.
e. an ace or not a diamond.
f. not a jack or not a diamond.

g. not an ace and not a club.
h. not a spade and not a 2, 3, 4 or 5.
i. neither red nor a 2.
j. not a queen or not a picture card.
k. not an ace or not a picture card.
l. a picture card but not a jack.

6.5 Independent Events

Introduction

The calculation of the probability of two events, A and B, denoted $P(A \cap B)$, can be complicated. Thus far, we have determined $P(A \cap B)$ by listing all of the elements in the sample space. This calculation can be done more easily if the two events are independent.

Two events, E_1 and E_2, are said to be *independent* if the occurrence of one of the events does not alter the probability that the other event will occur. As an example, if a coin is tossed into the air and at the same time, a die is rolled, the result of the coin toss does not affect the outcome of the die roll.

The Counting Principle

Given two activities like tossing a coin and rolling a die, it is often important to determine how many outcomes are possible when *both* activities are performed. The *Counting Principle* provides a way to to do this. The counting principle states that if one activity can occur in m ways, and a second activity can occur in n ways, then there are $m \times n$ ways for both activities to occur.

Example 1

A coin is tossed and a die is rolled. How many outcomes are possible?

Solution

The coin can land two ways (heads and tails) and the die can have one of six outcomes. The combined event can happen in $2 \times 6 = 12$ ways.

■

Example 2

A coin is tossed and a die is rolled. What is the probability that the coin will land "heads up" and the die will show a 5?

Solution

The outcome of the coin toss and the outcome of the die roll have no effect on one another, that is, the two events are independent. We can list the elements of the sample space as ordered pairs of events, using the notation (coin toss, die roll). If we use H to represent heads and T to represent tails, then our sample space is the set

$$S = \begin{Bmatrix} (H,1),(H,2),(H,3),(H,4),(H,5),(H,6), \\ (T,1),(T,2),(T,3),(T,4),(T,5),(T,6) \end{Bmatrix}.$$

There are exactly 12 outcomes for this coin toss and die roll. This is exactly what the counting principle guaranteed in Example 1. The combined event that concerns us, heads on the coin paired with 5 on the die, is represented by just one element in S, namely, $(H,5)$. Therefore,

$$\begin{aligned} P(H \cap 5) = P(H,5) &= \frac{\text{the number of ways the event can occur}}{\text{the number of elements in the sample space}} \\ &= \frac{1}{12}. \end{aligned}$$

■

Independent Probability Formula

There is another way to evaluate this type of "and" probability for independent events. If we consider just the coin in Example 2, $P(H) = \frac{1}{2}$, and if we consider just the die, $P(5) = \frac{1}{6}$. Observe that the product of $P(H)$ and $P(5)$ is $P(H) \times P(5) = \frac{1}{2} \times \frac{1}{6} = \frac{1}{12}$. This is the same result we obtained when we found $P((H,5))$ above. This suggests a generalized *independent probability formula. Given two independent events A and B,*

$$P(A \text{ and } B) = P(A \cap B) = P(A) \times P(B).$$

Example 3

Four chips, numbered 1, 2, 3 and 4, and four playing cards, a jack, a queen, a king and an ace, are all placed in a hat. A chip and a card are selected at random. What is the probability that an even-numbered chip and a picture card are selected?

Solution

We observe that the two events are independent. We first solve the problem by listing all of the members of the sample space, using J for jack, Q for queen, K for king and A for ace. Since there are 4 chips and 4 cards, there are $4\times4=16$ ways for this compound event to occur. Therefore, S will have 16 elements, as shown below.

$$S=\begin{Bmatrix}(1,J),(1,Q),(1,K),(1,A),(2,J),(2,Q),(2,K),(2,A),\\(3,J),(3,Q),(3,K),(3,A),(4,J),(4,Q),(4,K),(4,A)\end{Bmatrix}.$$

The set E containing ordered pairs of the form (even number, picture card) is the set

$$E=\{(2,J),(2,Q),(2,K),(4,J),(4,Q),(4,K)\}.$$

Then,

$$P(\text{even}\cap\text{picture card})=\frac{\text{number of ways the event can occur}}{\text{number of elements in the sample space}}$$

$$=\frac{6}{16}=\frac{3}{8}.$$

If we solve the problem using the independent probability formula, we obtain

$$P(\text{even}\cap\text{picture card})=P(\text{even})\times P(\text{picture card})=\frac{2}{4}\times\frac{3}{4}=\frac{6}{16}=\frac{3}{8}.$$

■

Example 4

A card is selected from an ordinary deck of 52 playing cards and a die is rolled. Find the probability that a diamond card is selected and an odd number is rolled.

Solution

The sample space for this problem is large. In fact, there are $52\times6=312$ outcomes. We will solve the problem using the independent probability formula. Since the events are independent,

$$P(\text{diamond}\cap\text{odd number})=P(\text{diamond})\times P(\text{odd number})$$

$$=\frac{13}{52}\times\frac{3}{6}=\frac{39}{312}=\frac{1}{8}.$$

■

The Probability of Several Events

Both the addition theorem for mutually exclusive events and the formula for independent probability can be extended to cases involving more than two events.

Suppose that $E_1, E_2, E_3, \ldots, E_n$ represent n events. If these events are mutually exclusive, then the probability of E_1 or E_2 or E_3 or...or E_n is found by adding the probabilities of each event. Symbolically, we write

$$P(E_1 \text{ or } E_2 \text{ or } E_3 \text{ or...or } E_n) = P(E_1 \cup E_2 \cup E_3 \cup \ldots \cup E_n)$$

$$= P(E_1) + P(E_2) + P(E_3) + \ldots + P(E_n).$$

Similarly, if these n events are *all* independent events, then the probability of E_1 and E_2 and E_3 and...and E_n is found by multiplying the probabilities of each event. Symbolically, we write

$$P(E_1 \text{ and } E_2 \text{ and } E_3 \text{ and...and } E_n) = P(E_1 \cap E_2 \cap E_3 \cap \ldots \cap E_n)$$

$$= P(E_1) \times P(E_2) \times P(E_3) \times \ldots \times P(E_n).$$

Example 5

The Lucky-Three Lottery requires the bettor to choose a three-digit number from 000 to 999. The winning number is picked at random. Determine

a. *P*(the winning number is 111 or 222 or 333 or 444 or 555 or 666 or 777 or 888 or 999),

b. *P*(the same three-digit number occurs five days in a row).

Solution

To answer part (a), we note that there are 1,000 possible elements in the sample space, one for each possible number. The probability that a particular number is the winning number is $\frac{1}{1{,}000}$. The events 111 or 222 or 333 or … or 999 are all mutually exclusive, and each has a probability of $\frac{1}{1{,}000}$ of occurring. Therefore,

$$P(111 \text{ or } 222 \text{ or...or } 999) = \frac{1}{1{,}000} + \frac{1}{1{,}000} + \ldots + \frac{1}{1{,}000} = \frac{9}{1{,}000}.$$

To answer part (b), we note that each day's winning number is independent of the previous day's number. For the same three-digit number to occur five days in a row, it does not matter which number wins on the first day. We need only calculate the probability that the same number reoccurs on the next four days. If we let E represent the event "the number that won on the first day wins again," the probability that the same number occurs on the next four days in a row is

$$P(E \text{ and } E \text{ and } E \text{ and } E) = P(E)\times P(E)\times P(E)\times P(E)$$

$$= \frac{1}{1,000}\times\frac{1}{1,000}\times\frac{1}{1,000}\times\frac{1}{1,000}$$

$$= \frac{1}{1,000,000,000,000}.$$

■

Example 6

Two dice are rolled. Determine

a. P(the sum of the two dice is 11),

b. P(the sum of the two dice is 5).

Solution

There are two ways that the dice can add up to 11. The first die can be 6 and the second 5, or the first die can be 5 and the second 6. Since the events are mutually exclusive,

$$P(5 \text{ and } 6) \text{ or } P(6 \text{ and } 5) = P(5 \text{ and } 6) + P(6 \text{ and } 5).$$

Also, since each die roll is independent of the other die roll, we have $P(5 \text{ and } 6) = P(5)\times P(6) = \frac{1}{6}\times\frac{1}{6} = \frac{1}{36}$. In a similar manner, we find that $P(6 \text{ and } 5) = P(6)\times P(5) = \frac{1}{36}$. Putting this all together, we obtain

$$P(\text{the sum is } 11) = P(5 \text{ and } 6) + P(6 \text{ and } 5)$$

$$= \frac{1}{36}+\frac{1}{36} = \frac{2}{36} = \frac{1}{18}.$$

We answer part (b) in a similar fashion. There are four different ways to produce the number 5 as a sum. We have

$$\begin{aligned} P(\text{the sum is 5}) &= P(1\text{ and }4) + P(2\text{ and }3) + P(3\text{ and }2) + P(4\text{ and }1) \\ &= \frac{1}{36} + \frac{1}{36} + \frac{1}{36} + \frac{1}{36} \\ &= \frac{4}{36} = \frac{1}{9}. \end{aligned}$$

■

Example 7

Alicia has three friends, Brittany, Carla and Dara. Find

a. *P*(none of Alicia's friends is born in the same month as Alicia)

b. *P*(at least one of Alicia's friends is born in a different month than Alicia)

Solution

The month in which Alicia was born does not matter. We let B represent the event "Brittany was born in Alicia's birth month," C represent the event "Carla was born in Alicia's birth month," and D represent the event "Dara was born in Alicia's birth month." Then,

$$P(B) = P(C) = P(D) = \frac{1}{12}, \text{ and } P(B') = P(C') = P(D') = \frac{11}{12}.$$

In part (a), we want to find the probability that all three friends were born in months different from Alicia's month. Therefore we want $P(B' \text{ and } C' \text{ and } D')$. Since the three events are independent,

$$\begin{aligned} P(B' \cap C' \cap D') &= P(B') \times P(C') \times P(D') \\ &= \frac{11}{12} \times \frac{11}{12} \times \frac{11}{12} = \frac{1331}{1728}. \end{aligned}$$

For part (b), the probability that at least one of Alicia's friends is born in a different month than hers is $P(B' \cup C' \cup D')$. These events are not mutually exclusive. It is possible, for instance, to have both Brittany and Carla born in months different from Alicia's. Thus, we cannot use the addition theorem for mutually exclusive events. But, we can use an extended form of DeMorgan's Law and the concept of complementary probability. Using our set theory notation

$$\begin{aligned} P(B' \cup C' \cup D') &= P((B \cap C \cap D)') \\ &= 1 - P(B \cap C \cap D) \\ &= 1 - [P(B) \times P(C) \times P(D)] \\ &= 1 - \left[\frac{1}{12} \times \frac{1}{12} \times \frac{1}{12}\right] = \frac{1727}{1728}. \end{aligned}$$

■

When we calculate the probability of event A and event B, it is critical to know whether the two events are independent. If they are independent, we can apply the independent probability formula. If not, we must list the members of the sample space in order to calculate the probability. This distinction is illustrated in the following two examples.

Example 8

Shira has three tiles lettered d, g, and o in a bag. She will use the tiles to form a word, using the first tile picked as the first letter, the second as the second letter and the remaining tile as the third letter. She does not replace the tiles in the bag. What is the probability that Shira will spell the word "dog"?

Solution

We cannot say that Shira's attempts are independent. Whichever letter she picks first will affect the choice of letters for the second and third attempts, so we cannot use the formula for independent events. One way to answer the question is to list the sample space of all possible outcomes. $S = \{\text{dgo, dog, gdo, god, odg, ogd}\}$. If we let D represent the event "the first letter is d," O represent the event "the second letter is o," and G represent the event "the third letter is g," then $P(\text{dog}) = P(D \cap O \cap G) = \frac{1}{6}$. The Venn diagram for the solution is shown in the following illustration.

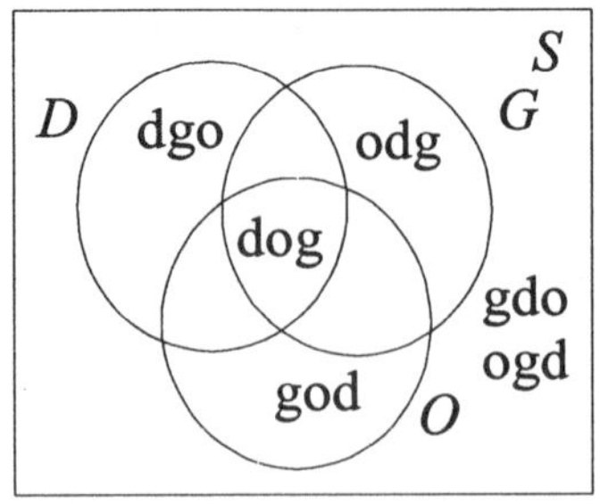

■

Example 9

Suppose Maxine took the bag containing the letters d, g and o from Shira. She picked one tile from the bag but she wrote down the letter. Then she placed that tile back in the bag and picked again. She then wrote down the second letter, replaced the tile, and picked a third time. What is the probability that she will spell the word "dog"?

Solution

Since Maxine replaced each tile after looking at the letter, each of her attempts to pick a tile is independent of every other attempt. For each attempt, $P(D) = P(O) = P(G) = \frac{1}{3}$. Then,

$$P(\text{dog}) = P(D \cap O \cap G) = P(D) \cdot P(O) \cdot P(G) = \frac{1}{3} \cdot \frac{1}{3} \cdot \frac{1}{3} = \frac{1}{27}.$$

The Venn diagram for the solution is shown below.

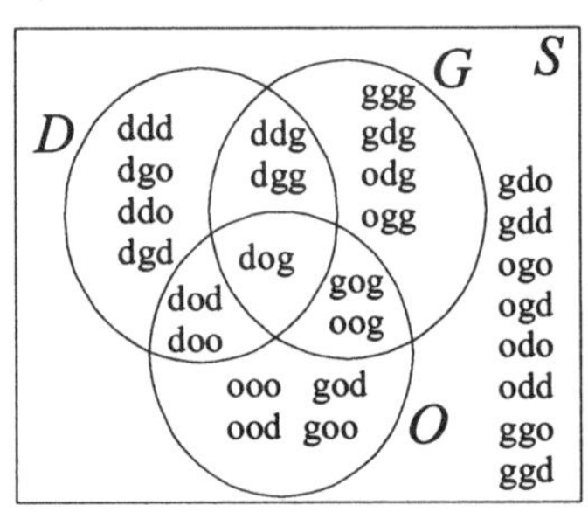

■

In-Class Exercises and Problems for Section 6.5

In-Class Exercises

For each pair of events in Exercises 1–2, answer the question "Are the two events independent?"

1. A coin is flipped and a die is rolled.
   a. heads on the coin; 5 on the die.
   b. heads on the coin; tails on the coin.
   c. 4 on the die; 2 on the die.
2. A slot machine consists of three spinning wheels with pictures of different objects (cherries, lemons, bells,...) on them. The gambler on the machine pulls a lever or pushes a button to activate the machine.
   a. The first wheel shows cherries; the second wheel shows cherries.
   b. The first wheel shows any object; the second wheel shows the same object as is shown on the first wheel.
   c. The third wheel shows a fruit; the third wheel shows a lemon.
   d. The first wheel shows a bell; the first wheel shows a fruit.

For Exercises 3–5, apply the rules of probability to answer each question.

3. Suppose H, J and K are three independent events with $P(H)=\frac{1}{3}$, $P(J)=\frac{1}{2}$, and $P(K)=\frac{3}{4}$. Find:
   a. $P(H \cap J)$.
   b. $P(H \cap K)$.
   c. $P(H \cap J \cap K)$.
   d. $P((J \cap K)')$.
   e. $P(J' \cup K')$.
   f. $P(J' \cap K')$.
   g. $P((H \cup J \cup K)')$.
   h. $P(H' \cup J' \cup K')$.

4. Ruven and Debbie are playing a game in which each calls out a random number from 1 to 5 simultaneously. Find the probability that:
   a. both call out 5.
   b. both call out 3.
   c. Ruven calls out 3 and Debbie calls out 5.
   d. either one of them calls out 3 and the other calls out 5.
   e. Debbie calls out 5 and Ruven a number less than 3.
   f. they call out the same number.
   g. each calls out an even number.
   h. neither calls out 2.
   i. either one of them calls out 4 while the other does not.
5. Two fair dice are rolled. Determine:
   a. P(the sum of the dice is 2).
   b. P(the sum of the dice is 12).
   c. P(the sum of the dice is 3).
   d. P(the sum of the dice is 8).
   e. P(the sum of the dice is 7).

Problems

6. Joshua's wardrobe consists of ten pairs of pants (five black, three green and two brown), eight shirts (four white, two green, one brown and one blue) and twelve pairs of socks (six black, four white and two brown). He selects one of each type of garment, but because of a power failure, he must select his clothes in the dark. Find the probability that Joshua selects:
   a. a black pair of pants and a white shirt.
   b. a green pair of pants and a blue shirt.
   c. a green shirt but not a black pair of pants.
   d. neither a brown pair of pants nor a brown shirt.
   e. brown pants and a brown shirt.
   f. green pants and a green shirt.
   g. the same color pants and shirt.
   h. the same color pants, shirt and socks.
   i. a white shirt and white socks.
   j. a white shirt or white socks.
   k. a white shirt or a brown shirt or a blue shirt.
   l. not a white shirt and not a brown shirt and not a blue shirt.

7. Four pills and three capsules are in a box. One of the pills is poisonous while one of the capsules is its antidote. Find the probability that a person who swallows a pill and a capsule will be poisoned and not saved.

8. Thirty percent of the women in a large investment firm earn over \$100,000 while forty percent of the men earn over \$100,000. A woman and a man are both randomly chosen to work together. Determine the probability that:
   a. both people earn over \$100,000.
   b. at least one of them earns over \$100,000.
   c. neither of them earns over \$100,000.
   d. only one of them earns over \$100,000.

9. A card is selected at random from an ordinary deck of playing cards. It is replaced in the deck, the deck is shuffled, and another card is selected. Determine the probability that:
   a. both cards are aces.
   b. the first card is an ace and the second card is a king.
   c. the first card is a picture card and the second is a club.
   d. the cards are a jack of diamonds and a queen of spades, in any order.
   e. the cards are an ace and a king, in any order.
   f. both cards are exactly the same.

10. A pair of dice has been altered so that for each die, $P(1) = P(2) = P(3) = \frac{1}{4}$ and $P(4) = P(5) = P(6) = \frac{1}{12}$. The two dice are rolled. Find:
    a. P(a 1 on the first die and a 1 on the second).
    b. P(a 2 on the first die or a 2 on the second).
    c. P(the sum of the two dice is 12).
    d. P(the sum of the two dice is 11).
    e. P(the sum of the two dice is 8).
    f. P(the sum of the two dice is 7).
    g. P(both dice show the same number).

6.6 Networks

Introduction

In Section 4.6, we saw how reducing circuits was an important application of logic. Studying the paths that current takes as it goes through a circuit is of concern to computer network designers. The circuitry that links various computers is referred to as a *network.* The switches are analogous to what are called *routers*. A router is a device that connects networks and routes information to designated networks. Routers can send the data in any number of directions. As our next examples illustrate, the concept of a network is not limited only to computer applications.

Continuous Paths

One very important efficiency concern is to insure that data does not go over the same path more than once as it travels through an entire network. If data does not go over the same pathway as it travels through a network, we say that the data has traveled a *continuous path* through the network.

Example 1

Suppose a network consists of five routers *A, B, C, D* and *E* as shown below. Data is required to be sent over the network, beginning at router *E*, to connect all of the routers. How should the data be sent through the network so that a continuous path is achieved?

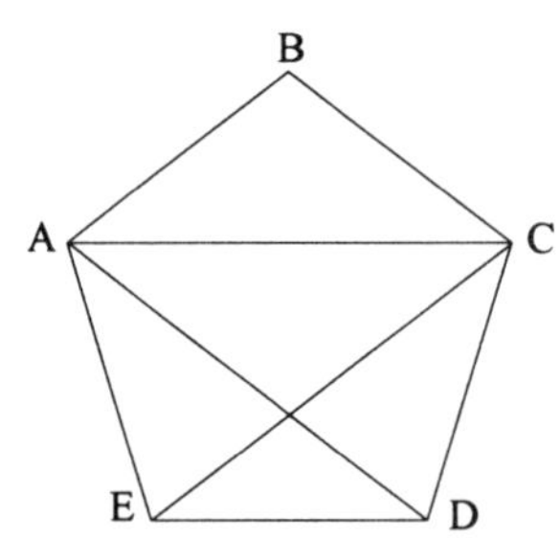

Solution

Starting at router *E* we proceed to *A*, then to *B*, then to *C*, to *D*, to *E*, to *C*, to *A*, and finally, to *D*. Thus, a continuous path is achieved. We abbreviate this path as *EABCDECAD*.

■

Example 2

Are there any other continuous paths that start from router *E*?

Solution

Starting from *E* and going in any direction (i.e. to *A*, *C* or *D*) and continuing along any unused path will ultimately result in a continuous path that ends at *D*. One such continuous path starting at *E* but different from the path in Example 1 is: *ECDEABCAD*

■

Example 3

Find a continuous path beginning at *D*.

Solution

The path *DCBAEDACE* is continuous and is a mirror image of the continuous path starting at *E* and ending at *D* as shown in Example 1.

■

Example 4

Find a continuous path starting at *A*.

Solution

Every attempt to trace a continuous path starting at *A* results in seven of the paths being covered. However, the eighth path can only be traced by traveling over a previously traveled path. Therefore, if we begin at router *A*, no continuous path exists.

■

Example 5

Which starting points result in continuous paths?

Solution

As we saw in Examples 1 and 2, continuous paths can be found starting at routers *D* and *E*. If we begin at any other router, no continuous path can be achieved.

■

Example 6

Using the network below, determine all starting routers such that a continuous path can be achieved.

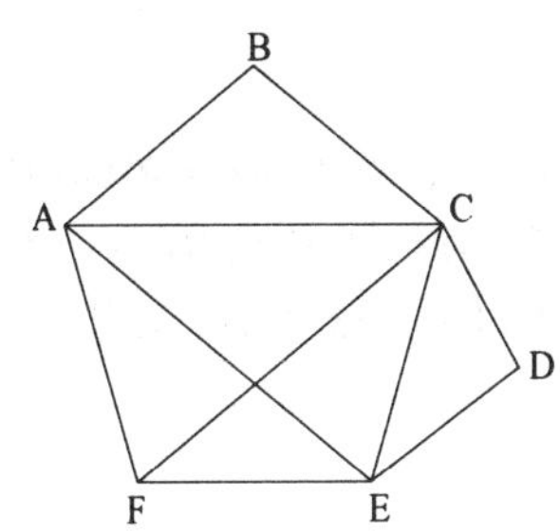

Solution

A continuous paths can be achieved by starting at router *F*. The path is: *FABCDEFCAEC*. There are ten pathways and they all have been covered with this arrangement. Another continuous path can be achieved by starting at *C*. The path is *CDEFABCAECF*. There are many other combinations for continuous paths starting from either *F* or *C*. However, no other starting routers will produce a continuous path. You should verify that no path starting at any of the remaining routers (*A, B, D* or *E*) will result in a continuous path.

■

Example 7

Using the network below, determine which starting routers lead to a continuous path and which ones do not.

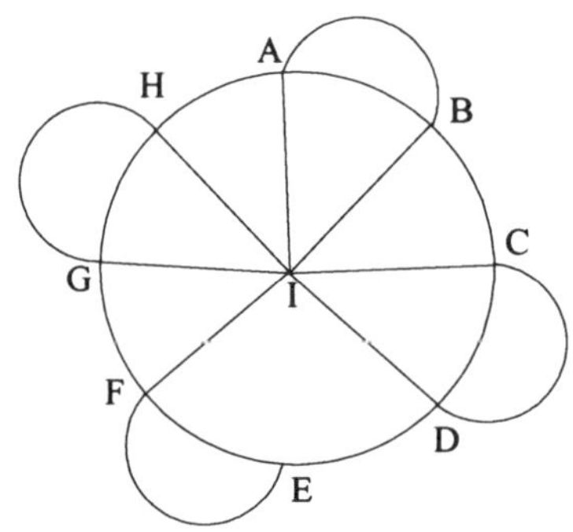

Solution

The starting routers that lead to a continuous path are *E* and *I*. An example of a continuous path that begins at *E* would be to first outline the complete circle and then to trace over the four semicircles and radii. This path is: *EFGHABCDEFIGHIABICDI*. In all there are nineteen paths to cover.

■

In our discussion of networks and routers we have seen that certain routers can serve as the starting points of continuous paths while others can not. What then are the requirements on a router such that it may be the starting point of a continuous path? To see if a pattern exists for starting routers, let us look at the three networks discussed and the number of paths meeting at each of these routers.

Network	Starting routers producing continuous paths	Number of paths meeting at the router
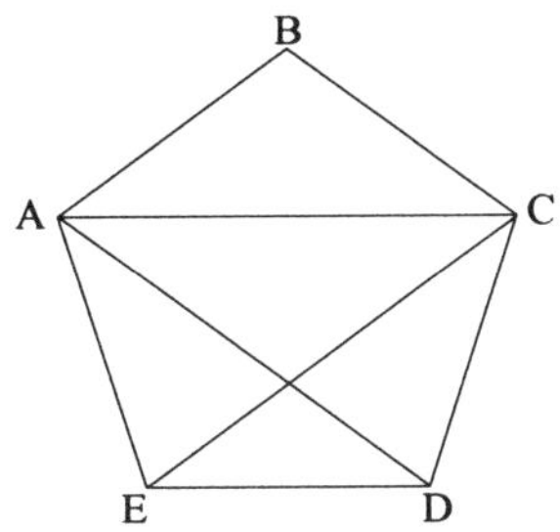	*E, D*	3, 3
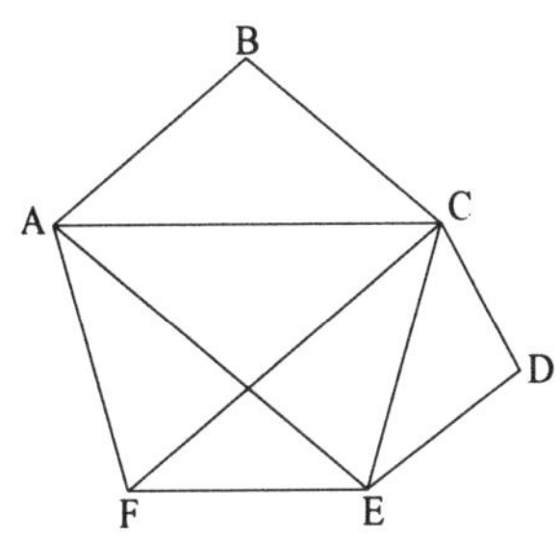	*F, C*	3, 5
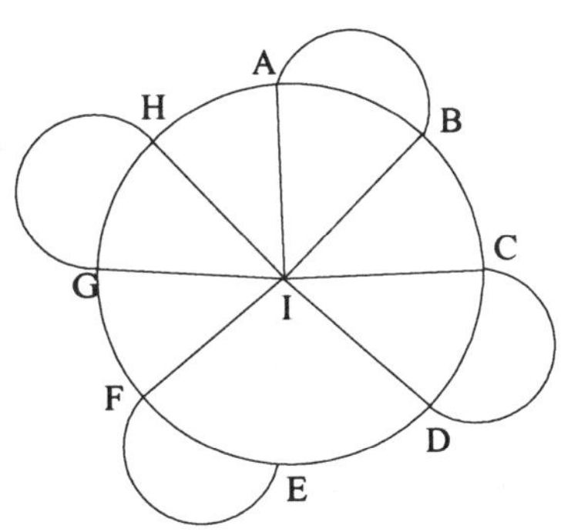	*E, I*	3, 7

We see that each starting router is the intersection point of an *odd* number of paths, while each router that does not produce a continuous path is the meeting point of an even number of paths. Observe that in the first network, *A, B* and *C* all have even numbers of paths. In the second network, the number of paths meeting at *A, B, D* and *E* are all even, as are the number of paths meeting at *A, B, C, D, F, G* and *H* in network three. This observation leads to a fundamental result about continuous paths for networks.

When a network has odd routers, it has a continuous path if and only if the number of odd routers is exactly two. The continuous path must begin at either odd router and must end at the other.

Many other laws relating to networks and continuous paths are studied by computer network designers in an attempt to produce more efficient and less costly networks.

Example 8

Determine if the following network has a continuous path.

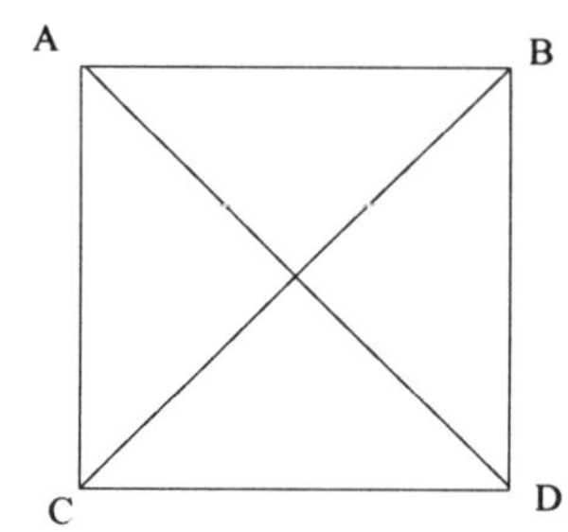

Solution

This network has no continuous path since the number of odd routers is not exactly two.

■

Example 9

Consider each of the following three networks. Determine if each has a continuous path. Determine the number of paths at each router and formulate a conclusion about such networks.

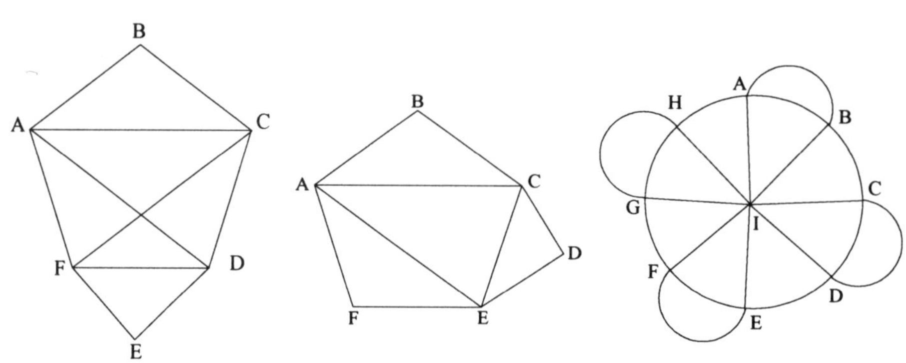

Solution

Each of the three networks has continuous paths and *any* router can serve as a starting point. As illustrations:

For the first network - *ABCDEFACFDA*

For the second network - *BCDEFACEAB*

For the third network - *IABCDEFGHABIFEIHGICDI* (twenty paths in all)

The conclusion we seem to come to is that *networks with no odd routers have continuous paths and any router may be used as a starting point.*

■

In-Class Exercises and Problems for Section 6.6

In-Class Exercises

In Exercises 1–4, determine which networks below have continuous paths.

1. 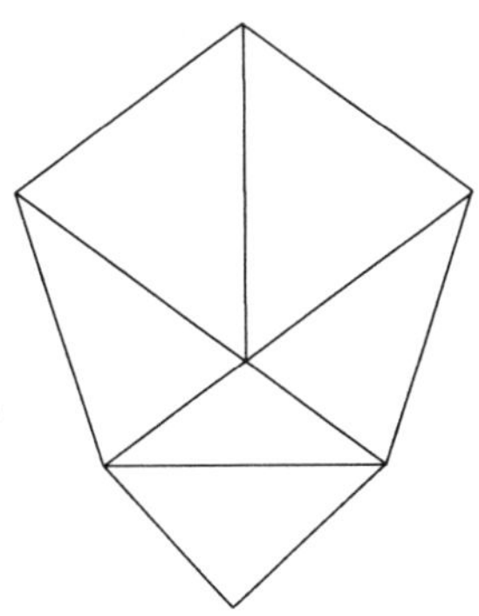

2. 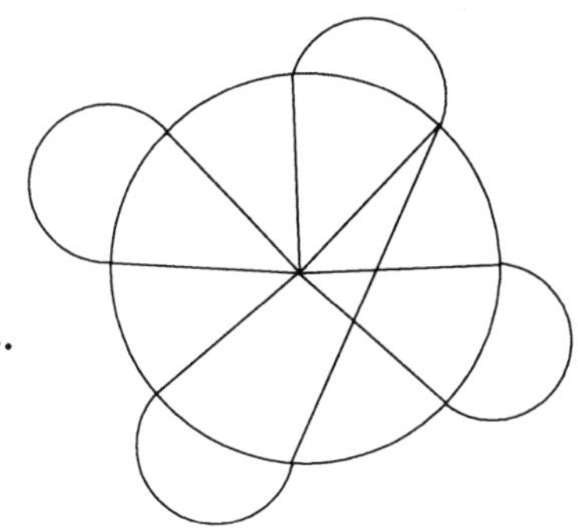

3. 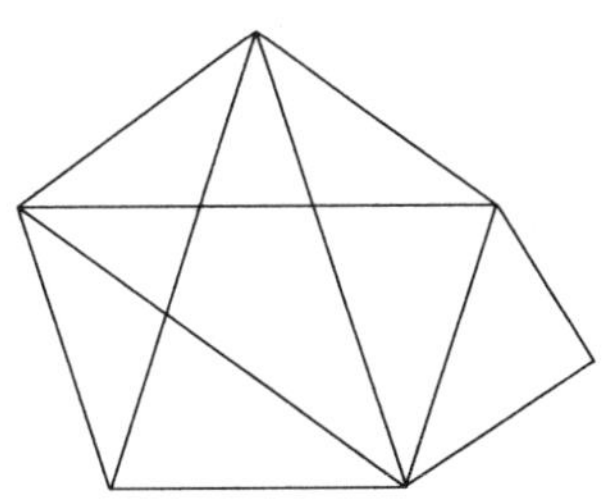

4. 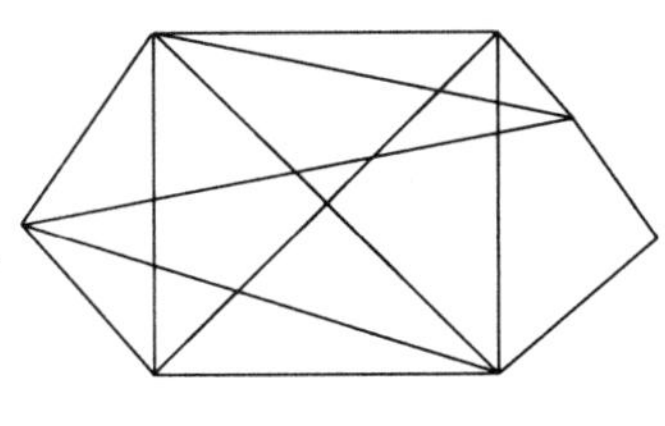

5. The three dimensional graph below depicts a cube with vertices at A-H. Is there a continuous path along the edges of this graph?

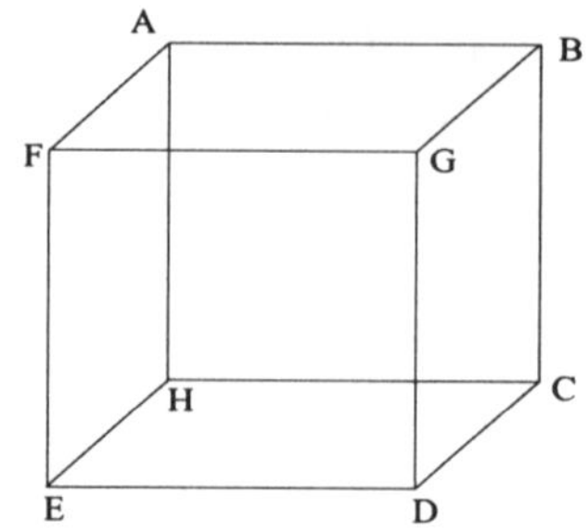

6. A letter carrier must deliver mail to the neighborhood shown below.

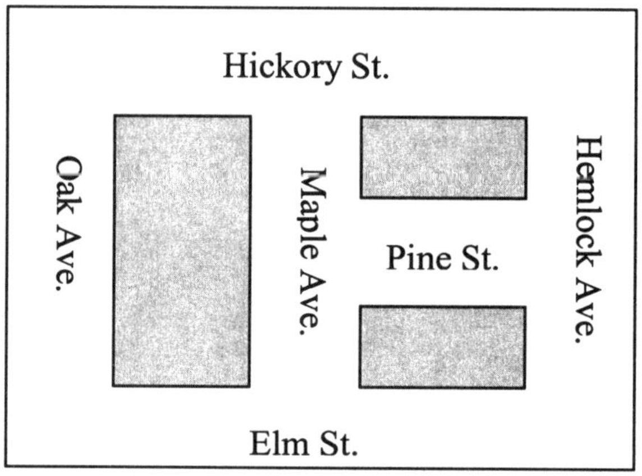

a. Represent the neighborhood as a network, making sure to include both sides of the street if there are buildings on both sides.

b. Determine if it is possible for the letter carrier to cover her route without walking the same sidewalk twice. If so, find a path that will accomplish this.

Problems

7. Identify which networks below have continuous paths:

(a)

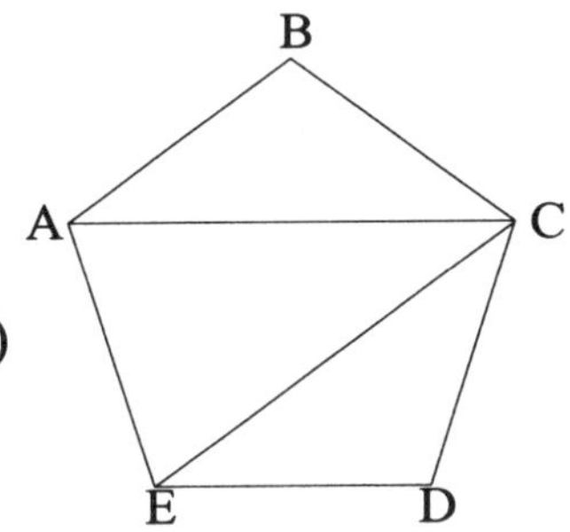

(b)

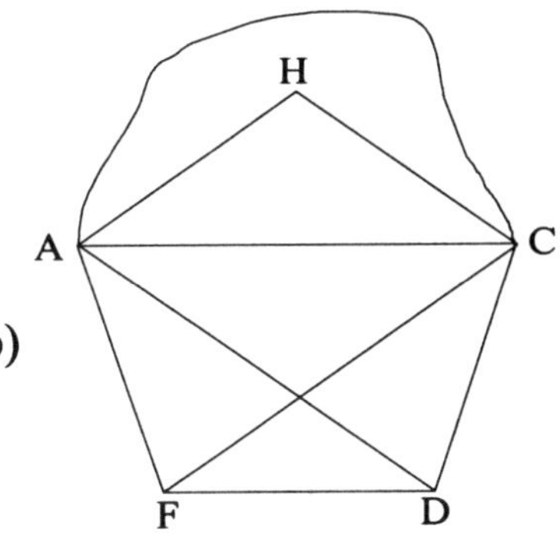

(c)

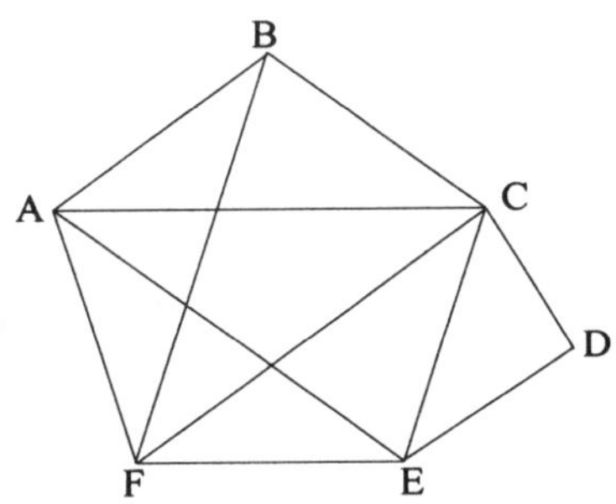

(d)

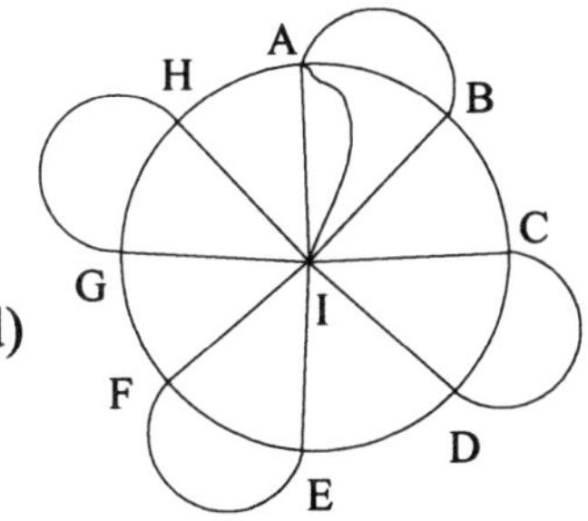

(e)

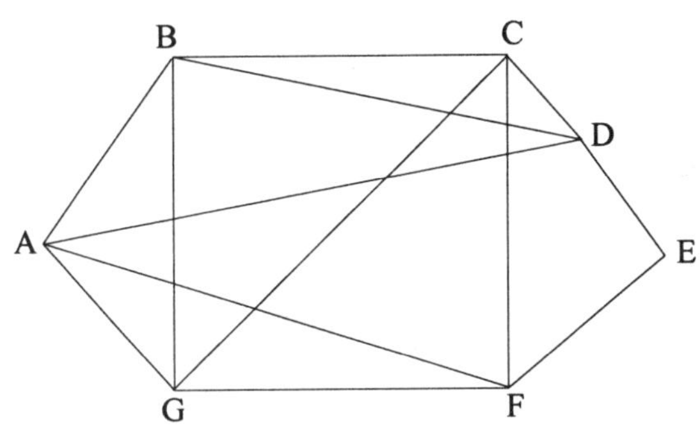

8. There are seven teams in a basketball league (A, B, C, D, E, F, and G) with required matches shown below. Because of geographic considerations, only one contest can be played at a time and the home team of each match plays the next match away. There are twelve matches in all. An example of the first three matches of a schedule might be: A travels to B, B travels to F, and F travels to G.

A against B and G
B against A, C, F and G
C against B, D, F and G
D against C, E and F
E against D and F
F against B, C, D, E, and G
G against A, B, C, and F

Provide a graph that shows these pairings.

9. A traveling salesman has a weekly route that takes him through the five towns of Scranton, Wilkes-Barre, Allentown, Reading and Harrisburg. The map below shows the entire route he must cover. Can he cover this route without traveling over any road more than once? What town should serve as his home base?

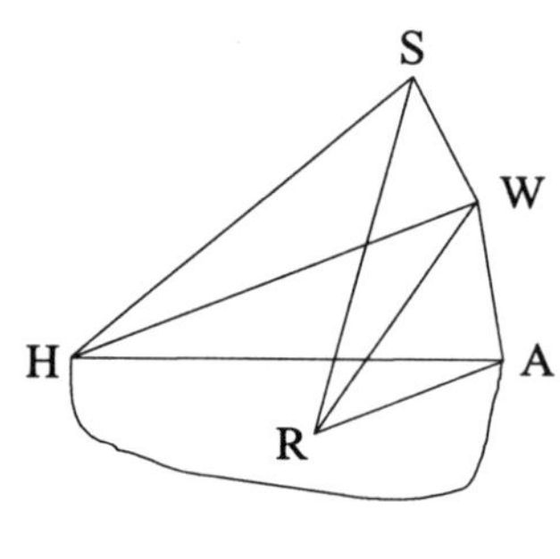

10. A famous problem in mathematics is called the Königsburg bridge problem. It is a question of determining whether it is possible to walk across seven bridges in Königsberg, Germany, without using the same bridge twice. The diagram that follows shows the four land locations at A, B, C, and D, the river and the seven bridges that cross the river. Is it possible to take such a walk?

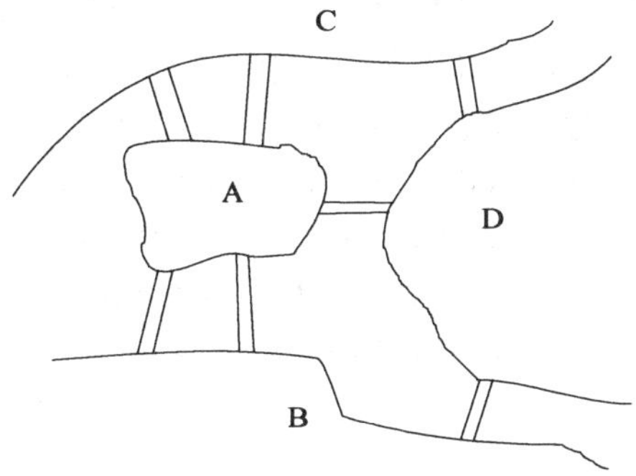

11. An ancient puzzle in mathematics is called the utility problem. The problem involves three houses and three wells that supply water to each of the houses. However, because of drought conditions, paths from the three houses to the three wells must be formed so as to allow the members of each household access to each well. The following diagram shows the arrangement.

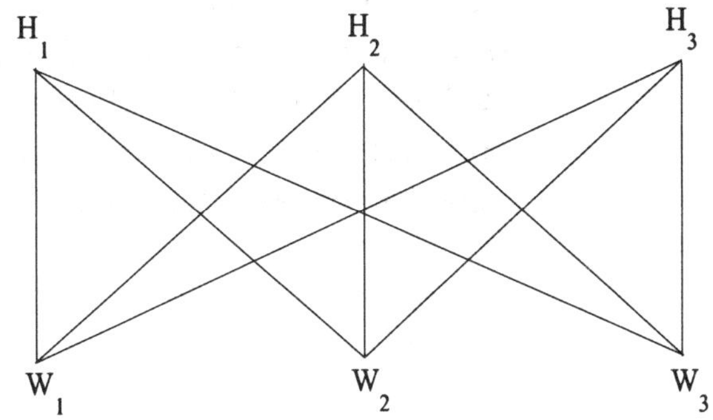

After a period of time, a feud developed among the three families. As a result, no family wished to meet another family while getting water from any well. Can such a network be constructed?

Chapter 6 Review

For questions 1–3, use a Venn diagram to answer each question.

1. In the census of a small town, 830 people claimed to be of European ancestry, 115 people claimed to be of Asian ancestry, 325 claimed to be of African ancestry, while 30 claimed to be of other ancestry. However, some of the people surveyed claimed dual ancestry. Ten claimed they were of European and Asian ancestry, 5 were of African and Asian ancestry while 20 were of African and European ancestry. No one claimed ancestry from all three regions.
   a. How many people in the town were surveyed?
   b. How many were of only Asian ancestry?
   c. How many were not of European and African ancestry?
   d. How many were of European or Asian ancestry, but not African?
   e. How many were not of European ancestry?

2. In a particular lab experiment, a scientist was working with some mice. Sixteen mice were male, 20 were trained and 18 were well-fed. Ten of the mice were trained, well-fed males. Twelve were trained males, 11 were well-fed males and 13 were trained and well-fed.
   a. How many mice were there?
   b. How many were females?
   c. How many were trained, starved female mice?

3. A restaurant owner wanted to know customers' fondness for three items on the dessert menu. Questioning a group of 50 people, he discovered that 20 people liked raspberry tarts, 15 people liked ice cream, and 25 people liked chocolate mousse. Three people indicated that they liked only raspberry tarts and ice cream, six people liked only raspberry tarts and chocolate mousse, while 10 people liked just raspberry tarts. Nine people said that they didn't like any of the three choices.
   a. How many people liked all three of these desserts?
   b. How many people liked only one of these desserts?
   c. How many people liked raspberry tarts or ice cream but not chocolate mousse?
   d. How many people like ice cream and chocolate mousse but not raspberry tarts?

For questions 4–5, use a Venn diagram to test the validity of each argument.

4.
$$\begin{array}{l} a \to (b \vee c) \\ \sim (b \wedge a) \\ \hline c \to \sim a \end{array}$$

5.
$$\begin{array}{l} \sim m \to \sim n \\ n \wedge w \\ m \vee \sim w \\ \hline m \to (n \to w) \end{array}$$

For questions 6-10, assume 100 M&M's are randomly chosen from a bag of M&M's with the following results: nine red, eleven blue, eleven yellow, fifteen orange, sixteen green, and thirty-eight brown. If Carmine randomly selects 3 M&M's, one at a time, from the 100 M&M's and replaces each one before making his next selection, find the probability that he selects:

6. a brown followed by a red followed by a green.
7. three green or three blue.
8. a yellow followed by two browns.
9. a green or an orange followed by a red and then a yellow.
10. no red ones.

For questions 11–14, assume that ninety percent of the cars manufactured by Speedy Motors have automatic transmissions, while 10% have standard transmissions. Ten percent have four-cylinder engines, 50% have six-cylinder engines and 40% have eight-cylinder engines. Assume any combination of transmission and engine is possible. If one of Speedy Motors' cars is selected at random, find the probability of it having

11. an automatic transmission and a four-cylinder.
12. a standard transmission or a eight-cylinder.
13. not an eight-cylinder.
14. neither a four-cylinder nor a standard transmission.
15. A salesman lives in city A, and must make stops at cities B, C and D before returning home. The distances between cities are:

A to B 70 miles;	A to C 90 miles;
A to D 130 miles;	B to C 20 miles;
B to D 50 miles;	C to D 60 miles

Find the shortest trip that he can make leaving from and returning to his home while making stops at B, C and D.

Sample Exam: Chapter 6

For questions 1–3, use a Venn diagram to help solve each problem.

1. A group of 47 co-workers planned a New Year's Eve party. Each person was surveyed about three dessert preferences: apple pie, cheesecake, and brownie sundae. The following data was compiled: 25 liked cheesecake, 24 liked brownie sundae, 4 liked only apple pie, 3 liked only brownie sundae, 6 like all three desserts, 9 like apple pie and cheesecake but not brownie sundae, and 14 like cheesecake and brownie sundae.
   a. How many people liked apple pie?
   b. How many people didn't like any of these desserts?
   c. How many people liked apple pie but not cheesecake?
   d. How many people liked cheesecake or brownie sundae but not both?
   e. How many people did not like brownie sundae?

2. A group of students were surveyed about their favorite musical artists and the following information was gathered: 32 liked Alanis Morissette, 32 liked Alicia Keys, 28 liked Britney Spears, 11 liked Alanis and Alicia, 10 liked only Alanis and Britney, 12 liked only Alicia, 5 liked all three artists, and 7 did not like any of these artists.
   a. How many students were surveyed?
   b. How many students did not like both Britney Spears and Alicia Keys?
   c. How many students liked exactly two artists?

3. Twenty percent of the population of Springhill has been to Italy at least once in the last five years and thirty percent of the population of Springhill has been to France at least once in the last five years. Therefore, half of the population of Springhill has been to Europe at least once in the last five years. The argument is faulty because it ignores the possibility that:
   a. Some of the population of Springhill has been neither to Italy nor to France in the last five years.
   b. Some of the population of Springhill may have been both to Italy and to France in the last five years.
   c. Some of the population of Springhill has been either to Italy or to France in the last five years, but not to both.

For questions 4–5, use a Venn diagram to test the validity of each argument.

4. $\sim(r \rightarrow \sim s)$
$a \vee \sim s$
r
$\overline{a \wedge r}$

5. $\sim(p \wedge \sim r)$
$\sim(r \vee s)$
$s \rightarrow p$
$\overline{p}$

For questions 6–9, consider a deck of playing cards and a group of 26 checkers, sixteen of which are red and ten of which are black. Two selections are made. Either two cards are selected with replacement, two checkers are selected with replacement, or a card and a checker may be selected. Determine the probability that:

6. if a card is selected and then a checker is selected, the card is black and the checker is black.
7. two red items are chosen.
8. if a card is selected and then a checker is selected, both are the same color.
9. a three or a black card and a red checker is selected.
10. A cleaning crew must vacuum the rooms and hallway in the house shown below.

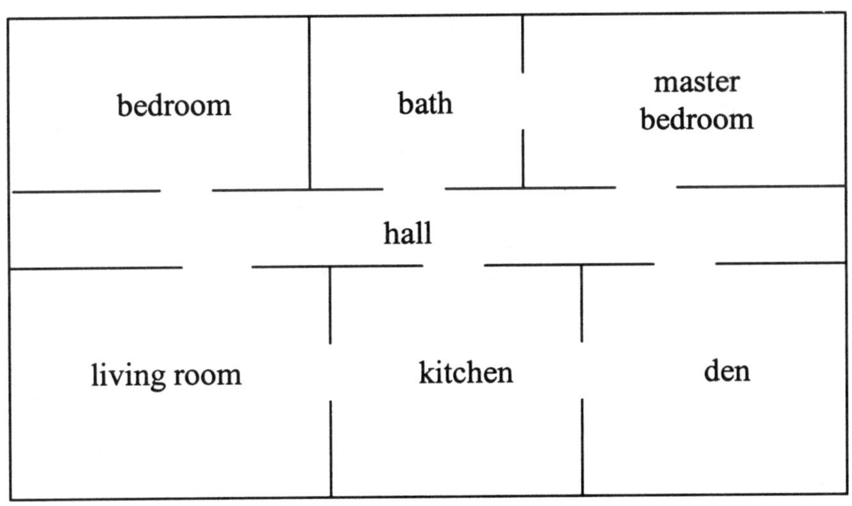

Represent the house as a network. Then determine if the cleaning crew can vacuum all the rooms and the hallway without passing through the same doorway twice? If so, in which rooms should they begin and end?

ANSWERS

Answers for Chapter 1

In-Class Excercises and Problems for Section 1.1

In-Class Exercises

1. $s \wedge b$
2. $s \rightarrow m$
3. $r \vee l$
4. $c \leftrightarrow (b \wedge v)$
5. $\sim h$
6. $c \rightarrow (s \wedge m)$
7. $\sim (d \wedge l)$
8. $s \wedge (w \rightarrow c)$
9. $(s \wedge c) \rightarrow w$
10. $c \wedge e$
11. $(b \rightarrow \sim w) \wedge (s \rightarrow d)$
12. $(s \wedge h) \rightarrow (l \wedge e)$
13. $(s \vee r) \rightarrow i$
14. $(c \vee s) \wedge (e \rightarrow \sim h)$
15. c
16. She wears glasses but she isn't tall.
17. It is not true that if she plays the flute then wears glasses.
18. She doesn't have red hair or she is tall.
19. If she is wears glasses then she has red hair and plays the flute.
20. She has red hair if and only if she wears glasses and is tall.
21. She plays the flute or if she has red hair she wears glasses.

22. F 23. T 24. F 25. CBD 26. F 27. F 28. T 29. T

30. T 31. CBD 32. T 33. T 34. CBD 35. T 36. T

Problems

37. tf; b
39. biconditional; $h \leftrightarrow \sim s$
41. conditional; $s \rightarrow m$
43. conjunction; $e \wedge b$
45. tf; $\sim c$
47. conditional; $\sim s \rightarrow \sim n$
49. conditional; $\sim n \rightarrow e$
51. conjunction; $(s \wedge a) \wedge \sim r$
53. $s \rightarrow (p \vee g)$
55. $\sim (s \rightarrow r)$
57. $o \leftrightarrow (j \wedge d)$
59. $(c \rightarrow w) \wedge [b \rightarrow (l \vee \sim j)]$
61. $[(h \wedge \sim t) \vee o] \rightarrow a$

63. T, F 65. F 67. F 69. F 71. T
73. T 75. T 77. F 79. F 81. T

In-Class Excercises and Problems for Section 1.2

In-Class Exercises

1. T 2. F 3. T 4. T 5. T 6. F 7. T 8. T
9. T 10. T 11. CBD 12. T 13. F 14. T 15. T 16. F
17. T 18. CBD 19. T 20. F 21. CBD 22. T 23. T 24. F
25. T

Problems

27. F 29. F 31. T 33. T 35. T 37. F
39. CBD 41. F 43. T 45. CBD 47. T 49. CBD
51. F 53. F 55. F

In-Class Excercises and Problems for Section 1.3

In-Class Exercises

1. Not a tautology - FTFF
2. Not a tautology - TFTT
3. Not a tautology - FTTF
4. Tautology
5. Tautology
6. Tautology
7. Tautology
8. Tautology
9. Not a tautology - FTFF
10. Tautology

Problems

11. Tautology
13. Tautology
15. Tautology
17. Not a tautology - TTFT
19. Not a tautology - TTFT
21. Not a tautology - TFFT
23. Tautology
25. Not a tautology - TTFT
27. Not a tautology - TTFF
29. Tautology

In-Class Excercises and Problems for Section 1.4

In-Class Exercises

1. Tautology
2. Not a tautology - TTTTFTFT
3. Tautology
4. Tautology
5. Tautology
6. Tautology
7. Tautology
8. Not a tautology - TTTTTFTF
9. Not a tautology - FTFFTTTT
10. Tautology

Problems

11. Tautology
13. Tautology
15. Tautology
17. Not a tautology - TTTTTFTF
19. Not a tautology - TFTTTFTT
21. 8
23. 2, F
25. 8 entries of F,2,1
27. $(a \wedge b) \vee (a \wedge c)$
29. $(m \vee l) \wedge (m \vee \sim w)$

In-Class Excercises and Problems for Section 1.5

In-Class Exercises

1. p
2. $\sim r \wedge w$
3. $(p \wedge q) \vee (p \wedge r)$
4. $\sim p \rightarrow s$
5. $(g \vee s) \vee \sim p$
6. $\sim c \vee (d \wedge \sim b)$
7. $p \wedge (s \rightarrow r)$
8. $\sim a \wedge (b \wedge \sim c)$
9. $(\sim b \wedge \sim p) \vee (\sim b \wedge q)$
10. $\sim [\sim (\sim w)]$
11. $\sim p \wedge q$
12. $\sim w \vee s$ 13. $p \wedge s$
14. $\sim (\sim b \vee r)$
15. $\sim (a \vee b)$
16. $\sim (\sim q \wedge s)$
17. $\sim m \wedge \sim (a \rightarrow b)$
18. $\sim (p \leftrightarrow q) \vee r$
19. $\sim [\sim p \wedge (a \rightarrow c)]$
20. $(w \wedge r) \vee \sim (p \rightarrow q)$
21. I will not take piano lessons or go to class.
22. Parallel parking and going around the block doesn't a driver make.
23. We can paint the town red or blue and we can paint it red or green.
24. It is not true that the swan's long neck didn't glistened in the noon day sun.
25. It is not true that I will not have milk or I want any cookies.

Problems

27. Distributive Equivalence
29. DeMorgan's Equivalence
31. Distributive Equivalence
33. DeMorgan's Equivalence
35. Double Negation Equivalence
37. DeMorgan's Equivalence
39. e
41. b
43. a
45. I will either take you out for dinner or buy you flowers.
47. In the winter either I go skiing, or I go ice skating or tobogganing.
49. It is not true that we will go out for dinner or stay home and cook.
51. The banana is ripe and the raspberry is plump.
53. Doctors advise us to eat right and exercise.

55.

p	q	r	$(p \wedge q) \wedge r$	$p \wedge (q \wedge r)$
T	T	T	T	T
T	T	F	F	F
T	F	T	F	F
T	F	F	F	F
F	T	T	F	F
F	T	F	F	F
F	F	T	F	F
F	F	F	F	F

57.

p	q	$\sim(p \wedge q)$	$\sim p \vee \sim q$
T	T	F	F
T	F	T	T
F	T	T	T
F	F	T	T

In-Class Excercises and Problems for Section 1.6

In-Class Exercises

1. $\sim c \vee k$
2. $\sim p \to \sim r$
3. $p \vee \sim s$
4. $w \to \sim z$
5. $b \vee a$
6. $m \vee (a \wedge b)$
7. $\sim p \to (c \wedge d)$
8. $(p \wedge r) \vee k$
9. $a \vee (r \wedge s)$
10. $(w \wedge h) \to (a \wedge \sim c)$
11. $h \to \sim a$
12. $\sim s \to c$
13. $q \to p$
14. $\sim (w \wedge a) \to \sim p$
15. $\sim b \to (m \vee n)$
16. $m \wedge \sim b$
17. $\sim s \wedge \sim w$
18. $\sim (a \to d)$
19. $\sim (\sim k \to \sim c)$
20. $(r \vee s) \wedge \sim w$
21. $\sim [p \to \sim (w \vee q)]$
22. g
23. a
24. e
25. f
26. c
27. b
28. d
29. If I don't do homework I don't pass.
30. If I don't quit my job I didn't win the lottery.
31. You won't go to the Chinese restaurant or you'll get a fortune cookie.
32. If the bird doesn't get up early it doesn't catch the worm.
33. If I don't fit into my jeans, I didn't go on a diet.

Problems

35. d
37. g
39. i
41. a
43. e
45. Contrapositive
47. Conditional
49. Conditional Negation
51. Conditional Negation
53. Contrapositive

55. a If I don't post your picture on the web I won't buy a digital camera.
    b. I won't buy a digital camera or I will post your picture on the web.
    c. I bought a digital camera and I didn't post your picture on the web.
57. a. If you don't feel better then you didn't take your medicine.
    b. You didn't take your medicine or you will feel better.
    c. You took your medicine and you didn't feel better.

In-Class Excercises and Problems for Section 1.7

In-Class Exercises

1. $\sim s \to r$ 2. $\sim p \to \sim q$ 3. $p \to \sim q$ 4. $\sim m \to r$
5. $\sim s \to (m \vee p)$ 6. $\sim d \to (b \wedge p)$ 7. $(c \vee d) \to \sim (p \leftrightarrow q)$
8. $\sim (p \wedge a) \to (p \vee \sim q)$ 9. $(p \leftrightarrow q) \to \sim (c \vee d)$ 10. $\sim (r \wedge \sim s) \to (w \vee c)$
11. $(p \wedge r) \vee (\sim r \wedge \sim p)$ 12. $(\sim w \wedge \sim s) \vee (s \wedge w)$ 13. $(\sim m \wedge n) \vee (\sim n \wedge m)$
14. If you don't live in a glass house then you should throw stones.
    If you shouldn't throw stones then you live in a glass house.
15. If you don't go to the birthday party you shouldn't bring a present.
    If you bring a present you should go to the birthday party.
16. If you can stand the heat, then you don't stay out of the kitchen.
    If you stay out of the kitchen then you can't stand the heat.
17. If it's raining take your umbrella.
    If you don't take your umbrella it's not raining.
18. If the sun is not shining the stars can be seen.
    If the stars can't be seen the sun is shining.
19. You should buy a new computer and have not owned it for more than two years or you have owned a computer for more than two years and should not buy a new one.
20. I will buy you a dozen roses and I get fired or I don't get fired and I don't buy you a dozen roses.
21. You don't fall off the horse and you don't take riding lessons or you take riding lessons and you fall off the horse.
22. We play bridge and we don't have four players or we have four players and we don't play bridge.
23. You can see the stars and the sun is shining or the sun is not shining and you can't see the stars.

Problems

25. $s \to \sim d, \sim s \to d, \sim d \to s$. If she is not too sleepy, Melissa can drive. If Melissa cannot drive, then she is too sleepy.
27. $\sim t \to \sim o, t \to o, \sim o \to \sim t$. If Sam does get time and a half he does work overtime. If he doesn't work overtime Sam doesn't get time and a half.
29. $a \to f, \sim a \to \sim f, f \to a$. If you don't live alone your food bills aren't low. If your food bills are low you live alone.
31. $s \to w, \sim s \to \sim w, w \to s$. If you don't swim, then you don't like water. If you like water then you swim.
33. $\sim m \to \sim s, m \to s, \sim s \to \sim m$. If you do well on the MCATs you will get into medical school. If you didn't get into medical school you didn't do well on the MCATs.
35. If weeds don't grow quickly they are attended.
37. The waiter will get a tip and the service is inadequate or the service is adequate and the waiter didn't get a tip.
39. We will use the motor and pick up the oars, or we will not pick up the oars and not use the motor

Chapter 1 Review

1. f 2. d 3. c 4. a 5. i 6. b 7. j 8. g 9. h 10. e
11. Commutative 12. Associative 13. Distributive 14. Biconditional

15. Conditional Negation
16. DeMorgan's
17. Conditional
18. DeMorgan's
19. Conditional
20. Distributive
21. Double Negation
22. Contrapositve
23. Commutative
24. I don't play field hockey or I play ice hockey.
25. Latesha is happy and she won't smile.
26. Margorie will play the drums and Juan isn't ill or Juan is ill and Margorie won't play the drums.
27. I won't take statistics and I won't take calculus.
28. We will play poker and I have a business trip or I don't have a business trip and we will not play poker.
29. The computer isn't cheap or the flat screen monitor isn't expensive.
30. The answer to part three is true but the answer to part four is false.
31. I do a chemistry experiment and it smells like rotten eggs but I didn't use sulfur dioxide.
32. The lawn will grow and it is not seeded or it was not watered, or the lawn is seeded and watered but it will not grow.
33. I had the salad and main course and I will be able to eat dessert.
34. It is a sentence and it doesn't have subject or it doesn't have a verb.
35. You don't stay in school and you become an accountant.
36. Computers are not fast or not accurate or they can think.
37. In her psychology class, she will neither study Skinner nor Freud.
38. This is a simple statement and therefore cannot be expressed as the negation of either a conjunction or a disjunction.
39. I'm not going to stay in NY but I will leave and not relocate to LA.
40. I'll stay in school and either I won't study chemistry or I won't work in the lab.
41. Native Americans didn't hunt buffalo for their skins or they didn't hunt buffalo for their meat, and they were not cold or they were not hungry.
42. Working after school is fun or I don't make money.
43. A flag is waved and a parade is in progress and a band is not playing.
44. If she doesn't want to be a doctor then she wants to be a lawyer.
45. They will not visit or not stay two weeks.
46. If she doesn't have a pedicure then she won't have a manicure.
47. You don't feed the dog or she will not bite.
48. Money is both an evil and a necessity.
49. It is not true that calculators have not made statistics easier or you do not have to do the homework.
50. If the number is not two then the number is either not even or not prime.
51. The enemy is not supplied with missiles and they are not defeated.
52. Either you use the Internet and your research skills will be enhanced, or you use the Internet and you will learn a great deal.
53. If I go to the play then the tickets are not expensive and if the tickets are not expensive then I will go to the play.
54. $\sim (h \wedge g) \rightarrow \sim c$ If I don't wear my hat and gloves it isn't cold outside.
55. $\sim c \vee (h \wedge g)$ It isn't cold outside or I wear my hat and gloves.
56. $c \wedge \sim (h \wedge g)$ It is cold outside but I did not wear my hat and gloves.
57. $\sim c \rightarrow \sim (h \wedge g)$ If it isn't cold outside I will not wear my hat and gloves .
58. $(h \wedge g) \rightarrow c$ If I wear my hat and gloves it is cold outside .

Sample Exam: Chapter 1

1. e	2. b	3. c	4. b	5. a	6. d	7. b	8. c	9. a	10. b
11. a	12. d	13. c	14. e	15. a	16. b	17. b	18. a	19. a	20. c

Answers for Chapter 2

In-Class Excercises and Problems for Section 2.1

In-Class Exercises

1. Invalid
2. Invalid
3. Invalid
4. Valid
5. Invalid
6. Invalid
7. Valid
8. Valid
9. Valid
10. Invalid

Problems

11. Valid
13. Valid
15. Invalid
17. Valid
19. Invalid

In-Class Excercises and Problems for Section 2.2

In-Class Excercises

1. Valid

2. Invalid

p	q	r
T	T	T
F	T	T

3. Invalid

r	s	w
F	T	F

4. Invalid

p	r	s
T	T	F

5. Valid

6. Invalid

p	q	r
F	F	T

7. Invalid

a	b	c
F	T	F

8. Valid

9. Invalid

a	b	c	d
F	T	T	F

10. Valid

Problems

11. Invalid

r	s
F	T

13. Invalid

c	k	p
F	F	F

15. Valid

17. Valid

19. Valid

21. Valid

23. Invalid

p	q	s
T	T	F

25. Valid

In-Class Excercises and Problems for Section 2.3

In-Class Exercises

1. T 2. CBD 3 T 4. F 5. T 6. T

7. F 8. F 9. F 10. T 11. T

12. Valid

13. Valid

14. Invalid

a	b	c
F	F	T

15. Invalid

c	m	n	r
F	T	F	F

16. Invalid

p	q	r	s
F	T	T	F

17. Valid

18. Invalid

e	p	s	w
F	T	T	T

19. Valid

20. Valid

21. Invalid

c	p	r	s	w
F	T	F	T	T

22. Valid

$$\begin{array}{l} \sim w \\ r \rightarrow w \\ \hline \sim r \end{array}$$

23. Invalid

$$\begin{array}{l} r \rightarrow p \\ r \vee i \\ \sim i \\ \hline \sim p \end{array}$$

24. Valid

$$\begin{array}{l} q \rightarrow \sim i \\ q \vee s \\ \sim s \\ \hline \sim i \end{array}$$

25. Invalid

$$\begin{array}{l} b \leftrightarrow f \\ \sim f \wedge m \\ \hline m \rightarrow b \end{array}$$

26. Valid

$$\begin{array}{l} m \vee o \\ o \rightarrow (s \wedge e) \\ \sim s \\ \hline m \end{array}$$

Problems

27. T 29. F 31. CBD 33. T 35. T 37. No

39. Yes 41. No 43. No 45. No

47. Valid

49. Valid

51. Invalid

a	b	p	w
T	F	T	F

53. Valid

55. $s \to m$ Valid

$\sim s \to e$

$w \wedge \sim m$

e

57. $\sim (h \wedge s) \to o$ Valid

$\sim h$

o

59. $\sim a \vee o$ Invalid

$o \to i$

$\sim i$

a

a	*i*	*o*
F	*F*	*F*

In-Class Excercises and Problems for Section 2.4

In-Class Exercises

1. Valid

2. Invalid

p	*q*	*r*	*s*
T	*T*	*T*	*T*
T	*T*	*F*	*T*
T	*F*	*T*	*T*

3. Invalid

q	*r*	*s*	*w*
F	*T*	*T*	*T*
T	*T*	*T*	*T*

4. Valid

5. Invalid

m	*n*	*p*	*q*	*r*
F	*T*	*F*	*F*	*T*

6. Valid

7. Valid

8. Valid

9. Valid

10. Valid

Problems

11. Valid

13. Valid

15. Invalid

b	*c*	*d*	*e*
T	*F*	*T*	*T*
T	*F*	*T*	*F*

17. Valid

19. Invalid

m	*q*	*r*	*s*	*w*
T	*F*	*F*	*T*	*T*

In-Class Excercises and Problems for Section 2.5

In-Class Exercises

1. Valid

2. Valid

3. Invalid

b	s	w
F	T	F

4. Valid

5. Valid

6. Valid

7. Invalid

p	q	r	s
T	T	F	T

8. Valid

9. Valid

10. Valid

11. Invalid

d	m	n	p	r	s
F	T	F	T	T	F

12. Invalid

a	b	c	d	p	x	z
T	F	F	T	F	T	T

13. $m \to h$

$(a \vee b) \to j$

$\sim w \to \sim j$

$h \to a$

$m \to w$

Valid

14. $\sim (s \vee c)$

$(c \vee b) \to (r \to s)$

$b \to r$

Invalid

b	c	r	s
T	F	F	F

15. $w \to (h \wedge m)$

$\sim e \to \sim h$

$w \vee \sim p$

$p \to (e \vee o)$

Valid

16. $\sim c \vee (s \vee w)$

$s \to p$

$\sim w$

$c \to \sim p$

Invalid

c	p	s	w
T	T	T	F

Problems

17. Valid

19. Valid

21. Invalid

a	d	j	m
F	T	F	F

23. Valid

25. Invalid

a	c	d	m	p
T	T	T	T	T

27. Valid

29. $c \to (j \vee d)$

$d \to \sim x$

$m \vee x$

$\sim m$

$c \to \sim j$

Invalid

c	d	j	m	x
T	F	T	F	T

31. $\sim s \to \sim n$

$(s \wedge p) \to r$

$(p \vee g) \to n$

$p \to r$

Valid

In-Class Excercises and Problems for Section 2.6

In-Class Exercises

1. Invalid

m	p	q	r
F	F	F	F

2. Valid

3. Valid

4. Invalid

a	b	c	p
F	T	F	T

5. Invalid

p	q	r	s	w
T	T	T	T	F
T	F	T	T	F

6. Invalid

a	b	c	d	g
T	T	F	F	F

7. Valid 8. Valid 9. Valid 10. Valid

11. $b \to \sim r$
$o \to (p \wedge s)$
$b \vee o$
———
$r \to s$

Valid

12. $e \to q$
$\sim q \vee (p \wedge c)$
$\sim m \to p$
———
$e \to \sim m$

Invalid

c	e	m	p	q
T	T	T	T	T

13. $d \to b$
$b \to \sim s$
$l \to (s \wedge d)$
———
$h \vee \sim l$

Valid

Problems

15. Valid 17. Valid 19. Valid

23. $c \to (l \to a)$
$c \vee p$
$\sim (l \wedge p)$
———
$a \vee \sim l$

Valid

25. $b \to c$
$c \to (p \wedge e)$
$r \to (\sim e \vee w)$
$(b \wedge w) \vee (\sim b \wedge r)$
———
$p \vee w$

Invalid

b	c	e	p	r	w
F	F	F	F	T	F

Chapter 2 Review

1. T 2. T 3. F 4. F 5. CBD

6. T 7. T 8. F 9. CBD 10. F

11. Valid

12. Invalid

a	b	p	q
F	F	F	T

13. Invalid

p	s	w
F	F	T

14. Valid

15. Valid

16. Invalid

a	m	n	p
F	F	F	F

17. Invalid

a	b	p	q	s	w
F	F	T	T	F	F

18. Valid

19. Valid

20. Invalid

a	b	m	r	s	w
T	T	F	T	T	T
T	F	F	T	T	T

21. Valid

22. Valid

23. Invalid

b	p	q	s	w
T	T	F	F	F
T	T	F	F	T
T	F	F	F	T

24. Invalid

a	b	c	p
T	F	T	F
F	F	T	F
F	F	F	F

25. Invalid

a	b	c	p	w
T	F	T	F	F
T	F	T	T	F

26. Valid

27. Valid

28. Valid

29. Invalid

p	q	s	w
F	F	T	T

30. Valid

31. Invalid

p	q	r	s
T	T	T	T

32. Valid

33. Valid

34. Valid

35. Valid

36. Invalid

a	b	c	p	q	r	w
T	T	F	F	T	T	F

37. $d \vee v$ valid

$d \rightarrow s$

$v \rightarrow l$

$\sim l$

---

s

38. $g \wedge w$ Invalid

$w \rightarrow [(d \wedge p) \vee r]$

$p \rightarrow \sim l$

$r \rightarrow \sim l$

d	g	l	p	r	w
F	T	T	F	T	T
T	T	T	F	T	T

39. $r \rightarrow \sim w$ Valid

$(\sim s \vee e) \rightarrow w$

$\sim (s \wedge \sim e)$

$\sim r$

40. $\sim (p \rightarrow s)$ Valid

$b \rightarrow p$

$b \wedge a$

$a \rightarrow \sim s$

41. $s \vee r$ Invalid

$(g \vee b) \rightarrow s$

$\sim g \rightarrow b$

$\sim r$

b	g	r	s
T	F	T	T
F	T	T	T
T	T	T	T

42. $n \rightarrow (a \vee b)$ Invalid

$o \rightarrow \sim a$

$o \vee (n \wedge \sim l)$

$b \rightarrow l$

a	b	l	n	o
F	T	F	T	T
F	T	F	F	T

43. $t \rightarrow p$ Invalid

$d \wedge \sim t$

$p \vee \sim d$

d	p	t
T	F	F

44. $\sim w \vee \sim (e \rightarrow j)$ Valid

$d \rightarrow w$

$\sim e \rightarrow \sim d$

45. $c \rightarrow (d \wedge s)$ Valid

$d \rightarrow (g \wedge \sim b)$

$c \vee e$

$\sim e$

---

g

46. $w \rightarrow b$ Invalid

$(w \wedge r) \rightarrow p$

$p \rightarrow b$

$r \vee \sim b$

$\sim w$

b	p	r	w
T	T	T	T

47. $p \to (g \wedge n)$ Invalid

$n \to t$

$t \to (h \vee d)$

$p \to h$

d	g	h	n	p	t
T	T	F	T	T	T

48. $p \vee i$ Invalid

$p \to (f \wedge s)$

$i \to \sim c$

$\sim c \wedge \sim f$

---

s

c	f	i	p	s
F	F	T	F	F

49. $d \vee r$ Valid

$r \to (l \wedge t)$

$\sim l$

---

d

Sample Exam: Chapter 2

1. T 2. F 3. CBD 4. F 5. T

6.

a	b	c	d	e	g
T	T	F	T	T	F

7.

a	b	p	q	r	s
T	F	F	F	F	T

8.

a	p	q	r	s
F	T	F	T	F

9.

p	q	r	s	w
T	T	F	T	F
F	T	F	T	F

10. Valid 11. Valid 12. Invalid

m	n	p	q	y	z
F	T	T	F	T	F

13. Invalid

p	q	r	s	w
T	T	T	F	F
T	T	F	F	F

14. Valid 15. Invalid

p	r	s	w
F	F	T	F

16. Valid 17. Invalid

a	b	c	d
F	T	T	T
T	T	T	T

18. Valid 19. Invalid

c	d	p	q	r	s	w
T	F	F	F	T	T	T
T	F	T	F	T	F	T
T	F	F	F	T	F	T

Answers for Chapter 3

In-Class Excercises and Problems for Section 3.1

In-Class Exercises

1. MT 2. MP 3. MT 4. MT 5. MP 6. MP 7. MT 8. MT

9.

S	R
1. $\sim a$	P
2. $\sim a \to b$	P
3. b	MP(1)(2)
4. $c \to \sim b$	P
5. $\sim c$	MT(3)(4)

10.

S	R
1. $\sim w$	P
2. $s \to w$	P
3. $\sim s$	MT(1)(2)
4. $\sim r \to s$	P
5. r	MT(3)(4)
6. $r \to p$	P
7. p	MP(5)(6)

11.

a	b	r	w
T	T	F	F
T	F	F	F
F	T	F	F

12.

S	R
1. $\sim (p \vee r)$	P
2. $s \to (p \vee r)$	P
3. $\sim s$	MT(1)(2)
4. $\sim q \to s$	P
5. q	MT(3)(4)

13.

S	R
1. $p \to q$	P
2. $(p \to q) \to (r \to s)$	P
3. $r \to s$	MP(1)(2)
4. $(r \to s) \to (m \to b)$	P
5. $m \to b$	MP(3)(4)

14.

k	p	s	w
F	T	T	F
F	T	F	T
F	T	F	F

15.

c	p	r	s	w
F	F	F	F	T

16.

S	R
1. $\sim p$	P
2. $(r \vee \sim s) \to p$	P
3. $\sim (r \vee \sim s)$	MT(1)(2)
4. $c \to (r \vee \sim s)$	P
5. $\sim c$	MT(3)(4)
6. $\sim c \to d$	P
7. d	MP(5)(6)

17.

S	R
1. $\sim (a \to c)$	P
2. $(p \vee r) \to (a \to c)$	P
3. $\sim (p \vee r)$	MT(1)(2)
4. $\sim (p \vee r) \to s$	P
5. s	MP(3)(4)
6. $s \to (k \wedge g)$	P
7. $k \wedge g$	MP(5)(6)

18.

k	l	p	q	r	w
F	F	F	T	T	F

Problems

19. Reason

1. P
2. P
3. MP(1)(2)
4. P
5. MP(3)(4)

21. Reason

1. P
2. P
3. MT(1)(2)
4. P
5. MP(3)(4)

23. Reason

1. P
2. P
3. MT(1)(2)
4. P
5. MP(3)(4)
6. P
7. MP(5)(6)

25.

S	R
1. $\sim s$	P
2. $\sim r \to s$	P
3. r	MT(1)(2)

27.

S	R
1. $\sim (p \to r)$	P
2. $s \to (p \to r)$	P
3. $\sim s$	MT(1)(2)

29.

S	R
1. $\sim q$	P
2. $r \to q$	P
3. $\sim r$	MT(1)(2)
4. $p \to r$	P
5. $\sim p$	MT(3)(4)

31.

p	q	r
F	F	F

33.

S	R
1. $\sim (p \wedge a)$	P
2. $w \to (p \wedge a)$	P
3. $\sim w$	MT(1)(2)
4. $\sim w \to k$	P
5. k	MP(3)(4)

35.

a	d	g	m	r
T	T	T	F	F
T	T	F	F	F
T	F	T	F	F

37.

S	R
1. $c \vee d$	P
2. $(c \vee d) \rightarrow q$	P
3. q	MP(1)(2)
4. $q \rightarrow \sim r$	P
5. $\sim r$	MP(3)(4)
6. $a \rightarrow r$	P
7. $\sim a$	MT(5)(6)
8. $b \rightarrow a$	P
9. $\sim b$	MT(7)(8)

39.

a	b	m	n	p	q	w
T	F	F	F	T	T	F

41. $(b \vee c) \rightarrow k$

$k \rightarrow \sim d$

$p \rightarrow d$

$\sim p \rightarrow (m \wedge s)$

$b \vee c$

$m \wedge s$

S	R
1. $b \vee c$	P
2. $(b \vee c) \rightarrow k$	P
3. k	MP(1)(2)
4. $k \rightarrow \sim d$	P
5. $\sim d$	MP(3)(4)
6. $p \rightarrow d$	P
7. $\sim p$	MT(5)(6)
8. $\sim p \rightarrow (m \wedge s)$	P
9. $m \wedge s$	MP(7)(8)

43. $h \rightarrow r$

$h \rightarrow w$

$r \rightarrow i$

$\sim i$

$\sim w$

h	i	r	w
F	F	F	T

In-Class Excercises and Problems for Section 3.2

In-Class Excercises

1. DS 2. DS 3. DA 4. DA 5. DS 6. DA 7. MT 8. DS 9. MP 10. DA

11.

S	R
1. a	P
2. $\sim a \vee b$	P
3. b	DS(1)(2)
4. $b \rightarrow c$	P
5. c	MP(3)(4)

12.

S	R
1. r	P
2. $r \rightarrow (s \vee q)$	P
3. $s \vee q$	MP(1)(2)
4. $\sim s$	P
5. q	DS(3)(4)

13.

p	q	r
F	T	F

14.

S	R
1. w	P
2. $w \to c$	P
3. c	MP(1)(2)
4. $c \vee p$	DA(3)
5. $(c \vee p) \to b$	P
6. b	MP(4)(5)

15.

S	R
1. $\sim (q \wedge r)$	P
2. $p \vee (q \wedge r)$	P
3. p	DS(1)(2)
4. $p \to s$	P
5. s	MP(3)(4)

16.

S	R
1. $\sim w$	P
2. $p \to w$	P
3. $\sim p$	MT(1)(2)
4. $\sim p \vee s$	DA(3)
5. $(\sim p \vee s) \to k$	P
6. k	MP(4)(5)

17.

S	R
1. $\sim c$	P
2. $(a \to b) \vee c$	P
3. $a \to b$	DS(1)(2)
4. $\sim b$	P
5. $\sim a$	MT(3)(4)
6. $\sim a \vee p$	DA(5)

18.

l	m	p	w
T	T	F	T
F	T	F	T

19.

S	R
1. b	P
2. $a \to \sim b$	P
3. $\sim a$	MT(1)(2)
4. $c \to a$	P
5. $\sim c$	MT(3)(4)
6. $c \vee d$	P
7. d	DS(5)(6)

20.

S	R
1. p	P
2. $\sim p \vee r$	P
3. r	DS(1)(2)
4. $s \to \sim r$	P
5. $\sim s$	MT(3)(4)
6. $\sim s \to w$	P
7. w	MP(5)(6)

21.

a	d	p	s
F	F	F	T

22.

S	R
1. $\sim q$	P
2. $p \to q$	P
3. $\sim p$	MT(1)(2)
4. $\sim p \to w$	P
5. w	MP(3)(4)
6. $w \vee z$	DA(5)
7. $(w \vee z) \to k$	P
8. k	MP(6)(7)

23.

S	R
1. $\sim q$	P
2. $d \vee q$	P
3. d	DS(1)(2)
4. $\sim c \rightarrow \sim d$	P
5. c	MT(3)(4)
6. $c \vee (p \rightarrow s)$	DA(5)

24.

S	R
1. $r \rightarrow w$	P
2. $(r \rightarrow w) \rightarrow (s \vee g)$	P
3. $s \vee g$	MP(1)(2)
4. $\sim s$	P
5. g	DS(3)(4)
6. $g \vee p$	DA(5)

25.

a	b	c	d
T	T	F	F

26.

S	R
1. s	P
2. $m \vee \sim s$	P
3. m	DS(1)(2)
4. $m \vee q$	DA(3)
5. $n \rightarrow \sim (m \vee q)$	P
6. $\sim n$	MT(4)(5)

27.

S	R
1. s	P
2. $r \vee \sim s$	P
3. r	DS(1)(2)
4. $r \vee w$	DA(3)
5. $(r \vee w) \rightarrow p$	P
6. p	MP(4)(5)
7. $q \rightarrow \sim p$	P
8. $\sim q$	MT(6)(7)

28.

S	R
1. $\sim p \rightarrow q$	P
2. $\sim (\sim p \rightarrow q) \vee [r \rightarrow (s \vee c)]$	P
3. $r \rightarrow (s \vee c)$	DS(1)(2)
4. $\sim (s \vee c)$	P
5. $\sim r$	MT(3)(4)
6. $\sim r \vee w$	DA(5)
7. $(\sim r \vee w) \rightarrow d$	P
8. d	MP(6)(7)

29.

$w \to i$

$(i \vee p) \to \sim s$

$s \vee c$

w

---

c

S	R
1. w	P
2. $w \to i$	P
3. i	MP(1)(2)
4. $i \vee p$	DA(3)
5. $(i \vee p) \to \sim s$	P
6. $\sim s$	MP(4)(5)
7. $s \vee c$	P
8. c	DS(6)(7)

31.

$e \to c$

$c \to \sim w$

$w \vee l$

$\sim l$

$\sim e$

S	R
1. $\sim l$	P
2. $w \vee l$	P
3. w	DS(1)(2)
4. $c \to \sim w$	P
5. $\sim c$	MT(3)(4)
6. $e \to c$	P
7. $\sim e$	MT(5)(6)

Problems

33.

Reason
1. P
2. P
3. MP(1)(2)
4. DA(3)
5. P
6. MP(4)(5)

35.

Reason
1. P
2. P
3. MP(1)(2)
4. P
5. MT(3)(4)
6. P
7. DS (5)(6)
8. DA (7)

37.

S	R
1. $\sim q$	P
2. $q \vee \sim p$	P
3. $\sim p$	DS(1)(2)
4. $\sim r \to p$	P
5. r	MT(3)(4)

39.

S	R
1. c	P
2. $c \to m$	P
3. m	MP(1)(2)
4. $m \vee p$	DA(3)
5. $(m \vee p) \to s$	P
6. s	MP(4)(5)

41.

S	R
1. s	P
2. $s \to a$	P
3. a	MP(1)(2)
4. $a \vee b$	DA(3)
5. $(a \vee b) \to c$	P
6. c	MP(4)(5)

43.

S	R
1. w	P
2. $w \to (p \to q)$	P
3. $p \to q$	MP(1)(2)
4. p	P
5. q	MP(3)(4)
6. $q \vee r$	DA(5)

45.

S	R
1. n	P
2. $m \vee \sim n$	P
3. m	DS(1)(2)
4. $m \to (p \vee r)$	P
5. $p \vee r$	MP(3)(4)
6. $\sim r$	P
7. p	DS(5)(6)

47.

S	R
1. a	P
2. $a \vee p$	DA(1)
3. $(a \vee p) \to k$	P
4. k	MP(2)(3)
5. $k \to s$	P
6. s	MP(4)(5)
7. s $\vee q$	DA(6)
8. $(s \vee q) \to w$	P
9. w	MP(7)(8)

49.

S	R
1. $\sim a$	P
2. $(b \to c) \vee a$	P
3. $b \to c$	DS(1)(2)
4. b	P
5. c	MP(3)(4)
6. $c \vee w$	DA(5)
7. $(c \vee w) \to p$	P
8. p	MP(6)(7)

51.

S	R
1. $\sim s$	P
2. $s \vee w$	P
3. w	DS(1)(2)
4. $w \to a$	P
5. a	MP(3)(4)
6. $a \vee b$	DA(5)
7. $(a \vee b) \to p$	P
8. p	MP(6)(7)
9. $p \vee q$	DA(8)

53.

S	R
1. $\sim s$	P
2. $b \to s$	P
3. $\sim b$	MT(1)(2)
4. $\sim p$	P
5. $\sim p \to (a \vee b)$	P
6. $a \vee b$	MP(4)(5)
7. a	DS(3)(6)
8. $a \to q$	P
9. q	MP(7)(8)

55.

S	R
1. $\sim (m \leftrightarrow n)$	P
2. $(m \leftrightarrow n) \vee r$	P
3. r	DS(1)(2)
4. $r \to p$	P
5. p	MP(3)(4)
6. $\sim (a \to m) \to \sim p$	P
7. $a \to m$	MT(5)(6)
8. a	P
9. m	MP(7)(8)

57. $b \to w$
$s \vee \sim w$
$s \vee i$
b
———
i

S	R
1. b	P
2. $b \to w$	P
3. w	MP(1)(2)
4. $s \vee \sim w$	P
5. $\sim s$	DS(3)(4)
6. $s \vee i$	P
7. i	DS(5)(6)

59. $h \vee \sim p$
$p \to s$
$\sim s$
———
$\sim h$

h	p	s
T	F	F

61. $m \to (n \vee c)$
$n \to g$
$\sim g$
m
———
$c \vee s$

S	R
1. m	P
2. $m \to (n \vee c)$	P
3. $n \vee c$	MP(1)(2)
4. $\sim g$	P
5. $n \to g$	P
6. $\sim n$	MT(4)(5)
7. c	DS(3)(6)
8. $c \vee s$	DA(7)

In-Class Excercises and Problems for Section 3.3

In-Class Exercises

1. CS 2. CA 3. CA 4. CS 5. DA 6. CA 7. DS 8. MT 9. CS 10. DA 11. MP 12. MT

13.

S	R
1. p	P
2. $\sim p \vee s$	P
3. s	DS(1)(2)
4. $p \wedge s$	CA(1)(3)
5. $(p \wedge s) \to q$	P
6. q	MP(4)(5)

14.

b	p	q	r
T	T	F	T

15.

S	R
1. $m \wedge n$	P
2. m	CS(1)
3. $\sim m \vee p$	P
4. p	DS(2)(3)
5. $p \to w$	P
6. w	MP(4)(5)

16.

S	R
1. $s \wedge g$	P
2. s	CS(1)
3. $s \to (a \vee w)$	P
4. $a \vee w$	MP(2)(3)
5. $d \to \sim (a \vee w)$	P
6. $\sim d$	MT(4)(5)

17.

S	R
1. p	P
2. $\sim p \vee (\sim s \to \sim w)$	P
3. $\sim s \to \sim w$	DS(1)(2)
4. $s \to \sim p$	P
5. $\sim s$	MT(1)(4)
6. $\sim w$	MP(3)(5)
7. $\sim s \wedge \sim w$	CA(5)(6)

18.

S	R
1. $a \wedge b$	P
2. a	CS(1)
3. $a \to p$	P
4. p	MP(2)(3)
5. b	CS(1)
6. $b \wedge p$	CA(4)(5)
7. $(b \wedge p) \to s$	P
8. s	MP(6)(7)

19.

p	r	s	w
T	T	T	T

20.

S	R
1. $a \wedge k$	P
2. k	CS(1)
3. $k \to \sim g$	P
4. $\sim g$	MP(2)(3)
5. $g \vee \sim h$	P
6. $\sim h$	DS(4)(5)
7. a	CS(1)
8. $a \wedge \sim h$	CA(6)(7)

21.

S	R
1. $\sim a \wedge b$	P
2. $\sim a$	CS(1)
3. $a \vee c$	P
4. c	DS(2)(3)
5. $d \to \sim c$	P
6. $\sim d$	MT(4)(5)
7. $\sim d \to e$	P
8. e	MP(6)(7)

22.

S	R
1. $\sim b$	P
2. $s \rightarrow b$	P
3. $\sim s$	MT(1)(2)
4. $s \vee c$	P
5. c	DS(3)(4)
6. $c \wedge \sim s$	CA(3)(5)
7. $(c \wedge \sim s) \rightarrow k$	P
8. k	MP(6)(7)

23.

S	R
1. s	P
2. $r \vee \sim s$	P
3. r	DS(1)(2)
4. $r \rightarrow (p \wedge q)$	P
5. $p \wedge q$	MP(3)(4)
6. p	CS(5)
7. $\sim w \rightarrow \sim p$	P
8. w	MT(6)(7)
9. $w \vee c$	DA(8)

24.

S	R
1. $c \wedge g$	P
2. c	CS(1)
3. $c \rightarrow \sim j$	P
4. $\sim j$	MP(2)(3)
5. $j \vee p$	P
6. p	DS(4)(5)
7. $p \rightarrow s$	P
8. s	MP(6)(7)
9. g	CS(1)
10. $s \wedge g$	CA(8)(9)

25.

S	R
1. k	P
2. $c \vee \sim k$	P
3. c	DS(1)(2)
4. $c \rightarrow (a \wedge b)$	P
5. $a \wedge b$	MP(3)(4)
6. b	CS(5)
7. $b \vee z$	DA(6)
8. $(b \vee z) \rightarrow m$	P
9. m	MP(7)(8)

26.

S	R
1. $p \wedge b$	P
2. p	CS(1)
3. $p \rightarrow q$	P
4. q	MP(2)(3)
5. $r \rightarrow \sim q$	P
6. $\sim r$	MT(4)(5)
7. $s \vee r$	P
8. s	DS(6)(7)
9. $s \vee w$	DA(8)
10. $(s \vee w) \rightarrow m$	P
11. m	MP(9)(10)

27.

S	R
1. $\sim p$	P
2. $(\sim a \wedge s) \vee p$	P
3. $\sim a \wedge s$	DS(1)(2)
4. $\sim a$	CS(3)
5. $\sim a \rightarrow j$	P
6. j	MP(4)(5)
7. $j \vee c$	DA(6)
8. $(j \vee c) \rightarrow d$	P
9. d	MP(7)(8)
10. $d \rightarrow k$	P
11. k	MP(9)(10)
12. s	CS(3)
13. $k \wedge s$	CA(11)(12)

28.

a	b	c	d	e	p
T	F	T	T	T	T

29. $c \vee s$

$s \rightarrow b$

$(c \wedge \sim s) \rightarrow m$

$\sim b$

---

m

S	R
1. $\sim b$	P
2. $s \rightarrow b$	P
3. $\sim s$	MT(1)(2)
4. $c \vee s$	P
5. c	DS(3)(4)
6. $c \wedge \sim s$	CA(3)(5)
7. $(c \wedge \sim s) \rightarrow m$	P
8. m	MP(6)(7)

30. $t \rightarrow p$

$p \rightarrow (c \wedge d)$

$d \rightarrow (r \vee u)$

$t \wedge \sim u$

$c \wedge r$

S	R
1. $t \wedge \sim u$	P
2. t	CS(1)
3. $t \rightarrow p$	P
4. p	MP(2)(3)
5. $p \rightarrow (c \wedge d)$	P
6. $c \wedge d$	MP(4)(5)
7. d	CS(6)
8. $d \rightarrow (r \vee u)$	P
9. $r \vee u$	MP(7)(8)
10. $\sim u$	CS(1)
11. r	DS(9)(10)
12. c	CS(6)
13. $c \wedge r$	CA(11)(12)

31. $(g \vee m) \wedge \sim e$

$\sim h \rightarrow \sim m$

$s \rightarrow h$

$\sim g$

---

s

e	g	h	m	s
F	F	T	T	F

Problems

33. Reason

1. P
2. P
3. MP(1)(2)
4. CS(3)
5. CS(3)
6. P
7. DS(4)(6)
8. CA(5)(7)

35. Reason

1. P
2. P
3. DS(1)(2)
4. CS(3)
5. CS(3)
6. P
7. MT(5)(6)
8. CA(4)(7)
9. P
10. MP(8)(9)
11. CA(4)(10)

37.

S	R
1. m	P
2. n	P
3. $m \wedge n$	CA(1)(2)
4. $\sim (m \wedge n) \vee q$	P
5. q	DS(3)(4)

39.

S	R
1. $p \wedge q$	P
2. q	CS(1)
3. $q \rightarrow s$	P
4. s	MP(2)(3)
5. $w \rightarrow \sim s$	P
6. $\sim w$	MT(4)(5)

41.

S	R
1. w	P
2. $w \to r$	P
3. r	MP(1)(2)
4. $\sim (a \wedge c) \to \sim r$	P
5. $a \wedge c$	MT(3)(4)
6. a	CS(5)

43.

S	R
1. s	P
2. $s \vee q$	DA(1)
3. $(s \vee q) \to p$	P
4. p	MP(2)(3)
5. $p \wedge s$	CA(1)(4)
6. $\sim (p \wedge s) \vee a$	P
7. a	DS(5)(6)

45.

S	R
1. $(\sim p \to \sim q) \wedge (\sim r \vee s)$	P
2. $\sim p \to \sim q$	CS(1)
3. q	P
4. p	MT(2)(3)
5. $\sim r \vee s$	CS(1)
6. r	P
7. s	DS(5)(6)
8. $p \wedge s$	CA(4)(7)

47.

a	p	q	s
T	T	T	T

49.

S	R
1. $(p \vee q) \wedge w$	P
2. $p \vee q$	CS(1)
3. $\sim p$	P
4. q	DS(2)(3)
5. $r \to \sim q$	P
6. $\sim r$	MT(4)(5)
7. $\sim r \to z$	P
8. z	MP(6)(7)

51.

S	R
1. $q \wedge r$	P
2. r	CS(1)
3. $\sim p \to \sim r$	P
4. p	MT(2)(3)
5. $p \vee s$	DA(4)
6. $(p \vee s) \to w$	P
7. w	MP(5)(6)

53.

S	R
1. $w \wedge s$	P
2. w	CS(1)
3. $\sim w \vee (a \wedge b)$	P
4. $a \wedge b$	DS(2)(3)
5. b	CS(4)
6. s	CS(1)
7. $b \wedge s$	CA(5)(6)
8. $(b \wedge s) \to q$	P
9. q	MP(7)(8)
10. $q \vee z$	DA(9)
11. $(q \vee z) \to c$	P
12. c	MP(10)(11)

55.

S	R
1. $\sim p \wedge r$	P
2. $\sim p$	CS(1)
3. $\sim p \to q$	P
4. q	MP(2)(3)
5. $\sim q \vee s$	P
6. s	DS(4)(5)
7. r	CS(1)
8. $s \wedge r$	CA(6)(7)
9. $(s \wedge r) \to w$	P
10. w	MP(8)(9)
11. $w \vee b$	DA(10)
12. $(w \vee b) \to c$	P
13. c	MP(11)(12)

57.

Argument	S	R
$j \to (c \wedge \sim t)$		
$j \vee a$	1. $\sim a$	P
$c \to l$	2. $j \vee a$	P
$\sim a$	3. j	DS(1)(2)
——— $l \vee s$	4. $j \to (c \wedge \sim t)$	P
	5. $c \wedge \sim t$	MP(3)(4)
	6. c	CS(5)
	7. $c \to l$	P
	8. l	MP(6)(7)
	9. $l \vee s$	DA(8)

59.

Argument	S	R
$d \to c$		
$d \vee h$	1. $s \wedge i$	P
$s \to \sim h$	2. s	CS(1)
$s \wedge i$	3. $s \to \sim h$	P
——— $i \wedge c$	4. $\sim h$	MP(2)(3)
	5. $d \vee h$	P
	6. d	DS(4)(5)
	7. $d \to c$	P
	8. c	MP(6)(7)
	9. i	CS(1)
	10. $i \wedge c$	CA(8)(9)

In-Class Exercises and Problems for Section 3.4
In-Class Exercises

1. CN	2. CE	3. DM	4. DA	5. CP	6. MT	7. DE
8. DS	9. DM	10. MP	11. CS	12. DM	13. CE	14. CN

15.

S	R
1. $\sim(p \vee q)$	P
2. $\sim p \wedge \sim q$	DM(1)
3. $\sim p$	CS(2)
4. $\sim p \to s$	P
5. s	MP(3)(4)

16.

S	R
1. b	P
2. $b \to (a \to c)$	P
3. $a \to c$	MP(1)(2)
4. $\sim a \vee c$	CE(3)
5. $\sim(\sim a \vee c) \vee n$	P
6. n	DS(4)(5)

17.

S	R
1. $\sim(s \to w)$	P
2. $s \wedge \sim w$	CN(1)
3. $\sim w$	CS(2)
4. $w \vee p$	P
5. p	DS(3)(4)
6. $a \to \sim p$	P
7. $\sim a$	MT(5)(6)

18.

S	R
1. $a \vee \sim b$	P
2. $\sim(\sim a \wedge b)$	DM(1)
3. $s \to (\sim a \wedge b)$	P
4. $\sim s$	MT(2)(3)
5. $\sim s \to c$	P
6. c	MP(4)(5)

19.

S	R
1. $\sim p \to m$	P
2. $\sim m \to p$	CP(1)
3. $(\sim m \to p) \to q$	P
4. q	MP(2)(3)
5. $a \vee \sim q$	P
6. a	DS(4)(5)
7. $a \vee b$	DA(6)

20.

S	R
1. $d \vee c$	P
2. $\sim d \to c$	CE(1)
3. $s \to \sim(\sim d \to c)$	P
4. $\sim s$	MT(2)(3)
5. $\sim s \vee p$	DA(4)
6. $(\sim s \vee p) \to k$	P
7. k	MP(5)(6)

21.

S	R
1. $\sim(w \to s)$	P
2. $w \wedge \sim s$	CN(1)
3. $\sim s$	CS(2)
4. $s \vee p$	P
5. p	DS(3)(4)
6. $p \to (a \to \sim b)$	p
7. $a \to \sim b$	MP(5)(6)
8. $b \to \sim a$	CP(7)
9. $r \to \sim (b \to \sim a)$	P
10. $\sim r$	MT(8)(9)

22.

S	R
1. $\sim (p \vee \sim w)$	P
2. $\sim p \wedge w$	DM(1)
3. w	CS(2)
4. $w \to s$	P
5. s	MP(3)(4)
6. $s \vee b$	DA(5)
7. $m \to \sim (s \vee b)$	P
8. $\sim m$	MT(6)(7)
9. $m \vee (q \to d)$	P
10. $q \to d$	DS(8)(9)
11. $\sim q \vee d$	CE(10)

23.

S	R
1. $(\sim p \vee q) \wedge s$	P
2. $\sim p \vee q$	CS(1)
3. $p \to q$	CE(2)
4. $(p \to q) \to \sim r$	P
5. $\sim r$	MP(3)(4)
6. $r \vee c$	P
7. c	DS(5)(6)

24.

a	b	p	w
T	T	T	F

25.

S	R
1. $\sim p \vee a$	P
2. $\sim (p \wedge \sim a)$	DM(1)
3. $r \to (p \wedge \sim a)$	P
4. $\sim r$	MT(2)(3)
5. $\sim r \to q$	P
6. q	MP(4)(5)

26.

S	R
1. $r \wedge (\sim q \to k)$	P
2. $\sim q \to k$	CS(1)
3. $\sim k \to q$	CP(2)
4. $\sim (\sim k \to q) \vee p$	P
5. p	DS(3)(4)
6. $p \vee w$	DA(5)
7. $(p \vee w) \to n$	P
8. n	MP(6)(7)
9. r	CS(1)
10. $r \wedge n$	CA(8)(9)

27.

S	R
1. $\sim (\sim s \to r)$	P
2. $\sim s \wedge \sim r$	CN(1)
3. $\sim r$	CS(2)
4. $r \vee (m \to n)$	P
5. $m \to n$	DS(3)(4)
6. $\sim (d \vee n)$	P
7. $\sim d \wedge \sim n$	DM(6)
8. $\sim n$	CS(7)
9. $\sim m$	MT(5)(8)

28.

S	R
1. $c \wedge (\sim p \vee w)$	P
2. $\sim p \vee w$	CS(1)
3. $p \to w$	CE(2)
4. $(p \to w) \to s$	P
5. s	MP(3)(4)
6. c	CS(1)
7. $s \wedge c$	CA(5)(6)
8. $\sim r \to \sim (s \wedge c)$	P
9. r	MT(7)(8)

29.

S	R
1. $g \to \sim k$	P
2. $\sim g \vee \sim k$	CE(1)
3. $(\sim k \vee \sim g) \to p$	P
4. p	MP(2)(3)
5. $(w \vee \sim s) \to \sim p$	P
6. $\sim (w \vee \sim s)$	MT(4)(5)
7. $\sim w \wedge s$	DM(6)
8. s	CS(7)

30.

a	b	p	r	w
F	F	T	F	T
F	F	F	F	T

31.

S	R
1. $p \to q$	P
2. $\sim p \vee q$	CE(1)
3. $(\sim p \vee q) \to w$	P
4. w	MP(2)(3)
5. $(s \vee \sim r) \to \sim w$	P
6. $\sim (s \vee \sim r)$	MT(4)(5)
7. $\sim s \wedge r$	DM(6)
8. r	CS(7)
9. $\sim r \vee c$	P
10. c	DS(8)(9)

32.

S	R
1. $\sim (w \rightarrow \sim s)$	P
2. $w \wedge s$	CN(1)
3. w	CS(2)
4. $\sim w \vee p$	P
5. p	DS(3)(4)
6. s	CS(2)
7. $p \wedge s$	CA(5)(6)
8. $(p \wedge s) \rightarrow q$	P
9. q	MP(7)(8)
10. $q \rightarrow (\sim a \vee b)$	P
11. $\sim a \vee b$	MP(9)(10)
12. $a \rightarrow b$	CE(11)

33.

a	b	c	h	p	s
T	T	T	F	T	T
T	T	F	F	T	T

34.

S	R
1. $(a \wedge \sim b) \vee (a \wedge c)$	P
2. $a \wedge (\sim b \vee c)$	DE(1)
3. a	CS(2)
4. $\sim b \vee c$	CS(2)
5. $b \rightarrow c$	CE(4)
6. $(b \rightarrow c) \rightarrow q$	P
7. q	MP(5)(6)
8. $(w \vee \sim s) \rightarrow \sim q$	P
9. $\sim (w \vee \sim s)$	MT(7)(8)
10. $\sim w \wedge s$	DM(9)
11. s	CS(10)
12. $s \wedge a$	CA(3)(11)
13. $(s \wedge a) \rightarrow p$	P
14. p	MP(12)(13)

35. $\sim (h \rightarrow \sim s)$

$(h \wedge s) \rightarrow c$

$\sim c \vee i$

---

i

S	R
1. $\sim (h \rightarrow \sim s)$	P
2. $h \wedge s$	CN(1)
3. $(h \wedge s) \rightarrow c$	P
4. c	MP(2)(3)
5. $\sim c \vee i$	P
6. i	DS(4)(5)

36.

$r \to (n \wedge \sim b)$
$\sim r \to (o \vee m)$
$\sim(\sim m \to o)$
―――――
$n \vee d$

	S	R
1.	$\sim(\sim m \to o)$	P
2.	$\sim m \wedge \sim o$	CN(1)
3.	$\sim(m \vee o)$	DM(2)
4.	$\sim r \to (o \vee m)$	P
5.	r	MT(3)(4)
6.	$r \to (n \wedge \sim b)$	P
7.	$n \wedge \sim b$	MP(5)(6)
8.	n	CS(7)
9.	$n \vee d$	DA(8)

37.

$o \to (n \wedge s)$
$\sim(r \to b)$
$r \to n$
$b \vee s$
―――――
o

b	n	o	r	s
F	T	F	T	T

Problems

39.

Reason
1. P
2. CN(1)
3. CS(2)
4. CS(2)
5. P
6. DS(4)(5)
7.CA(3)(6)
8.P
9.MP(7)(8)

41.

Reason
1. P
2. CP(1)
3.P
4. MT(2)(3)
5. P
6. DS(4)(5)
7.CE(6)

43.

Reason
1. P
2. CE(1)
3. CP(2)
4.P
5. MP(3)(4)
6. CN(5)
7.CS(6)
8. CS(6)
9. DA(7)
10.P
11.DS(9)(10)
12.CA(8)(11)

45.

	S	R
1.	$m \vee q$	P
2.	$\sim m \to q$	CE(1)
3.	$(\sim m \to q) \to p$	P
4.	p	MP(2)(3)
5.	$n \to \sim p$	P
6.	$\sim n$	MT(4)(5)

47.

	S	R
1.	$q \to \sim p$	P
2.	$p \to \sim q$	CP(1)
3.	$\sim(p \to \sim q) \vee w$	P
4.	w	DS(2)(3)
5.	$\sim r \to \sim w$	P
6.	r	MT(4)(5)

49.

	S	R
1.	$\sim p \vee \sim w$	P
2.	$\sim(p \wedge w)$	DM(1)
3.	$r \to (p \wedge w)$	P
4.	$\sim r$	MT(2)(3)
5.	$r \vee s$	P
6.	s	DS(4)(5)

51.

S	R
1. $g \to \sim a$	P
2. $a \to \sim g$	CP(1)
3. $(p \vee \sim r) \to \sim (a \to \sim g)$	P
4. $\sim (p \vee \sim r)$	MT(2)(3)
5. $\sim p \wedge r$	DM(4)
6. r	CS(5)
7. $r \to s$	P
8. s	MP(6)(7)
9. $s \vee w$	DA(8)
10. $(s \vee w) \to m$	P
11. m	MP(9)(10)

53.

S	R
1. $\sim (a \to b)$	P
2. $a \wedge \sim b$	CN(1)
3. $\sim (\sim a \vee b)$	DM(2)
4. $w \vee (\sim a \vee b)$	P
5. w	DS(3)(4)
6. $w \vee p$	DA(5)
7. $(w \vee p) \to c$	P
8. c	MP(6)(7)

55.

S	R
1. $\sim p \vee s$	P
2. $\sim (p \wedge \sim s)$	DM(1)
3. $m \to (p \wedge \sim s)$	P
4. $\sim m$	MT(2)(3)
5. $\sim m \vee a$	DA(4)
6. $(\sim m \vee a) \to w$	P
7. w	MP(5)(6)

57.

S	R
1. $d \to \sim n$	P
2. $n \to \sim d$	CP(1)
3. $\sim (n \to \sim d) \vee w$	P
4. w	DS(2)(3)
5. $w \to (a \wedge b)$	P
6. $a \wedge b$	MP(4)(5)
7. b	CS(6)
8. $b \vee c$	DA(7)

59.

a	q	r	s	w
F	T	T	T	F
T	T	T	T	F
T	F	T	T	F

61.

S	R
1. $s \wedge \sim w$	P
2. $\sim (s \to w)$	CN(1)
3. $\sim (c \wedge p) \to (s \to w)$	P
4. $c \wedge p$	MT(2)(3)
5. c	CS(4)
6. s	CS(1)
7. $c \wedge s$	CA(5)(6)
8. $(c \wedge s) \to n$	P
9. n	MP(7)(8)

63.

S	R
1. $\sim r \to s$	P
2. $r \vee s$	CE(1)
3. $(r \vee s) \to (w \to \sim a)$	P
4. $w \to \sim a$	MP(2)(3)
5. $a \to \sim w$	CP(4)
6. $(a \to \sim w) \to \sim b$	P
7. $\sim b$	MP(5)(6)
8. $p \wedge q$	P
9. p	CS(8)
10. $p \wedge \sim b$	CA(7)(9)

65.

S	R
1. $\sim (a \to b)$	P
2. $a \wedge \sim b$	CN(1)
3. $(a \wedge \sim b) \to c$	P
4. c	MP(2)(3)
5. $c \vee q$	DA(4)
6. $(c \vee q) \to (w \to \sim s)$	P
7. $w \to \sim s$	MP(5)(6)
8. s	P
9. $\sim w$	MT(7)(8)
10. a	CS(2)
11. $a \wedge \sim w$	CA(9)(10)

67. $r \vee h$

$\sim (m \rightarrow \sim v)$

$(m \vee p) \rightarrow r$

---

h

h	m	p	r	v
F	T	T	T	T
F	T	F	T	T

69. $(b \wedge g) \vee (b \wedge h)$

$b \rightarrow (i \rightarrow e)$

$i \wedge \sim h$

$e \wedge g$

S	R
1. $(b \wedge g) \vee (b \wedge h)$	P
2. $b \wedge (g \vee h)$	DE(1)
3. b	CS(2)
4. $b \rightarrow (i \rightarrow e)$	P
5. $i \rightarrow e$	MP(3)(4)
6. $i \wedge \sim h$	P
7. i	CS(6)
8. e	MP(5)(7)
9. $\sim h$	CS(6)
10. $g \vee h$	CS(2)
11. g	DS(9)(10)
12. $e \wedge g$	CA(8)(11)

In-Class Exercises and Problems for Section 3.5
In-Class Exercises

1.

S	R
1. c	ACP
2. $c \rightarrow (d \wedge g)$	P
3. $d \wedge g$	MP(1)(2)
4. g	CS(3)
5. $c \rightarrow g$	CP(1)(4)

2.

S	R
1. r	ACP
2. $\sim p \rightarrow \sim r$	P
3. p	MT(1)(2)
4. $\sim p \vee q$	P
5. q	DS(3)(4)
6. $r \rightarrow q$	CP(1)(5)

3.

S	R
1. $\sim r$	ACP
2. $r \vee s$	P
3. s	DS(1)(2)
4. $\sim a \vee \sim s$	P
5. $\sim a$	DS(3)(4)
6. $\sim r \rightarrow \sim a$	CP(1)(5)

4.

h	s	w
F	F	T

5.

S	R
1. $\sim k$	ACP
2. $\sim k \rightarrow (p \wedge b)$	P
3. $p \wedge b$	MP(1)(2)
4. b	CS(3)
5. $\sim b \vee a$	P
6. a	DS(4)(5)
7. $\sim k \rightarrow a$	CP(1)(6)

6.

S	R
1. h	ACP
2. $h \vee p$	DA(1)
3. $\sim (h \vee p) \vee l$	P
4. l	DS(2)(3)
5. $s \rightarrow \sim l$	P
6. $\sim s$	MT(4)(5)
7. $h \rightarrow \sim s$	CP(1)(6)

7.

S	R
1. a	ACP
2. p	P
3. $p \wedge a$	CA(1)(2)
4. $(p \wedge a) \rightarrow n$	P
5. n	MP(3)(4)
6. $\sim n \vee (c \wedge d)$	P
7. $c \wedge d$	DS(5)(6)
8. c	CS(7)
9. $a \rightarrow c$	CP(1)(8)

8.

p	q	r	s
T	T	F	T

9.

S	R
1. r	ACP
2. $r \vee s$	DA(1)
3. $(r \vee s) \rightarrow q$	P
4. q	MP(2)(3)
5. $\sim p \rightarrow \sim q$	P
6. p	MT(4)(5)
7. $p \wedge r$	CA(1)(6)
8. $(p \wedge r) \rightarrow w$	P
9. w	MP(7)(8)
10. $r \rightarrow w$	CP(1)(9)

10.

S	R
1. p	ACP
2. $p \vee q$	DA(1)
3. $(p \vee q) \rightarrow w$	P
4. w	MP(2)(3)
5. a	P
6. $a \rightarrow (w \rightarrow s)$	P
7. $w \rightarrow s$	MP(5)(6)
8. s	MP(4)(7)
9. $\sim s \vee r$	P
10. r	DS(8)(9)
11. $p \rightarrow r$	CP(1)(10)

11.

S	R
1. d	P
2. $\sim d \vee p$	P
3. p	DS(1)(2)
4. $p \rightarrow (\sim m \rightarrow n)$	P
5. $\sim m \rightarrow n$	MP(3)(4)
6. $\sim n$	ACP
7. m	MT(5)(6)
8. $m \wedge d$	CA(1)(7)
9. $(m \wedge d) \rightarrow r$	P
10. r	MP(8)(9)
11. $r \vee a$	DA(10)
12. $\sim n \rightarrow (r \vee a)$	CP(6)(11)

12.

a	b	c	d	p	s	w
T	F	F	T	F	T	T
T	F	F	T	F	F	T

Problems

13.

Reason
1. ACP
2. P
3. DS (1) (2)
4. P
5. MT (3) (4)
6. P
7. MP(5)(6)
8. CP (1)(7)

15.

S	R
1. p	ACP
2. $c \vee p$	DA(1)
3. $(c \vee p) \rightarrow s$	P
4. s	MP(2)(3)
5. $s \vee b$	DA(4)
6. $p \rightarrow (s \vee b)$	CP (1)(5)

17.

S	R
1. w	ACP
2. $\sim a \rightarrow \sim w$	P
3. a	MT(1)(2)
4. $a \vee d$	DA(3)
5. $(a \vee d) \rightarrow \sim p$	P
6. $\sim p$	MP(4)(5)
7. $w \rightarrow \sim p$	CP (1)(6)

19.

S	R
1. $\sim p$	P
2. s	ACP
3. $\sim p \wedge s$	CA(1)(2)
4. $(\sim p \wedge s) \rightarrow q$	P
5. q	MP(3)(4)
6. $\sim q \vee \sim r$	P
7. $\sim r$	DS(5)(6)
8. $s \rightarrow \sim r$	CP (2)(7)

21.

S	R
1. r	ACP
2. $r \rightarrow (a \wedge \sim b)$	P
3. $a \wedge \sim b$	MP(1)(2)
4. a	CS(3)
5. $d \rightarrow \sim a$	P
6. $\sim d$	MT(4)(5)
7. $d \vee \sim q$	P
8. $\sim q$	DS(6)(7)
9. $r \rightarrow \sim q$	CP (1)(8)

23.

S	R
1. $\sim d$	P
2. $d \vee \sim s$	P
3. $\sim s$	DS(1)(2)
4. $\sim s \to (\sim c \vee q)$	P
5. $\sim c \vee q$	MP(3)(4)
6. c	ACP
7. q	DS(5)(6)
8. $q \to p$	P
9. p	MP(7)(8)
10. $c \to p$	CP (6)(9)

25.

a	b	c	p	w
T	F	T	T	F

27.

S	R
1. $\sim (a \wedge \sim p)$	P
2. $\sim a \vee p$	DM(1)
3. a	ACP
4. p	DS(2)(3)
5. $\sim w \to \sim a$	P
6. w	MT(3)(5)
7. $w \wedge p$	CA(4)(6)
8. $\sim (w \wedge p) \vee s$	P
9. s	DS(7)(8)
10. $s \to b$	P
11. b	MP(9)(10)
12. $a \to b$	CP (3)(11)

29.

S	R
1. $\sim (a \to b)$	P
2. $a \wedge \sim b$	CN(1)
3. $(a \wedge \sim b) \to (c \vee d)$	P
4. $c \vee d$	MP(2)(3)
5. g	ACP
6. $g \to (w \wedge \sim d)$	P
7. $w \wedge \sim d$	MP(5)(6)
8. $\sim d$	CS(7)
9. c	DS(4)(8)
10. $\sim c \vee z$	P
11. z	DS(9)(10)
12. $g \to z$	CP(5)(11)

31. j

$\sim l \to \sim j$

$e \to \sim a$

$l \to (a \vee k)$

$\sim k \to (\sim e \vee r)$

S	R
1. j	P
2. $\sim l \to \sim j$	P
3. l	MT(1)(2)
4. $l \to (a \vee k)$	P
5. $a \vee k$	MP(3)(4)
6. $\sim k$	ACP
7. a	DS(5)(6)
8. $e \to \sim a$	P
9. $\sim e$	MT(7)(8)
10. $\sim e \vee r$	DA(9)
11. $\sim k \to (\sim e \vee r)$	CP (6)(10)

33. $c \vee d$

$c \to \sim b$

$m \to (\sim b \vee \sim h)$

$\sim d \to m$

b	c	d	h	m
F	T	F	T	F
F	T	F	F	F

In-Class Exercises and Problems for Section 3.6
In-Class Exercises

1.

S	R
1. $\sim n$	AIP
2. $m \vee n$	P
3. m	DS(1)(2)
4. $m \to n$	P
5. n	MP(3)(4)
6. n	CD(1)(5)

2.

S	R
1. $\sim [(a \wedge c) \to r]$	AIP
2. $(a \wedge c) \wedge \sim r$	CN(1)
3. $a \wedge c$	CS(2)
4. a	CS(3)
5. $a \to (c \to r)$	P
6. $c \to r$	MP(4)(5)
7. c	CS(3)
8. r	MP(6)(7)
9. $\sim r$	CS(2)
10. $(a \wedge c) \to r$	CD(8)(9)

3.

S	R
1. $\sim (g \vee m)$	AIP
2. $\sim g \wedge \sim m$	DM(1)
3. $\sim g$	CS(2)
4. $c \to g$	P
5. $\sim c$	MT(3)(4)
6. $c \vee l$	P
7. l	DS(5)(6)
8. $l \to m$	P
9. m	MP(7)(8)
10. $\sim m$	CS(2)
11. $g \vee m$	CD(9)(10)

4.

d	k	p	r	s
F	F	F	F	F

5.

a	b	c	g	n	p
F	F	T	F	F	F
F	F	F	F	F	F

6.

S	R
1. $\sim(\sim w \rightarrow \sim r)$	AIP
2. $\sim w \wedge r$	CN(1)
3. r	CS(2)
4. $r \rightarrow \sim p$	P
5. $\sim p$	MP(3)(4)
6. $\sim(s \vee \sim n)$	P
7. $\sim s \wedge n$	DM(6)
8. $\sim s$	CS(7)
9. $\sim s \rightarrow (\sim p \rightarrow w)$	P
10. $\sim p \rightarrow w$	MP(8)(9)
11. w	MP(5)(10)
12. $\sim w$	CS(2)
13. $\sim w \rightarrow \sim r$	CD(11)(12)

7.

S	R
1. $\sim(p \vee s)$	AIP
2. $\sim p \wedge \sim s$	DM(1)
3. $\sim s$	CS(2)
4. $d \rightarrow s$	P
5. $\sim d$	MT(3)(4)
6. $\sim a \vee w$	P
7. $a \rightarrow w$	CE(6)
8. $(a \rightarrow w) \rightarrow (p \vee d)$	P
9. $p \vee d$	MP(7)(8)
10. p	DS(5)(9)
11. $\sim p$	CS(2)
12. $p \vee s$	CD(10)(11)

8.

S	R
1. $r \wedge s$	AIP
2. r	CS(1)
3. $r \vee p$	DA(2)
4. $(r \vee p) \rightarrow m$	P
5. m	MP(3)(4)
6. $\sim m \vee (w \rightarrow \sim s)$	P
7. $w \rightarrow \sim s$	DS(5)(6)
8. s	CS(1)
9. $\sim w$	MT(7)(8)
10. $(p \vee s) \rightarrow w$	P
11. $\sim(p \vee s)$	MT(9)(10)
12. $\sim p \wedge \sim s$	DM(11)
13. $\sim s$	CS(12)
14. $\sim(r \wedge s)$	CD(8)(13)

9.

S	R
1. $\sim(c \rightarrow \sim d)$	AIP
2. $c \wedge d$	CN(1)
3. d	CS(2)
4. $d \rightarrow a$	P
5. a	MP(3)(4)
6. $\sim h$	P
7. $\sim s \vee h$	P
8. $\sim s$	DS(6)(7)
9. $\sim s \rightarrow (a \rightarrow \sim c)$	P
10. $a \rightarrow \sim c$	MP(8)(9)
11. $\sim c$	MP(5)(10)
12. c	CS(2)
13. $c \rightarrow \sim d$	CD(11)(12)

10.

a	b	c	p	q
F	F	F	T	F

11.

S	R
1. k	P
2. $\sim z \rightarrow \sim k$	P
3. z	MT(1)(2)
4. $\sim z \vee (\sim s \rightarrow w)$	P
5. $\sim s \rightarrow w$	DS(3)(4)
6. $\sim (w \vee \sim a)$	AIP
7. $\sim w \wedge a$	DM(6)
8. $\sim w$	CS(7)
9. s	MT(5)(8)
10. $a \rightarrow \sim s$	P
11. $\sim a$	MT(9)(10)
12. a	CS(7)
13. $w \vee \sim a$	CD(11)(12)

12.

S	R
1. r	P
2. $\sim (c \rightarrow \sim p)$	AIP
3. $c \wedge p$	CN(2)
4. p	CS(3)
5. $p \wedge r$	CA(1)(4)
6. $\sim (p \wedge r) \vee s$	P
7. s	DS(5)(6)
8. $s \rightarrow (c \rightarrow w)$	P
9. $c \rightarrow w$	MP(7)(8)
10. c	CS(3)
11. w	MP(9)(10)
12. $(k \vee c) \rightarrow \sim w$	P
13. $\sim (k \vee c)$	MT(11)(12)
14. $\sim k \wedge \sim c$	DM(13)
15. $\sim c$	CS(14)
16. $c \rightarrow \sim p$	CD(10)(15)

Problems

13.

Reason
1. AIP
2. P
3. DS(1)(2)
4. CS(3)
5. CS(3)
6. P
7. MP(5)(6)
8. P
9. MT(7)(8)
10. DM(9)
11. CS(10)
12. CD(4)(11)

15.

Reason
1. AIP
2. DM(1)
3. CS(2)
4. CS(2)
5. P
6. DS(3)(5)
7. CS(6)
8. CS(6)
9. P
10. MP(8)(9)
11. DA(4)
12. P
13. MP(11)(12)
14. MP(10)(13)
15. CD(7)(14)

17.

S	R
1. $\sim k$	AIP
2. $\sim k \rightarrow h$	P
3. h	MP(1)(2)
4. $\sim h \vee k$	P
5. k	DS(3)(4)
6. k	CD(1)(5)

19.

S	R
1. $\sim(\sim p \rightarrow r)$	AIP
2. $\sim p \wedge \sim r$	CN(1)
3. $\sim r$	CS(2)
4. $\sim k \vee r$	P
5. $\sim k$	DS(3)(4)
6. $\sim s \vee k$	P
7. $\sim s$	DS(5)(6)
8. $(p \rightarrow q) \rightarrow s$	P
9. $\sim(p \rightarrow q)$	MT(7)(8)
10. $p \wedge \sim q$	CN(9)
11. p	CS(10)
12. $\sim p$	CS(2)
13. $\sim p \rightarrow r$	CD(11)(12)

21.

S	R
1. $\sim(\sim q \vee s)$	AIP
2. $q \wedge \sim s$	DM(1)
3. q	CS(2)
4. $\sim w \vee \sim q$	P
5. $\sim w$	DS(3)(4)
6. $(n \vee b) \rightarrow w$	P
7. $\sim(n \vee b)$	MT(5)(6)
8. $\sim n \wedge \sim b$	DM(7)
9. $\sim n$	CS(8)
10. $\sim n \rightarrow s$	P
11. s	MP(9)(10)
12. $\sim s$	CS(2)
13. $\sim q \vee s$	CD(11)(12)

23.

S	R
1. $\sim(\sim p \vee a)$	AIP
2. $p \wedge \sim a$	DM(1)
3. $\sim a$	CS(2)
4. $\sim a \rightarrow (\sim p \vee c)$	P
5. $\sim p \vee c$	MP(3)(4)
6. $a \vee (c \rightarrow \sim p)$	P
7. $c \rightarrow \sim p$	DS(3)(6)
8. p	CS(2)
9. $\sim c$	MT(7)(8)
10. $\sim p$	DS(5)(9)
11. $\sim p \vee a$	CD(8)(10)

25.

c	d	e	z
T	F	F	F

27.

a	c	d	p	q	s
T	F	T	T	T	T
F	F	T	T	T	T

29.

S	R
1. $\sim (d \rightarrow l)$	P
2. $d \wedge \sim l$	CN(1)
3. d	CS(2)
4. $d \rightarrow (b \vee s)$	P
5. $b \vee s$	MP(3)(4)
6. $\sim (b \vee \sim c)$	AIP
7. $\sim b \wedge c$	DM(6)
8. $\sim b$	CS(7)
9. s	DS(5)(8)
10. $s \rightarrow (m \wedge \sim c)$	P
11. $m \wedge \sim c$	MP (9)(10)
12. $\sim c$	CS(11)
13. c	CS(7)
14. $b \vee \sim c$	CD(12)(13)

31.

S	R
1. $\sim r$	P
2. $s \rightarrow r$	P
3. $\sim s$	MT(1)(2)
4. $\sim s \rightarrow (d \rightarrow l)$	P
5. $d \rightarrow l$	MP(3)(4)
6. $\sim (\sim l \rightarrow c)$	AIP
7. $\sim l \wedge \sim c$	CN(6)
8. $\sim l$	CS(7)
9. $\sim d$	MT(5)(8)
10. $\sim c$	CS(7)
11. $c \vee d$	P
12. d	DS(10)(11)
13. $\sim l \rightarrow c$	CD(9)(12)

33. $c \vee \sim v$

$(f \vee d) \rightarrow \sim c$

$\sim f \rightarrow b$

$v \rightarrow b$

S	R
1. $\sim (v \rightarrow b)$	AIP
2. $v \wedge \sim b$	CN(1)
3. $\sim b$	CS(2)
4. $\sim f \rightarrow b$	P
5. f	MT(3)(4)
6. $f \vee d$	DA(5)
7. $(f \vee d) \rightarrow \sim c$	P
8. $\sim c$	MP(6)(7)
9. $c \vee \sim v$	P
10. $\sim v$	DS(8)(9)
11. v	CS(2)
12. $v \rightarrow b$	CD(10)(11)

35. $p \vee c$

$g \rightarrow (\sim j \vee p)$

$(g \wedge j) \vee \sim c$

---

p

S	R
1. $\sim p$	AIP
2. $p \vee c$	P
3. c	DS(1)(2)
4. $(g \wedge j) \vee \sim c$	P
5. $g \wedge j$	DS(3)(4)
6. g	CS(5)
7. $g \rightarrow (\sim j \vee p)$	P
8. $\sim j \vee p$	MP(6)(7)
9. j	CS(5)
10. p	DS(8)(9)
11. p	CD(1)(10)

Chapter 3 Review

1. d 2. g 3. b 4. c 5. i 6. h 7. f 8. a 9. j 10. e

11. Reason

1. P
2. CS(1)
3. CS(1)
4. DA(2)
5. P
6. MP(4)(5)
7. P
8. MT(6)(7)
9. DM(8)
10. CS(9)
11. P
12. MP(10)(11)
13. DS(3)(12)

12. Reason

1. ACP
2. P
3. MP(1)(2)
4. P
5. DS(3)(4)
6. P
7. MT(5)(6)
8. P
9. MP(7)(8)
10. CS(9)
11. CP(1)(10)

13. Reason

1. AIP
2. DM(1)
3. CS(2)
4. CS(2)
5. P
6. MP(3)(5)
7. P
8. MT(4)(7)
9. CN(8)
10. CS(9)
11. CS(9)
12. DA(10)
13. P
14. MP(12)(13)
15. MT(6)(11)
16. CD(14)(15)

14.

S	R
1. $\sim p$	P
2. $\sim p \vee q$	DA(1)
3. $(\sim p \vee q) \rightarrow r$	P
4. r	MP(2)(3)
5. $a \rightarrow \sim r$	P
6. $\sim a$	MT(4)(5)

15.

S	R
1. $(c \vee d) \wedge a$	P
2. $c \vee d$	CS(1)
3. $b \rightarrow \sim (c \vee d)$	P
4. $\sim b$	MT(2)(3)
5. $b \vee \sim s$	P
6. $\sim s$	DS(4)(5)

16.

S	R
1. $\sim p$	AIP
2. $\sim p \rightarrow \sim d$	P
3. $\sim d$	MP(1)(2)
4. $d \vee \sim s$	P
5. $\sim s$	DS(3)(4)
6. $(\sim d \vee a) \rightarrow s$	P
7. $\sim (\sim d \vee a)$	MT(5)(6)
8. $d \wedge \sim a$	DM(7)
9. d	CS(8)
10. p	CD(3)(9)

17.

S	R
1. $a \wedge \sim q$	P
2. a	CS(1)
3. $a \to (\sim s \vee \sim c)$	P
4. $\sim s \vee \sim c$	MP(2)(3)
5. $\sim q$	CS(1)
6. $\sim (p \wedge s) \to q$	P
7. $p \wedge s$	MT(5)(6)
8. p	CS(7)
9. s	CS(7)
10. $\sim c$	DS(4)(9)
11. $p \wedge \sim c$	CA(8)(10)

18.

S	R
1. $\sim c$	P
2. $b \to c$	P
3. $\sim b$	MT(1)(2)
4. $\sim b \vee \sim p$	DA(3)
5. $(\sim b \vee \sim p) \to s$	P
6. s	MP(4)(5)
7. $s \vee q$	DA(6)

19.

S	R
1. $\sim d \to c$	P
2. $d \vee c$	CE(1)
3. $(d \vee c) \to k$	P
4. k	MP(2)(3)
5. $k \to p$	P
6. p	MP(4)(5)

20.

S	R
1. $m \wedge p$	P
2. m	CS(1)
3. $m \to \sim n$	P
4. $\sim n$	MP(2)(3)
5. $\sim n \to (a \wedge b)$	P
6. $a \wedge b$	MP(4)(5)
7. b	CS(6)
8. $\sim b \vee s$	P
9. s	DS(7)(8)

21.

S	R
1. $\sim l$	ACP
2. $l \vee a$	P
3. a	DS(1)(2)
4. $a \to (p \to \sim s)$	P
5. $p \to \sim s$	MP(3)(4)
6. p	P
7. $\sim s$	MP(5)(6)
8. $\sim s \vee k$	DA(7)
9. $(\sim s \vee k) \to b$	P
10. b	MP(8)(9)
11. $\sim l \to b$	CP(1)(10)

22.

a	g
T	F

23.

S	R
1. $\sim(p \vee a) \wedge \sim s$	P
2. $\sim(p \vee a)$	CS(1)
3. $(\sim p \wedge \sim a)$	DM(2)
4. $\sim p$	CS(3)
5. $p \vee (\sim r \rightarrow s)$	P
6. $\sim r \rightarrow s$	DS(4)(5)
7. $\sim s$	CS(1)
8. r	MT(6)(7)

24.

S	R
1. $\sim k$	ACP
2. $k \vee \sim(a \vee \sim b)$	P
3. $\sim(a \vee \sim b)$	DS(1)(2)
4. $\sim a \wedge b$	DM(3)
5. b	CS(4)
6. $b \rightarrow c$	P
7. c	MP(5)(6)
8. $c \vee n$	DA(7)
9. $\sim k \rightarrow (c \vee n)$	CP(1)8)

25.

a	b	p	s	w
T	T	F	T	F

26.

S	R
1. $\sim l$	P
2. $(a \vee g) \rightarrow l$	P
3. $\sim(a \vee g)$	MT(1)(2)
4. $\sim a \wedge \sim g$	DM(3)
5. $\sim g$	CS(4)
6. $g \vee (p \vee q)$	P
7. $p \vee q$	DS(5)(6)
8. $\sim p \rightarrow q$	CE(7)

27.

a	g	l	m	w
T	F	T	T	F

28.

S	R
1. $a \wedge \sim b$	P
2. a	CS(1)
3. $a \vee s$	DA(2)
4. $(a \vee s) \rightarrow n$	P
5. n	MP(3)(4)
6. $\sim n \vee \sim p$	P
7. $\sim p$	DS(5)(6)
8. $\sim b$	CS(1)
9. $\sim p \wedge \sim b$	CA(7)(8)
10. $\sim (p \vee b)$	DM(9)

29.

S	R
1. $\sim (s \rightarrow \sim q)$	P
2. $s \wedge q$	CN(1)
3. q	CS(2)
4. $\sim q \vee p$	P
5. p	DS(3)(4)
6. $p \vee d$	DA(5)
7. $(p \vee d) \rightarrow c$	P
8. c	MP(6)(7)

30.

S	R
1. $\sim a$	ACP
2. $a \vee \sim b$	P
3. $\sim b$	DS(1)(2)
4. $(\sim p \vee q) \rightarrow b$	P
5. $\sim (\sim p \vee q)$	MT(3)(4)
6. $p \wedge \sim q$	DM(5)
7. $\sim q$	CS(6)
8. $\sim s \rightarrow q$	P
9. s	MT(7)(8)
10. $\sim a \rightarrow s$	CP(1)(9)

31.

a	b	d	e	p	w
F	T	F	F	T	F
F	F	F	F	T	F

32.

S	R
1. a	P
2. $a \vee b$	DA(1)
3. $(a \vee b) \rightarrow r$	P
4. r	MP(2)(3)
5. $\sim r \vee w$	P
6. w	DS(4)(5)
7. $(m \vee n) \rightarrow \sim w$	P
8. $\sim (m \vee n)$	MT(6)(7)
9. $\sim m \wedge \sim n$	DM(8)
10. $\sim m$	CS(9)

33.

S	R
1. $\sim w$	AIP
2. $\sim w \vee q$	DA(1)
3. $(\sim w \vee q) \rightarrow \sim b$	P
4. $\sim b$	MP(2)(3)
5. $\sim a \vee b$	P
6. $\sim a$	DS(4)(5)
7. $(\sim r \rightarrow s) \rightarrow a$	P
8. $\sim (\sim r \rightarrow s)$	MT(6)(7)
9. $\sim r \wedge \sim s$	CN(8)
10. $\sim r$	CS(9)
11. $r \vee (s \wedge p)$	P
12. $s \wedge p$	DS(10)(11)
13. s	CS(12)
14. $\sim s$	CS(9)
15. w	CD(13)(14)

34.

S	R
1. a	ACP
2. $w \rightarrow \sim a$	P
3. $\sim w$	MT(1)(2)
4. $\sim (a \wedge \sim b)$	P
5. $\sim a \vee b$	DM(4)
6. b	DS(1)(5)
7. $\sim w \wedge b$	CA(3)(6)
8. $(\sim w \wedge b) \rightarrow c$	P
9. c	MP(7)(8)
10. $\sim c \vee p$	P
11. p	DS(9)(10)
12. $a \rightarrow p$	CP(1)(11)

35.

a	k	p	q	r	s	w
T	T	F	F	T	T	T

36. a, d   37. a, c

Sample Exam: Chapter 3

1. DS 2. MT 3. CS 4. CN 5. CE 6. DA 7. DM 8. MP

9. b 10. d 11. c 12. d 13. c 14. c

15. Reason

1. P
2. CN(1)
3. CS(2)
4. CS(2)
5. P
6. MP(3)(5)
7. MT(4)(6)
8. P
9. DS(7)(8)
10. AIP
11. DM(10)
12. CS(11)
13. CS(11)
14. P
15. MP(13)(14)
16. DS(9)(15)
17. CD(12)(16)

16.

S	R
1. d	P
2. $(\sim s \vee r) \rightarrow \sim d$	P
3. $\sim(\sim s \vee r)$	MT(1)(2)
4. $s \wedge \sim r$	DM(3)
5. $\sim r$	CS(4)
6. $r \vee \sim b$	P
7. $\sim b$	DS(5)(6)
8. $\sim b \vee n$	DA(7)

17.

S	R
1. d	ACP
2. $d \rightarrow (a \wedge \sim c)$	P
3. $a \wedge \sim c$	MP(1)(2)
4. a	CS(3)
5. $\sim s \rightarrow \sim a$	P
6. s	MT(4)(5)
7. $s \vee q$	DA(6)
8. $(s \vee q) \rightarrow r$	P
9. r	MP(7)(8)
10. $d \rightarrow r$	CP(1)(9)

18.

S	R
1. $\sim a \wedge p$	P
2. $\sim a$	CS(1)
3. $\sim a \to (k \to b)$	P
4. $k \to b$	MP(2)(3)
5. p	CS(1)
6. $(b \vee \sim c) \to \sim p$	P
7. $\sim (b \vee \sim c)$	MT(5)(6)
8. $\sim b \wedge c$	DM(7)
9. $\sim b$	CS(8)
10. $\sim k$	MT(4)(9)

19.

S	R
1. $g \to k$	P
2. $\sim g \vee k$	CE(1)
3. $(\sim g \vee k) \to p$	P
4. p	MP(2)(3)
5. $p \vee a$	DA(4)
6. $(p \vee a) \to c$	P
7. c	MP(5)(6)
8. $c \wedge p$	CA(4)(7)

20.

a	b	n	r	s	w
T	T	F	T	T	T

21.

a	b	k	p	s
F	F	F	T	F

22. $c \to p$

$\sim c \to g$

$e \to \sim p$

e

―――

g

S	R
1. e	P
2. $e \to \sim p$	P
3. $\sim p$	MP(1)(2)
4. $c \to p$	P
5. $\sim c$	MT(3)(4)
6. $\sim c \to g$	P
7. g	MP(5)(6)

23. $c \to g$

$g \to (f \vee s)$

$f \to (p \wedge l)$

―――

$c \to l$

c	f	g	l	p	s
T	F	T	F	T	T
T	F	T	F	F	T

Answers for Chapter 4

In-Class Excercises and Problems for Section 4.1

In-Class Exercises

1. My car is old. If my car is old and has 85,000 miles on it, then I will buy a new one soon. If my car is a Ford then I will buy a better model. Therefore I will buy a better model.
2. She is lazy or uncooperative. If she is uncooperative, she will not volunteer to help. If she is lazy then she will not do a good job. If she does not do a good job, then we will not ask her again.
3. Many correct answers.

4. a. 0.54 b. 0.17 c. 0.29 5 a.0.4 b. 0.2 c. 0.6 d. 1 e. 0 6. a. 0.8 b.0.6 c. 0.4 d. 0 e. 1

Problems

7. Fuzzy statement
9. Fuzzy statement
11. Not a fuzzy statement
13. Not a fuzzy statement
15. Fuzzy statement

17.-26. Many correct answers

	Statement	Degree of truthfulness
27.	Joshua understood Chapter 1.	0.75
	Maria understood Chapter 1.	0.4
	Davis understood Chapter 1.	1.00
29.	Autoland is expensive.	0.55
	Custom Auto Design is expensive.	1.00
	Ray's Wholesale is expensive.	0.0
	Tire World is expensive.	0.27
31.	The Clarkes	0.5
	The Davidsons	0.2
	The Martuccis	0.6
	The Waxmans	0.3
	The Jordans	1.00
	The Yeardleys	0.0

In-Class Excercises and Problems for Section 4.2

In-Class Exercises

1. 0.37 2. 0.81 3. 0.55 4. 0.55 5. 0.88 6. 0.81 7. 0.63 8. 0.55 9. 0.81 10. 0.63
11. a. 0.81 b. 0.81 c. same d. logically equivalent 12. a. 0.45 b. 0.55 c. degrees of truthfulness add to 1 d. negation

Problems

13. 0.3 15. 0.7 17. 0.3 19. 0.65 21. 0.6 23. 0.7 25. 0.4 27. 0.6

In-Class Excercises and Problems for Section 4.3

In-Class Exercises

1. a. Lilia should study 3 hours b. Ricky should study 6 hours c. Nancy should study 1hour d. Tom did well on the test. e. None of the above will need to study for 5 hours.
2. a. Adjust the headlights to brightness #4 b. Adjust the headlights to brightness #1 c. Adjust the headlights to brightness #3 d. Adjust the headlights to brightness #4
3. a. Quality Comfort b. Royal Comfort c. Restful Comfort d. EcoRest

Problems

5. a. Sleep 7 hours b. Sleep 4 hours c. Sleep 7 hours d. Sleep 12 hours

7. a. 18 days	b. 28 days	c. 8 days	d. 28 days	e. 24 days
9. a. 30 min	b. 20 min	c. 40 min	d. 40 min	e. 30 min

In-Class Excercises and Problems for Section 4.4

In-Class Exercises

1. Many correct answers
2. Many correct answers
3. How much is the car?
4. What are you doing today?
5. Input command, Find out when it will stop.
6. Simple Command, Be happy.
7. Simple Command, Take some medicine.
8. If the customer has a club card, charge $159.99; otherwise charge $179.99
9. If the person is an adult or a child 12 or over, add $13.95 to the bill; otherwise add $6.95 to the bill.
10. If it is Monday or Thursday, park elsewhere; otherwise park here.
11. If the shopper has 10 items or less, choose lane 1, if the shopper has 20 items or less, choose lane 2, if the shopper is using a credit card, choose lane 3; otherwise choose from lane 4 thuough lane 8.

Problems

13. Yes/No question
15. Input question
17. Yes/No question
19. Input command
21. Yes/No question
23. Input command
25. Simple command
27. Conditional command
29. Input command
31. Yes/No question
33. Input: Determine the starting and ending time of the show.
    Decisions: Is it time for the show to start?
    Is the VCR taping? Is it time for the show to end?
    Actions: If it is time for the show to start, then start taping, otherwise do not start taping.
    If the show is taping and the clock shows it is time for the show to end, then stops taping.
    Otherwise, don't change any setting.

In-Class Excercises and Problems for Section 4.5

In-Class Exercises

1. a. Many correct flowcharts b. Ask for alternate directions c. Take the Northern Parkway
   d. Take the service road.
2. a. Many correct flowcharts b. no c. no

Problems

3. a. weekends b. everyone makes their own dinner
5. a. $200 b. $100 c. $300 d. $1,000 and go to jail e. $450
7. many correct answers

In-Class Excercises and Problems for Section 4.6

In-Class Exercises

1-8. Current flows in all cases except cases 7 and 8. 9. Yes 10. No 11. Yes 12. Yes 13. Yes

14.
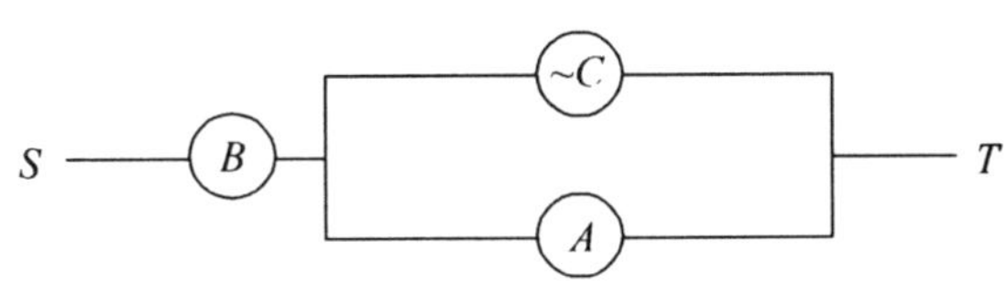

Problems

15. No
17. Yes
19. No
21. Yes
23. C
25. $F \cup (E \cap G)$
27. $A \cap B$
29. $A \cup B'$
31. No circuit
33. a. $A \cap B$, $A \cap C$, $B \cap C$ b. Union

c. One solution is $[A \cap (B \cup C)] \cup (B \cap C)$

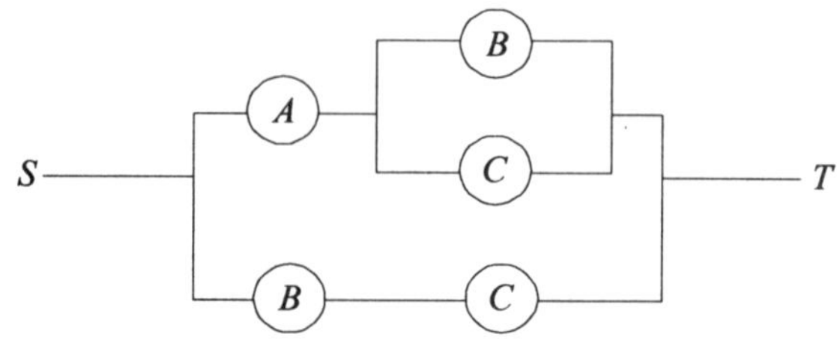

d. $A \cup (B \cap C)$

Chapter 4 Review

1. Either Candice will go to the Bahamas or she will stay home. She will not go to the Bahamas and she will take 5 vacation days next week. If she works 60 hours this week, then she will not take 5 vacation days next week. Therefore, Candice did not work 60 hours this week and she will stay home.

S	R
1. $\sim b \wedge v$	P
2. $\sim b$	CS (1)
3. $b \vee h$	P
4. h	DS (2) (3)
5. v	CS (1)
6. $w \rightarrow \sim v$	P
7. $\sim w$	MT (5) (6)
8. $\sim w \wedge h$	CA (4) (7)

2. a. 0.24 b. same as $\tau(q)$ c. $(p \vee q) \wedge q$ is logically equivalent to q
3. Take job offer 2
4. a. Many correct flowcharts b. Keep starter in one more inning c. Bring relief pitcher in
   d. Bring relief pitcher in
5. Did I pass? 6.-13. Current flows only in cases 6, 7, and 8.

14. 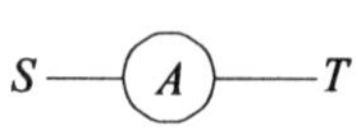

15. 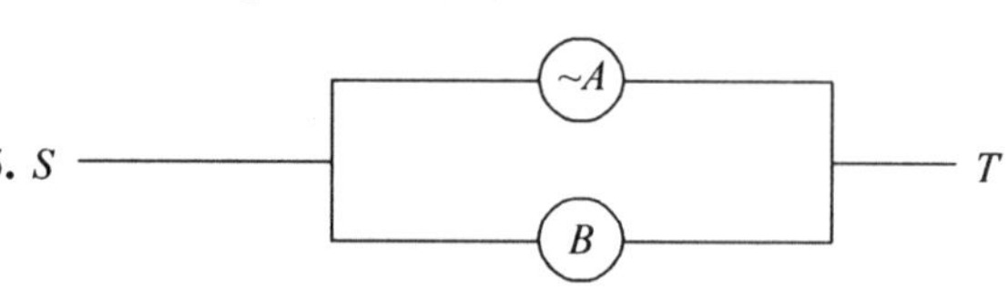

16. 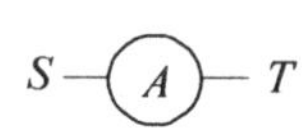

Sample Exam: Chapter 4

1. c 2. b 3. c 4. d 5. b 6. b 7. c 8. Many correct
9. Plastic disposable plates 10. Coated paper plates 11-18. Current flows only in cases 11 and 12.

19, 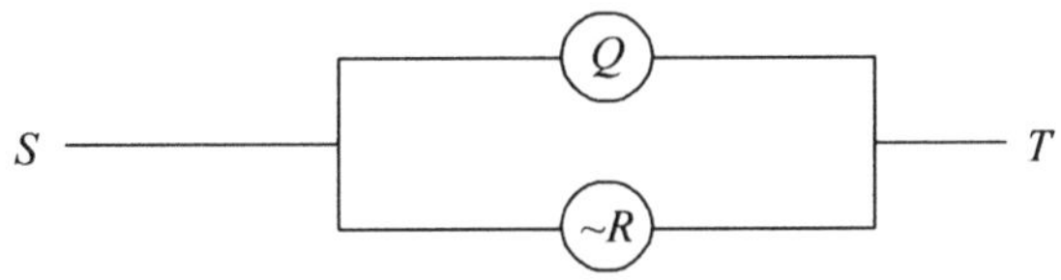

Answers for Chapter 5

In-Class Excercises and Problems for Section 5.1

In-Class Exercises

1. {0,4,8,12,16} 2. {20,25,30,35,40} 3. {on,no,one,eon} 4. ∅ 5. {Monday, Friday}
6. The even numbers from zero to ten 7.The odd numbers between 30 and 40
8. The first four letters of the alphabet
9. a. F b. T c. F d. F e. T f. T
g. F h. T i. T j. T k. F l. T
10. 5 11. 32 12. 31 13. 32 14. 64 15. 63 16. 16 17. 15 18. 32
19. {{2},{8},{10},{2,8},{2,10},{8,10},{2,8,10},∅ }

Problems

21. {4, 5, 6, 7, 8, 9}
23. {dog, god, cod, cog} 25. ∅ 27. The set of all odd numbers less than 22
29. The set of the first six integers squared
31. The set of all days of the week beginning with the letter z
33. True 35. False 37. True 39. True 41. False
43. 8 45. 63 47. Yes 49. False 51. False

In-Class Excercises and Problems for Section 5.2

In-Class Exercises

1. {1,2,3,4,5,6} 2. {2} 3. {1,5,6} 4.{3,4} 5. {1,2,4,5,6,7}
6. {3} 7. ∅ 8. {1,3,5,7} 9. {1,5,7} 10. {6}
11. {1,3,4,5,7} 12. {1,5} 13. {2,4,6} 14. U 15. {1,2,3,5,7}
16. {1,3,4,5,6,7} 17. {1,3,5,6,7} 18. {4,7} 19. {6} 20. {2,3,6,7}

Problems

21. {3, 7} 23. {1, 4, 5, 6} 25. {2, 5, 6, 9} 27. {1, 2, 3, 4, 8, 9}
29. {1, 2, 3, 4, 5, 6, 8, 9} 31. {April, September, November}
33. {May, June, July, August} 35. {May, July, August}
37. {May, June} 39. {May, June, July, August, September, October}

In-Class Excercises and Problems for Section 5.3

In-Class Exercises

1. 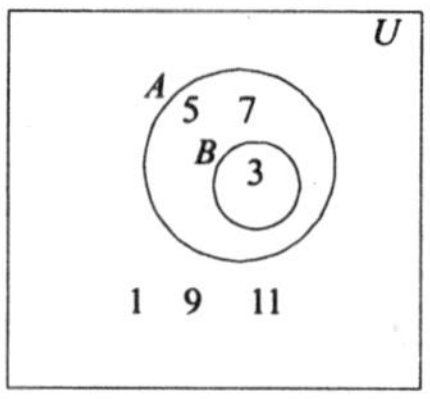

2. 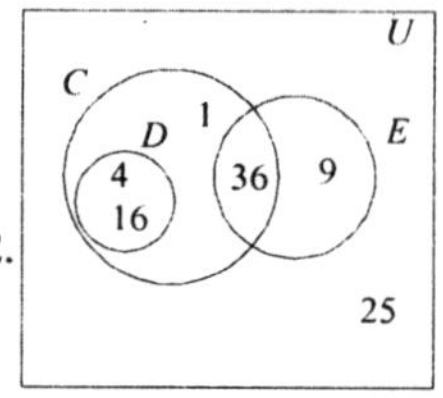

3. 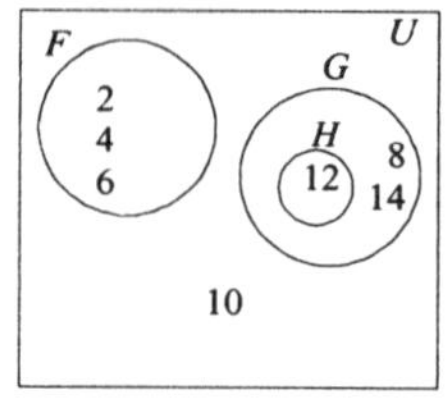

4. 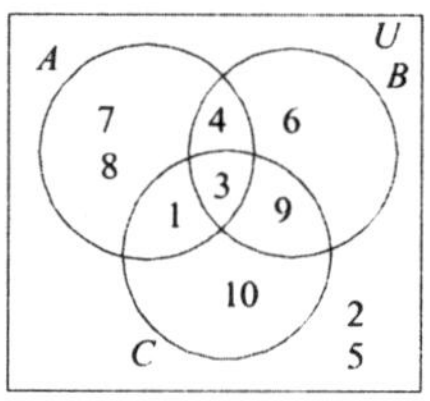

5. 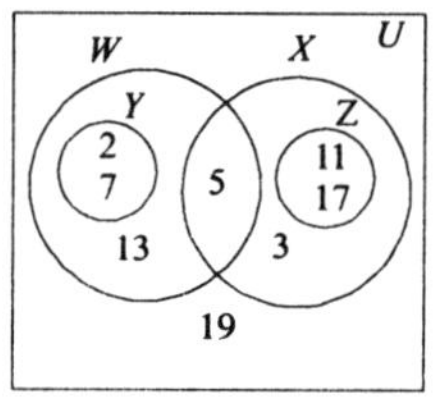

6. 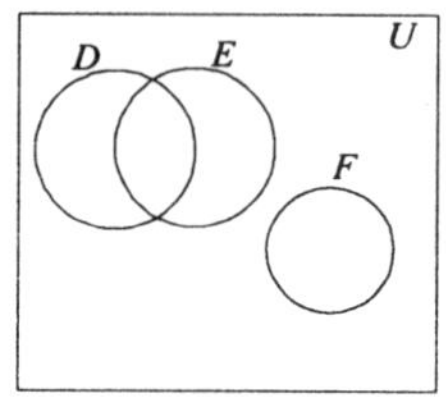

7. 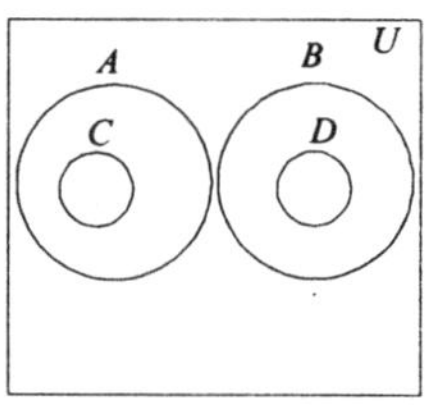

8. 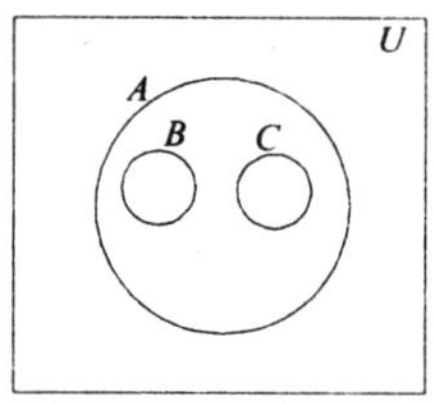

9. 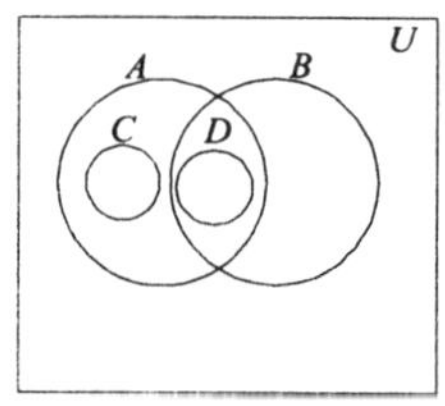

10. III, IV 11. I, III 12. I, II 13. II, IV 14. III 15. IV 16. I, II, IV 17. I, II, III
18. I, IV 19. II, IV 20. I, II,, V, VI 21. II, IV, VI, VII 22. I, V 23. I, III, V, VI, VII, VIII
24. II, VI 25. III 26. II, V, VI 27. III, IV, VI, VIII 28. III, V, VI, VII 29. I, II, IV, V, VI
30. III, IV, VI, VIII 31. IV 32. V, VI 33. V 34. II, IV, V, VII
35. $A \cap B \cap C$ 36. $A - B$ 37. $(B \cup C) - A$
38. $[A - (B \cup C)] \cup [B - (A \cup C)]$ or $[(A \cup B) - (A \cap B)] - C$

Problems

39. 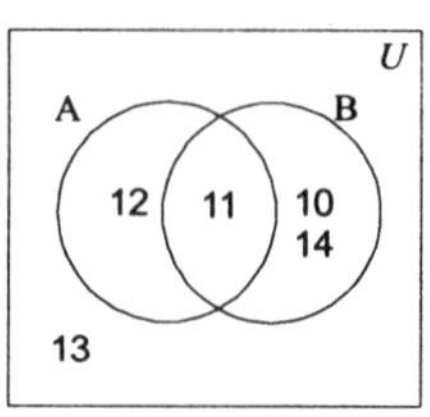

41. 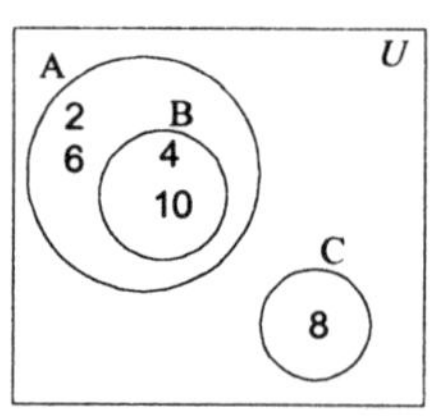

43. 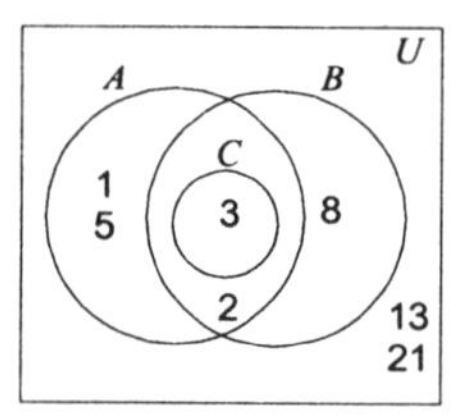

45. 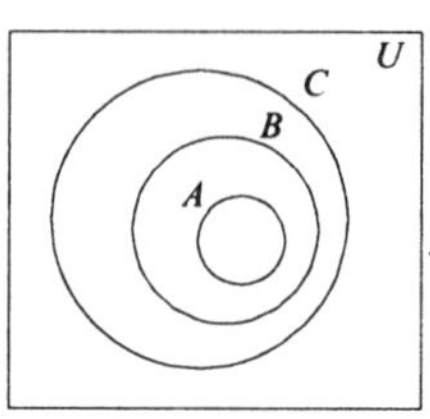

47. 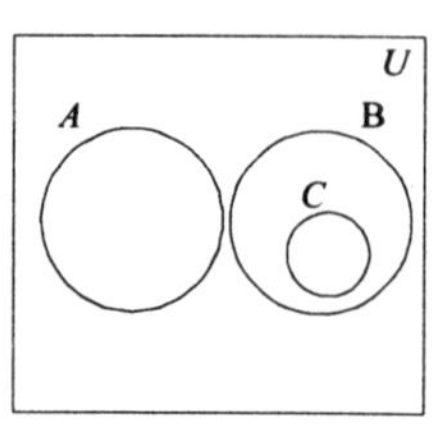

49. 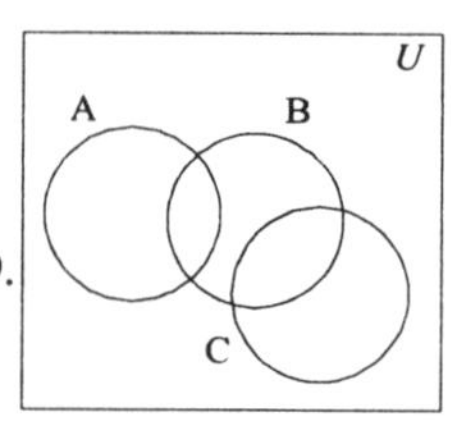

51. II 53. IV 55. II 57. I, III 59. I, II, III, IV
61. I, II 63. VII, VIII 65. I, II 67. I, III,V 69. III, IV, VII 71. IV
73. II, III, IV,V, VI, VII, VIII 75. $\varnothing$ 77. VII 79. VI 81. VIII 83. I, III, IV, V
85. $(A \cap B \cap C)'$ 87. $(B \cap C) - A$ 89. $(A \cup B \cup C)'$ 91. $A \cup (B \cap C)$

In-Class Excercises and Problems for Section 5.4

In-Class Exercises

1. a. Set Difference b. DeMorgan's Law c. Distributive Law d. Inverse Law e. Identity Law
2. a. Absorption Law b. Set Difference c. Distributive Law d. Idempotent Law
   e. Commutative Law f. Absorption Law

6. $B \cap C$ 7. A' 8. B' 9. A 10. U

Problems

11.

A	B	B'	$A \cap B$	$A - B'$
$\in$	$\in$	$\notin$	$\in$	$\in$
$\in$	$\notin$	$\in$	$\notin$	$\notin$
$\notin$	$\in$	$\notin$	$\notin$	$\notin$
$\notin$	$\notin$	$\in$	$\notin$	$\notin$

13.

A	B	C	$B \cup C$	$A \cap B$	$A \cap C$	$A \cap (B \cup C)$	$(A \cap B) \cup (A \cap C)$
$\in$	$\in$	$\in$	$\in$	$\in$	$\in$	$\in$	$\in$
$\in$	$\in$	$\notin$	$\in$	$\in$	$\notin$	$\in$	$\in$
$\in$	$\notin$	$\in$	$\in$	$\notin$	$\in$	$\in$	$\in$
$\in$	$\notin$	$\notin$	$\notin$	$\notin$	$\notin$	$\notin$	$\notin$
$\notin$	$\in$	$\in$	$\in$	$\notin$	$\notin$	$\notin$	$\notin$
$\notin$	$\in$	$\notin$	$\in$	$\notin$	$\notin$	$\notin$	$\notin$
$\notin$	$\notin$	$\in$	$\in$	$\notin$	$\notin$	$\notin$	$\notin$
$\notin$	$\notin$	$\notin$	$\notin$	$\notin$	$\notin$	$\notin$	$\notin$

15. III, IV 17. I, II, III, IV, V, VI, VII 19. I, II, III, IV

21. a. DeMorgan's Law b. Distributive Law c. Inverse Law d. Identity Law
    e. Set Difference

23. A' 25. $A \cap C$ 27. B' 29. C' 31. $P \cap R$

Chapter 5 Review

1. {y} 2. {Africa, Antarctica, Asia, Australia, Europe, North America, South America 3. $\varnothing$
4. {April, June, September, November}
5. {Mercury, Venus, Earth, Mars, Jupiter, Saturn, Uranus, Neptune, Pluto}
6. {5,9} 7. {5} 8. $\varnothing$ 9. $\varnothing$ 10. {1,2,3,4,5,7,9}
11. true 12. true 13. false 14. true 15. true 16. true 17. true 18. false

19.

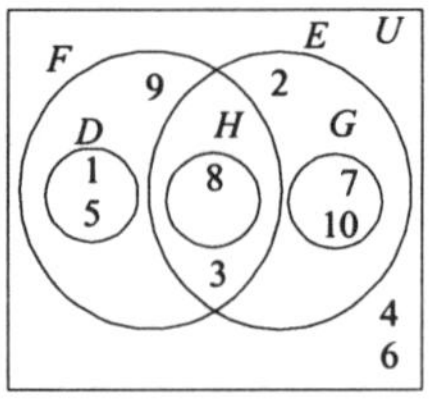

20.

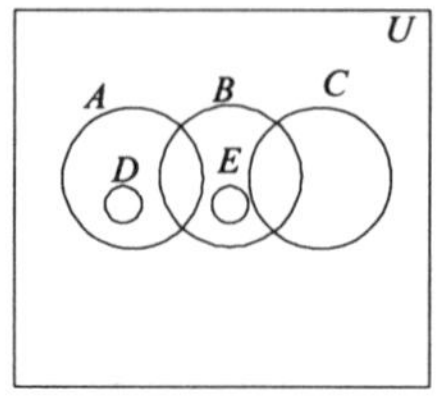

21. 4 22. 16 23. 128 24. 127 25. $\{\{h\},\varnothing\}$

26. Yes 27. Yes 28. No 29. No

30.

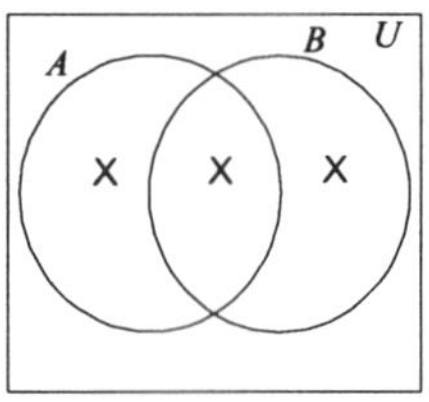

31.

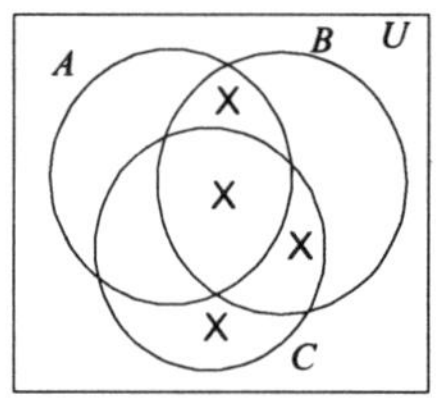

32. a. II b. II, IV c. VII, VIII d. V, VII

33. a. $[(A \cap B) - C] \cup [(B \cap C) - A]$ b. $B - (A \cup C)$

c. $(A \cap B \cap C) \cup (A - (B \cup C)) \cup (C - (A \cup B))$ d. $(A \cap C) - B$

e. $(C - (A \cup B)) \cup (A \cup B \cup C)'$

34. Yes, the compound statement is true.

35. *i.* d *ii.* c *iii.* a *iv.* e *v.* b

36. Sets A, B, C and D are proper subsets of the universal set. $C \subset A$, $D \subset A$, $C \cap D \neq \varnothing$ and neither C nor D is a subset of the other set, $A \cap B \neq \varnothing$. Neither set A nor set B is a subset of the other set. $B \cap C = \varnothing$, $B \cap D \neq \varnothing$ and neither B nor D is a subset of the other.

37. a. 2. Set Difference 3. DeMorgan's Law4. Set Difference 5. Commutative Law
6. Distributive Law 7. Inverse Law 8. Identity Law 9. Set Difference

b. 2. Set Difference 3. Absorption Law 4. Set Difference 5. DeMorgan's Law
6. Commutative Law7. Associative Law 8. Idempotent Law 9. Absorption Law

38. a. $(A - B) \cup (A - C') = (A \cap B') \cup (A \cap C) = A \cap (B' \cup C) = A - (B' \cup C)' = A - (B \cap C')$

b. $[P' \cap (P' \cup S)] - (P' \cap Q) = P' - (P' \cap Q) = P' \cap (P' \cap Q)' = P' \cap (P \cup Q')$

$= (P' \cap P) \cup (P' \cap Q') = \varnothing \cup (P' \cap Q') = P' \cap Q' = (P \cup Q)'$

c. $(A \cup C)' - [A' \cup (A' \cap B)] = (A \cup C)' - A' = (A \cup C)' \cap A = (A' \cap C') \cap A$

$= (C' \cap A') \cap A = C' \cap (A' \cap A) = C' \cap \varnothing = \varnothing$

Sample Exam: Chapter 5

1. $\{3,6,9\}$ 2. U 3. $\{1,3,10\}$ 4. $\{5,7,8\}$ 5. $\{1,7,10\}$ 6. 8 7. yes

8. no 9. b 10. d 11. true 12. true 13. true 14. false

15. false 16. false 17. false 18. true 19. true 20. false 21. c

22. 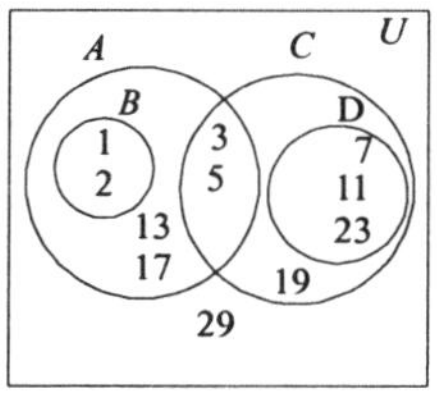

23. not true 24. d 25. Set Difference 26. Commutative Law

27. DeMorgan's Law 28. Commutative Law 29. Associative Law 30. Idempotent Law

31. Distributive Law 32. Inverse Law 33. Identity Law

34. $W \cap (S \cap W)' = W \cap (S' \cup W') = (W \cap S') \cup (W \cap W') = (W \cap S') \cup \varnothing = W \cap S' = W - S$

Answers for Chapter 6

In-Class Excercises and Problems for Section 6.1

In-Class Exercises

1. Invalid 2. Valid 3. Valid 4. Valid 5. Invalid 6. Invalid 7. Valid 8. Valid

Problems

9. Invalid 11. Valid 13. Valid 15. Invalid

In-Class Excercises and Problems for Section 6.2

In-Class Exercises

1. 4. 2. a 5 b. 10 3. a. 22 b. 24 c. 45 d. 35 4. a 960 b. 180 c. 720 d. 40 e. 580

5. a. 690 b. 10 c. 100 d. 350 e. 50 f. 360 g. 430 h.190 6. a. 23 b. 3 c. 2 d. 6 e. 5

7. a. 5 b. 15 c. 20

Problems

9. a. 4 b. 5 c. 3 11. a. 0 b. 3 c. 6 d. 25 13. a. 0 b. 200 c. 400 d.100

In-Class Excercises and Problems for Section 6.3

In-Class Exercises

1. a. $\frac{1}{26}$ b. $\frac{1}{26}$ c. $\frac{25}{26}$ d. $\frac{21}{26}$ e. $\frac{3}{26}$ f. $\frac{22}{26}$ g. $\frac{24}{26}$

2. a. $\frac{1}{6}$ b. 1 c. $\frac{2}{6}$ d. $\frac{3}{6}$ e. $\frac{5}{6}$ f. $\frac{5}{6}$ g. $\frac{4}{6}$ h. 0 i. $\frac{3}{6}$ j. $\frac{4}{6}$ k. $\frac{5}{6}$ l. 0

3. a. $\frac{5}{100}$ b. $\frac{35}{100}$ c. $\frac{50}{100}$ d. $\frac{50}{100}$

4. a. $\frac{13}{52}$ b. $\frac{39}{52}$ c. $\frac{26}{52}$ d. $\frac{48}{52}$ e. $\frac{1}{52}$ f. $\frac{8}{52}$ g. $\frac{44}{52}$ h. $\frac{16}{52}$

Problems

5. a. $\frac{2}{7}$ b. $\frac{4}{7}$ c. $\frac{1}{7}$ d. $\frac{5}{7}$ e. $\frac{6}{7}$ f. $\frac{2}{7}$ g. 0 h. $\frac{6}{7}$

7. a. $\frac{40}{100}$ b. $\frac{30}{100}$ c. $\frac{90}{100}$ d. $\frac{70}{100}$ e. $\frac{30}{100}$

9. a. $\frac{1}{12}$ b. $\frac{11}{12}$ c. $\frac{6}{12}$ d. 0 e. 1 f. $\frac{2}{12}$ g. $\frac{10}{12}$ h. $\frac{9}{12}$ i. $\frac{5}{12}$

In-Class Excercises and Problems for Section 6.4

In-Class Exercises

1. yes 2. no 3. yes 4. no 5. yes

6.

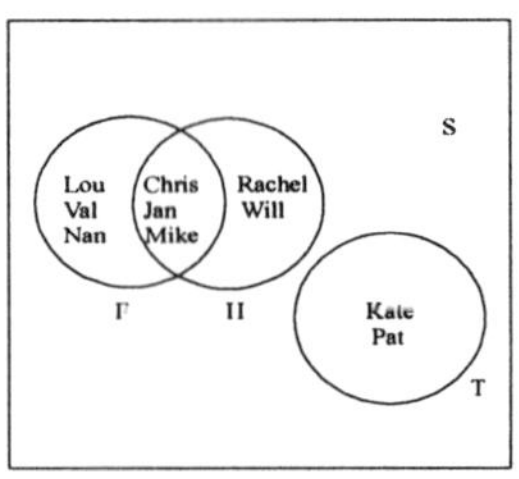

7. a. $\frac{6}{10}$ b. $\frac{5}{10}$ c. $\frac{3}{10}$ d. $\frac{8}{10}$ e. $\frac{2}{10}$

f. 0 g. $\frac{8}{10}$ h. $\frac{5}{10}$ i. $\frac{2}{10}$ j. $\frac{3}{10}$

8. $\frac{12}{17}$ 9. $\frac{13}{17}$ 10. $\frac{10}{17}$ 11. $\frac{14}{17}$ 12. $\frac{7}{17}$

13. $\frac{10}{17}$ 14. $\frac{16}{17}$ 15. $\frac{16}{17}$ 16. $\frac{5}{17}$ 17. $\frac{7}{17}$

18. $\frac{1}{75}$ 19. $\frac{29}{75}$ 20. $\frac{45}{75}$ 21. $\frac{74}{75}$ 22. $\frac{37}{75}$ 23. $\frac{38}{75}$

24. $\frac{9}{75}$ 25. $\frac{5}{75}$ 26. $\frac{42}{75}$ 27. $\frac{33}{75}$ 28. $\frac{30}{75}$ 29. $\frac{30}{75}$

30. $\frac{4}{26}$ 31. $\frac{6}{26}$ 32. $\frac{2}{26}$ 33. 1 34. 0

35. $\frac{8}{26}$ 36. $\frac{18}{26}$ 37. $\frac{3}{26}$ 38. $\frac{23}{26}$ 39. $\frac{10}{26}$

Problems

41. a. no b. no d. yes d. yes

43.

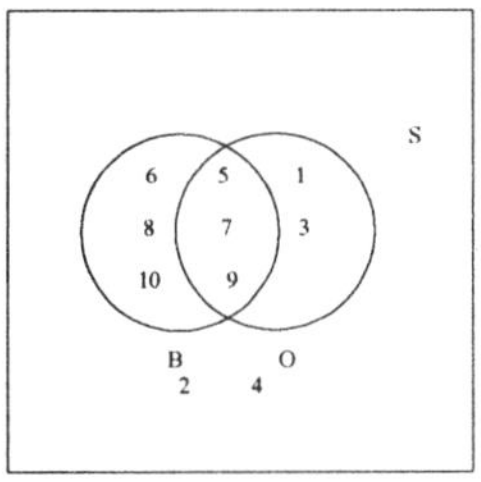

a. $\frac{6}{10}$ b. $\frac{5}{10}$ c. $\frac{3}{10}$ d. $\frac{8}{10}$

e. $\frac{2}{10}$ f. $\frac{7}{10}$ g. $\frac{7}{10}$ h. $\frac{2}{10}$

45. a. $\frac{26}{31}$ b. $\frac{24}{31}$ c. $\frac{9}{31}$ d. $\frac{22}{31}$ e. $\frac{28}{31}$ f. $\frac{28}{31}$ g. $\frac{22}{31}$

47. a. $\frac{39}{52}$ b. $\frac{32}{52}$ c. $\frac{16}{52}$ d. $\frac{24}{52}$ e. $\frac{40}{52}$ f. $\frac{51}{52}$

g. $\frac{36}{52}$ h. $\frac{27}{52}$ i. $\frac{24}{52}$ j. $\frac{48}{52}$ k. 1 l. $\frac{8}{52}$

In-Class Excercises and Problems for Section 6.5

In-Class Exercises

1. a. yes b. no c. no 2. a. yes b. yes c. no d. no

3. a. $\frac{1}{6}$ b. $\frac{3}{12}$ c. $\frac{3}{24}$ d. $\frac{5}{8}$ e. $\frac{5}{8}$ f. $\frac{1}{8}$ g. $\frac{2}{24}$ h. $\frac{21}{24}$

4. a. $\frac{1}{25}$ b. $\frac{1}{25}$ c. $\frac{1}{25}$ d. $\frac{2}{25}$ e. $\frac{2}{25}$

f. $\frac{1}{5}$ g. $\frac{4}{25}$ h. $\frac{16}{25}$ i. $\frac{8}{25}$

5. a. $\frac{1}{36}$ b. $\frac{1}{36}$ c. $\frac{2}{36}$ d. $\frac{5}{36}$ e. $\frac{6}{36}$

Problems

7. $\frac{2}{12}$ 9. a. $\frac{16}{2704}$ b. $\frac{16}{2704}$ c. $\frac{156}{2704}$ d. $\frac{2}{2704}$ e. $\frac{32}{2704}$ f. $\frac{1}{52}$

In-Class Excercises and Problems for Section 6.6

In-Class Exercises

1. No 2. Yes 3. Yes 4. Yes 5. No

6. a. 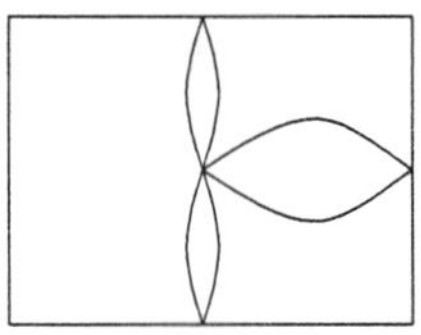

b. Since this can be thought of as a network with no odd routers, a continuous path may be found starting at any street corner.

Problems

7. a, c, d and e

9. A continuous route is possible starting at either Scranton or Reading.

11. Because all six locations would have an odd number (3) of paths connected to them, no such network can be constructed.

Chapter 6 Review

1. a. 1265 b. 100 c. 1245 d. 910 e. 435 2. a. 28 b. 12 c. 5 3. a. 1 b. 23 c. 16 d. 8

4. Invalid 5. Valid

6. $\frac{5,472}{1,000,000}$ 7. $\frac{5,427}{1,000,000}$ 8. $\frac{15,884}{1,000,000}$ 9. $\frac{3,096}{1,000,000}$ 10. $\frac{753,571}{1,000,000}$

11. 0.09 12. 0.46 13. 0.60 14. 0.81 15. ABDCA is the shortest route and covers 270 miles.

Sample Exam: Chapter 6

1. a. 26 b. 8 c. 11 d. 21 e.23 2. a. 64 b. 50 c. 25 3. b

4. Valid 5. Invalid 6. $\frac{260}{1,352}$ 7. $\frac{1,764}{6,084}$ 8. $\frac{676}{1,352}$ 9. $\frac{448}{1,352}$

10. 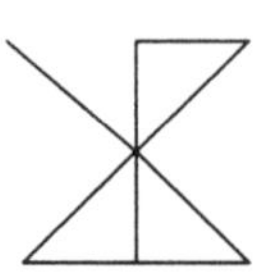

Because this may be thought of as a network with two odd routers, a continuous path is possible. The cleaning crew should begin in the kitchen and end in the bedroom or vice versa.

Index